Textile Reinforced Cement Composites

Textile Reinforced Cement Composites

New Insights in Structural and Material Engineering

Special Issue Editors

Jan Wastiels
Tine Tysmans

MDPI • Basel • Beijing • Wuhan • Barcelona • Belgrade • Manchester • Tokyo • Cluj • Tianjin

Special Issue Editors

Jan Wastiels
Vrije Universiteit Brussel
Belgium

Tine Tysmans
Vrije Universiteit Brussel
Belgium

Editorial Office
MDPI
St. Alban-Anlage 66
4052 Basel, Switzerland

This is a reprint of articles from the Special Issue published online in the open access journal *Applied Sciences* (ISSN 2076-3417) (available at: https://www.mdpi.com/journal/applsci/special_issues/textile_reinforced_cement).

For citation purposes, cite each article independently as indicated on the article page online and as indicated below:

LastName, A.A.; LastName, B.B.; LastName, C.C. Article Title. *Journal Name* **Year**, *Article Number*, Page Range.

ISBN 978-3-03928-330-9 (Pbk)
ISBN 978-3-03928-331-6 (PDF)

Cover image courtesy of Michael El Kadi.

Contents

About the Special Issue Editors . . . . **vii**

Tine Tysmans and Jan Wastiels
Editorial on Special Issue "Textile-Reinforced Cement Composites: New Insights into Structural
and Material Engineering"
Reprinted from: *Appl. Sci.* **2020**, *10*, 576, doi:10.3390/app10020576 . . . . **1**

Barzin Mobasher, Vikram Dey, Jacob Bauchmoyer, Himai Mehere and Steve Schaef
Reinforcing Efficiency of Micro and Macro Continuous Polypropylene Fibers in
Cementitious Composites
Reprinted from: *Appl. Sci.* **2019**, *9*, 2189, doi:10.3390/app9112189 . . . . **5**

Marco Carlo Rampini, Giulio Zani, Matteo Colombo and Marco di Prisco
Mechanical Behaviour of TRC Composites: Experimental and Analytical Approaches
Reprinted from: *Appl. Sci.* **2019**, *9*, 1492, doi:10.3390/app9071492 . . . . **23**

Ting Gong, Ali. A. Heravi, Ghaith Alsous, Iurie Curosu and Viktor Mechtcherine
The Impact-Tensile Behavior of Cementitious Composites Reinforced with Carbon Textile and
Short Polymer Fibers
Reprinted from: *Appl. Sci.* **2019**, *9*, 4048, doi:10.3390/app9194048 . . . . **47**

Arne Spelter, Sarah Bergmann, Jan Bielak and Josef Hegger
Long-Term Durability of Carbon-Reinforced Concrete: An Overview and
Experimental Investigations
Reprinted from: *Appl. Sci.* **2019**, *9*, 1651, doi:10.3390/app9081651 . . . . **67**

**Matthias De Munck, Tine Tysmans, Jan Wastiels, Panagiotis Kapsalis, Jolien Vervloet,
Michael El Kadi and Olivier Remy**
Fatigue Behaviour of Textile Reinforced Cementitious Composites and Their Application in
Sandwich Elements
Reprinted from: *Appl. Sci.* **2019**, *9*, 1293, doi:10.3390/app9071293 . . . . **81**

Juliane Wagner and Manfred Curbach
Bond Fatigue of TRC with Epoxy Impregnated Carbon Textiles
Reprinted from: *Appl. Sci.* **2019**, *9*, 1980, doi:10.3390/app9101980 . . . . **101**

Paraskevi D. Askouni, Catherine (Corina) G. Papanicolaou and Michael I. Kaffetzakis
The Effect of Elevated Temperatures on the TRM-to-Masonry Bond: Comparison of Normal
Weight and Lightweight Matrices
Reprinted from: *Appl. Sci.* **2019**, *9*, 2156, doi:10.3390/app9102156 . . . . **123**

**Panagiotis Kapsalis, Michael El Kadi, Jolien Vervloet, Matthias De Munck, Jan Wastiels,
Thanasis Triantafillou and Tine Tysmans**
Thermomechanical Behavior of Textile Reinforced Cementitious Composites Subjected to Fire
Reprinted from: *Appl. Sci.* **2019**, *9*, 747, doi:10.3390/app9040747 . . . . **137**

Mathias Flansbjer, Natalie Williams Portal and Daniel Vennetti
Verification of the Structural Performance of Textile Reinforced Reactive Powder Concrete
Sandwich Facade Elements
Reprinted from: *Appl. Sci.* **2019**, *9*, 2456, doi:10.3390/app9122456 . . . . **153**

Jolien Vervloet, Tine Tysmans, Michael El Kadi, Matthias De Munck, Panagiotis Kapsalis, Petra Van Itterbeeck, Jan Wastiels and Danny Van Hemelrijck
Validation of a Numerical Bending Model for Sandwich Beams with Textile-Reinforced Cement Faces by Means of Digital Image Correlation
Reprinted from: *Appl. Sci.* **2019**, *9*, 1253, doi:10.3390/app9061253 . . . . **179**

Jan Bielak, Viviane Adam, Josef Hegger and Martin Classen
Shear Capacity of Textile-Reinforced Concrete Slabs without Shear Reinforcement
Reprinted from: *Appl. Sci.* **2019**, *9*, 1382, doi:10.3390/app9071382 . . . . **195**

Silke Scheerer, Robert Zobel, Egbert Müller, Tilo Senckpiel-Peters, Angela Schmidt and Manfred Curbach
Flexural Strengthening of RC Structures with TRC—Experimental Observations, Design Approach and Application
Reprinted from: *Appl. Sci.* **2019**, *9*, 1322, doi:10.3390/app9071322 . . . . **215**

Rostislav Chudoba, Ehsan Sharei, Frank Schladitz andTilo Senckpiel-Peters
Numerical Modeling of Non-Uniformly ReinforcedCarbon Concrete Lightweight Ceiling Elements
Reprinted from: *Appl. Sci.* **2019**, *9*, 2348, doi:10.3390/app9112348 . . . . **233**

Amir Asgharzadeh and Michael Raupach
Damage Mechanisms of Polymer Impregnated Carbon Textiles Used as Anode Material for Cathodic Protection
Reprinted from: *Appl. Sci.* **2019**, *9*, 110, doi:10.3390/app9010110 . . . . **257**

About the Special Issue Editors

Jan Wastiels obtained the degree of Structural Engineer at Vrije Universiteit Brussel (VUB) in 1973, and received his PhD in Engineering Sciences from the same university in 1980 on the subject of the multiaxial behaviour of concrete. Since the 1980s, he has been active in the research and development of alternative cement materials for construction purposes, such as kaolinite-based mineral polymers, metakaolinite- and fly-ash-based alkaline activated materials, and inorganic phosphate cement. He has been heading and participating in national and international research projects on the subject, mostly in the context of fibre-reinforced cementitious composites. He has been the supervisor (promoter) of 16 doctoral theses (PhD) and over 100 master's theses, and is the author or co-author of over 280 scientific publications in international journals and books (of which 101 are Web of Science publications, mostly in Q1 journals, WoS h-index 18). He is emeritus Professor of VUB and Guest Professor of University of Ghent, and had teaching duties in concrete and steel design, building materials, and composite materials, amongst others. He was Vice-Dean and Dean of the Faculty of Engineering Sciences of VUB, and Head of the departments of Architectural Engineering (ARCH) and of Mechanics of Materials and Constructions (MeMC).

Tine Tysmans (°1983) graduated as a Civil Engineer in 2006 at Vrije Universiteit Brussel (VUB, Brussels, Belgium). In 2010 she received her PhD degree on the topic of thin shell structures in textile-reinforced cement composites. Shortly after obtaining her doctoral degree, Tine Tysmans was appointed as Professor at the department of Mechanics of Materials and Constructions (MeMC) at VUB (01/10/2011). Ever since, she has built out her research in the field of the development and analysis of new lightweight structures, mostly using cement matrix composite innovation. In a short period, she succeeded in expanding her research team significantly. She guided a total of 13 doctoral theses, nine of which are already successfully finished. Her research is driven by the search for innovative material-efficient lightweight structures such as sandwich panels or shells using cement composites. In doing so, she has built up expertise in the finite element modelling of cement composites, their durability, and their mechanical characterisation. In her search for better performance structures, she also performs research on form finding and structural optimisation. Her record shows a strongly growing number of scientific publications: 48 ISI Web of Science publications, cited more than 300 times (H-index of 11).

Editorial

Editorial on Special Issue "Textile-Reinforced Cement Composites: New Insights into Structural and Material Engineering"

Tine Tysmans * and Jan Wastiels

Department Mechanics of Materials and Constructions, Vrije Universiteit Brussel (VUB), Pleinlaan 2, 1050 Brussels, Belgium; jan.wastiels@vub.be
* Correspondence: Tine.Tysmans@vub.be

Received: 20 December 2019; Accepted: 23 December 2019; Published: 13 January 2020

Abstract: This special issue presents the latest advances in the field of Textile-Reinforced Cement Composites, including Textile-Reinforced Concrete (TRC), Textile-Reinforced Mortar (TRM), Fabric-Reinforced Cementitious Matrix (FRCM), etc. These composite materials distinguish themselves from other fibre reinforced concrete materials by their strain-hardening behaviour under tensile loading. This Special Issue is composed of 14 papers covering new insights in structural and material engineering. The papers include investigations on the level of the fibre reinforcement system as well as on the level of the composites, investigating their impact and fatigue behaviour, durability and fire behaviour. Both strengthening of existing structures and development of new structural systems such as lightweight sandwich systems are presented, and analysis and design methods are discussed. This Special Issue demonstrates the broadness and intensity of the ongoing advancements in the field of Textile-Reinforced Cement composites and the importance of several future research directions.

Keywords: cement composites; fibre; textile; textile-reinforced concrete; textile-reinforced mortar

---

1. Introduction

As sustainability is rising higher on the agenda, research towards new construction materials and systems has gained importance. One of the new material systems that has been investigated over the last few decades is textile-reinforced cement composites. These materials are known under different names, e.g., Textile-Reinforced Concrete (TRC) or Cement, Textile-Reinforced Mortar (TRM), and Fabric-Reinforced Cementitious Matrix (FRCM). All of these composite materials are characterized by a cementitious matrix material reinforced by textiles to provide for continuous fibre reinforcement in such a way that they show strain hardening behaviour under tensile loading.

Thanks to this tensile capacity provided by the textiles, traditional steel reinforcement can be omitted and the thickness of the concrete cover that is used to prevent the steel from corroding can be significantly reduced. As a result, thin concrete elements can be designed with a slenderness that cannot be achieved with traditional steel-reinforced concrete. As the dimensions of the concrete elements are reduced, the amount of concrete used, the weight of the elements, the emissions due to transport of the materials, etc., are also reduced. On top of this, the sustainability of the building elements can be increased because the flexible fibre reinforcement allows for the design of structural elements in optimal shapes, with tailored fibre textiles.

2. TRC in Structural and Material Engineering

The motivation for this Special Issue was the potential of TRC composites to contribute to a sustainable built environment. It presents the latest insights and results from the research community dealing with TRC composites on both the material level and the structural level. On the material level, much research has been carried out towards the optimization of the fibre reinforcement system. On the structural level, many researchers have investigated new structural systems that can benefit from the slenderness and low weight of TRCs, and methods to analyze loadbearing behaviour and design these new applications. Finally, the durability of the composites and their behaviour under an elevated temperature (including in fire) are important fields of investigation, both on the material level and the structural level. In total, 21 papers were submitted, and 14 were accepted.

Three papers focus on the reinforcement system. Although carbon or AR (alkali-resistant) glass fibres are used as the reinforcement material for TRCs in all other papers in this Special Issue due to their stiffness and strength properties, the paper by Mobasher, Dey, Bauchmoyer, Mehere, and Schaef [1] analyses the effect of high-toughness hydrophilic PP fibres on the cracking distribution associated with strengthening and toughening. The paper by Rampini, Zani, Colombo, and di Prisco [2] presents extensive experimental results of tensile tests on a wide set of AR glass fabrics and TRCs. The effect of the fabric coating and the addition of short fibres on the load capacity and the energy absorption is discussed. A hybrid fibre system is also used in the contribution by Gong, Heravi, Alsous, Curosu, and Mechtcherine [3], where the synergetic action between a carbon grid and short PE or PBO fibres is investigated in quasi-static uniaxial tensile tests, pull-out tests at different strain rates, and split Hopkinson impact tests.

Long-term durability is discussed and experimentally investigated in the paper by Spelter, Bergmann, Bielak, and Hegger [4], leading to the conclusion that it appears to be possible that no reduction due to external influences is necessary for the investigated type of carbon reinforcement. The effect of fatigue is treated in two contributions. The paper by De Munck, Tysmans, Wastiels, Kapsalis, Vervloet, El Kadi, and Remy [5] presents the effect of 100,000 loading cycles at serviceability load levels on the residual behaviour during uniaxial coupon tensile tests and four-point bending tests on sandwich panels reinforced with AR glass textiles. The paper by Wagner and Curbach [6] examines the bond fatigue of TRC with epoxy-impregnated carbon textiles as a function of the anchorage length and the load level, and presents the obtained results in S–N diagrams. The effect of an elevated temperature (120 °C and 200 °C) on the shear bond to masonry is treated in the feature paper by Askouni, Papanicolaou, and Kaffetzakis [7], using AR glass textiles with a normal weight and a lightweight mortar. The degradation of TRCs with SBR-coated carbon or AR glass textiles is treated in the paper by Kapsalis, El Kadi, Vervloet, De Munck, Wastiels, Triantafillou, and Tysmans [8].

The design of TRC structures is treated in four contributions. The paper by Flansbjer, Williams, Portal, and Vennetti [9] verifies the structural performance of non-load-bearing sandwich façade elements through a large-scale experimental program focused on anchorage and wind load tests, as well as through numerical modeling as validation. The paper by Vervloet, Tysmans, El Kadi, De Munck, Kapsalis, Van Itterbeeck, Wastiels, and Van Hemelrijck [10] focuses on the large-scale experimental validation of a numerical bending model for sandwich beams with TRC faces reinforced with two-dimensional (2D) and three-dimensional (3D) AR glass textiles. The shear capacity of carbon TRC slabs without shear reinforcement is investigated by Bielak, Adam, Hegger, and Classen [11] through three-point bending tests with a different shear slenderness. The results are compared to existing models and design provisions. The feature review paper by Scheerer, Zobel, Müller, Senckpiel-Peters, Schmidt, and Curbach [12] elaborates on design rules for the flexural strengthening of reinforced concrete structures with TRC, and highlights some practical applications.

The feature paper by Chudoba, Sharei, Senckpiel-Peters, and Schladitz [13] contributes to the discussion on macro-scale modeling methods for the structural analysis of thin-walled concrete shells reinforced with a layup of non-metallic fabrics. Large scale tests are performed on carbon-reinforced shell ceiling elements, and a smeared crack cross-section model is proposed, calibrated, and verified with a numerical analysis.

A final paper by Asgharzadeh and Raupach [14] proposes a specific application in which carbon textiles are, besides their use as a structural reinforcement, intended to act as an anode material for cathodic corrosion protection.

3. Conclusions

This Special Issue presents great advancements in the field of Textile-Reinforced Cement composites, as well as future research directions. Another essential step that we expect in the future is the translation of these research results into tangible design guidelines for the construction industry.

Author Contributions: Both authors have contributed to the conceptualization, writing, review and editing of this manuscript. Both authors have read and agreed to the published version of the manuscript.

Funding: This research received no external funding.

Acknowledgments: Thanks are due to all of the authors and peer reviewers for their valuable contributions to this Special Issue. The editorial team at MDPI has been of great help; special thanks go to Jennifer Li, Managing Editor for the Acoustics and Vibration Section.

Conflicts of Interest: The authors declare no conflict of interest.

References

1. Mobasher, B.; Dey, V.; Bauchmoyer, J.; Mehere, H.; Schaef, S. Reinforcing efficiency of micro and macro continuous polypropylene fibers in cementitious composites. *Appl. Sci.* **2019**, *9*, 2189. [CrossRef]
2. Rampini, M.C.; Zani, G.; Colombo, M.; di Prisco, M. Mechanical behaviour of TRC composites: Experimental and analytical approaches. *Appl. Sci.* **2019**, *9*, 1492. [CrossRef]
3. Gong, T.; Heravi, A.A.; Alsous, G.; Curosu, I.; Mechtcherine, V. Impact-tensile behavior of cementitious composites reinforced with carbon textile and short polymer fibers. *Appl. Sci.* **2019**, *9*, 4048. [CrossRef]
4. Spelter, A.; Bergmann, S.; Bielak, J.; Hegger, J. Long-term durability of carbon-reinforced concrete: An overview and experimental investigations. *Appl. Sci.* **2019**, *9*, 1651. [CrossRef]
5. De Munck, M.; Tysmans, T.; Wastiels, J.; Kapsalis, P.; Vervloet, J.; El Kadi, M.; Remy, O. Fatigue behaviour of textile reinforced cementitious composites and their application in sandwich elements. *Appl. Sci.* **2019**, *9*, 1293. [CrossRef]
6. Wagner, J.; Curbach, M. Bond fatigue of TRC with epoxy impregnated carbon textiles. *Appl. Sci.* **2019**, *9*, 1980. [CrossRef]
7. Askouni, P.D.; Papanicolaou, C.G.; Kaffetzakis, M.I. The effect of elevated temperatures on the TRM-to-masonry bond: Comparison of normal weight and lightweight matrices. *Appl. Sci.* **2019**, *9*, 2156. [CrossRef]
8. Kapsalis, P.; El Kadi, M.; Vervloet, J.; De Munck, M.; Wastiels, J.; Triantafillou, T.; Tysmans, T. Thermomechanical behavior of textile reinforced cementitious composites subjected to fire. *Appl. Sci.* **2019**, *9*, 747. [CrossRef]
9. Flansbjer, M.; Williams Portal, N.; Vennetti, D. Verification of the structural performance of textile reinforced reactive powder concrete sandwich façade elements. *Appl. Sci.* **2019**, *9*, 2456. [CrossRef]
10. Vervloet, J.; Tysmans, T.; El Kadi, M.; DE Munck, M.; Kapsalis, P.; Van Itterbeeck, P.; Wastiels, J.; Van Hemelrijck, D. Validation of a numerical bending model for sandwich beams with textile-reinforced cement faces by means of digital image correlation. *Appl. Sci.* **2019**, *9*, 1253. [CrossRef]
11. Bielak, J.; Adam, V.; Hegger, J.; Classen, M. Shear capacity of textile-reinforced concrete slabs without shear reinforcement. *Appl. Sci.* **2019**, *9*, 1382. [CrossRef]
12. Scheerer, S.; Zovel, R.; Müller, E.; Senckpiel-Peters, T.; Schmidt, A.; Curbach, M. Flexural strengthening of RC structures with TRC—Experimental observations, design approach and application. *Appl. Sci.* **2019**, *9*, 1322. [CrossRef]

13. Chudoba, R.; Sharei, E.; Senckpiel-Peters, T.; Schladitz, F. Numerical modeling of non-uniformly reinforced carbon concrete lightweight ceiling elements. *Appl. Sci.* **2019**, *9*, 2348. [CrossRef]
14. Asgharzadeh, A.; Raupach, M. Damage mechanisms of polymer impregnated carbon textiles used as anode material for cathodic protection. *Appl. Sci.* **2019**, *9*, 110. [CrossRef]

 *applied sciences*

Article

Reinforcing Efficiency of Micro and Macro Continuous Polypropylene Fibers in Cementitious Composites

Barzin Mobasher [1,*], Vikram Dey [2], Jacob Bauchmoyer [3], Himai Mehere [1] and Steve Schaef [4]

[1]   School of Sustainable Engineering and the Built Environment, Arizona State University, Tempe, AZ 85287, USA; hmehere@asu.edu
[2]   Structural Designer, PK Associates Structural Engineers, Scottsdale, AZ 85250, USA; vikram.dey@asu.edu
[3]   Structural Engineer, CDM Smith, Phoenix, AZ 85028, USA; jbauchmo@asu.edu
[4]   Materials Engineer, Development Admixture Systems, Beachwood, OH 44133, USA; steve.schaef@basf.com
*   Correspondence: barzin@asu.edu; Tel.: +1-480-965-0141; Fax: +1-480-965-0557

Received: 17 April 2019; Accepted: 18 May 2019; Published: 29 May 2019

Abstract: The effect of the microstructure of hydrophilic polypropylene (PP) fibers in the distribution of cracking associated with the strengthening and toughening mechanism of cement-based composites under tensile loading was studied. Using a filament winding system, continuous cement-based PP fiber composites were manufactured. The automated manufacturing system allows alignment of the fiber yarns in the longitudinal direction at various fiber contents. Composites with surface-modified hydrophilic macro-synthetic continuous polypropylene fibers and monofilament yarns with different diameters and surface structures were used. Samples were characterized using the tensile first cracking strength, post-crack stiffness, ultimate strength, and strain capacity. A range of volume fractions of 1–4% by volume of fibers was used, resulting in tensile first cracking strength in the range of 1–7 MPa, an ultimate strength of up to 22 MPa, and a strain capacity of 6%. The reinforcing efficiency based on crack spacing and width was documented as a function of the applied strain using digital image correlation (DIC). Quantitative analysis of crack width and spacing showed the sequential formation and gradual intermittent opening of several active and passive cracks as the key parameters in the toughening mechanism. Results are correlated with the tensile response and stiffness degradation. The mechanical properties, as well as crack spacing and composite stiffness, were significantly affected by the microstructure and dosage of continuous fibers.

Keywords: fiber-reinforced concrete; crack spacing; fiber; micro-fiber; tensile strength; toughness

---

1. Introduction

Development of strain-hardening cementitious composites (SHCC) using polypropylene (PP) fibers is a major breakthrough for a variety of applications in civil infrastructure systems. SHCC materials, such as textile reinforced concrete (TRC), exhibit high tensile strength, enhanced strain capacity, and ductility [1–3]. The superior mechanical properties offered by the polymeric based continuous fiber or textile system can be utilized as structural panels subjected to dynamic loads, such as impact and high speed, along with applications requiring blast resistance and fracture tolerance. SHCC systems could also be used as skin reinforcement laminates for the strengthening of unreinforced masonry walls, retrofit of existing structures, and beam–column connections [4–6].The tensile hardening behavior is attributed to the fiber bridging effect, which stabilizes crack growth and opening at the expense of the formation of multiple, parallel fine cracks. This cracking network gives rise to high energy absorption, both under quasi-static and dynamic loading conditions. The post-crack stiffness and the corresponding damage distribution may form with a variety of fiber systems and is governed by the fiber's ability to provide a sufficient degree of bond strength [7].

A class of SHCC materials made with polypropylene fibers with a high tensile ductility and stiffness retention over a large strain range is investigated in this study. Ductility enhancement is attractive from a cost point of view since polymeric fibers have a lower cost than steel, carbon, or other high-performance fibers; however, the efficiency of PP-based fibers is created in the form of developing composites with improved bond characteristics. Results are characterized by the improved bond characteristics of long and multifilament fibers, surface modifications, reduced diameter, and increased surface area of yarns. It is shown that proper mix proportioning results in excellent matrix properties [8–12].

Continuous unidirectional yarns were evaluated for two types of fiber compositions in this study. Effectiveness of the fiber–matrix bond interface in load transfer and distributed cracking in mechanical performance is addressed. Limitations in interfacial bond and low adhesion strength are major inefficiencies limiting the structural application of polymeric fibers in concrete materials. A combination of low organic–inorganic bond stiffness and strength limits the effectiveness of fiber–matrix stress transfer. Strength and toughness increases due to the increased aspect ratio in continuous fiber composites can be utilized in a variety of structural elements subjected to extreme loading conditions as discussed earlier [13].

Hydrophilic polymeric surfaces improve fiber performance and efficiency by affecting the bond stiffness and strength. Anchorage and bonding are also enhanced by geometrical modifications of the surface texture of the fiber [14]. Increasing the contact surface area by using small diameter filaments bundled into the form of yarn leads to additional bonding. In both these cases, the efficiency of the fiber performance is measured in the context of the fibers bridging over the cracks in the cementitious matrix, which subsequently de-bond and pullout, thus hindering the extension of cracks [15]. The fiber bridging and pullout force transmission reduces the crack tip stresses and increases toughness through energy dissipation [16]. The stress transmission through the bridging fibers is a major source of toughening and permits the initiation of new cracks, thus improving energy dissipation capacity of the composite [17].

Fiber length and orientation plays an important role in the mechanical response of cementitious composites. In order to eliminate the reduction factors due to length and orientation, unidirectional continuous composites were manufactured using a filament winding technique and tested in uniaxial tension. In the current study, two different polypropylene fiber types, namely macro-monofilaments and micro-multifilament yarns, at different dosages, are compared in terms of composite performance based on the tensile strength, crack spacing, and stiffness reduction as a function of measured strain. Matrix formulations consisted of blended cementitious matrices containing various proportions of Type II Ordinary Portland Cement (OPC), sand, and fly ash as a control matrix mix. Mechanical tests were performed under uniaxial tension, and three-dimensional digital image correlation (DIC) method and image analysis were used to quantify the damage mechanism and the non-uniform strain distributions. The distributed cracking mechanism was quantified by measuring the crack width and spacing and was further compared to the experimental stress–strain measures.

2. Experimental Program

Proprietary polypropylene yarns manufactured by BASF Construction Chemicals, Beachwood OH, USA were studied. A macrofiber labeled as MAC 2200CB (abbreviated as MAC in this study) is a commercially available monofilament macro-synthetic polypropylene fiber with an average diameter of 0.82 mm and pinched surface to improve the bond (see Figure 1a). It is used in cast-in-place concrete applications, such as slab-on-grade, pavements, bridge decks, and in precast concrete, mainly as a secondary reinforcement to restrain temperature cracking [18]. The second fiber evaluated in this study is a recently developed multifilament microfiber yarn with 500 thin filaments 40 microns in diameter and identified as MF 40 microfibers, as shown with two different magnifications in Figure 1b,c. The effective yarn diameters of MF/MAC measured from the SEM images represent a surface to volume ratio of about 20.

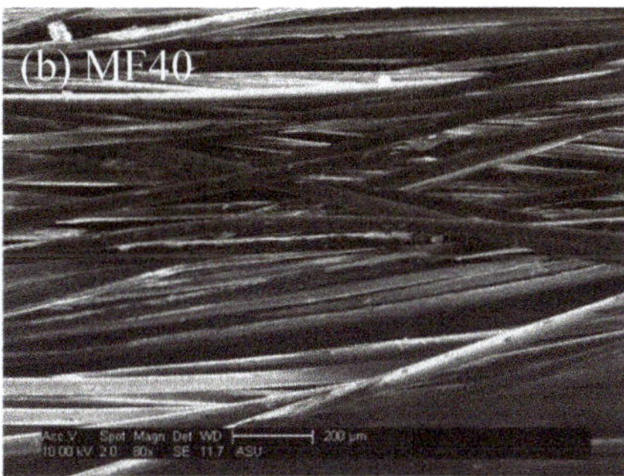

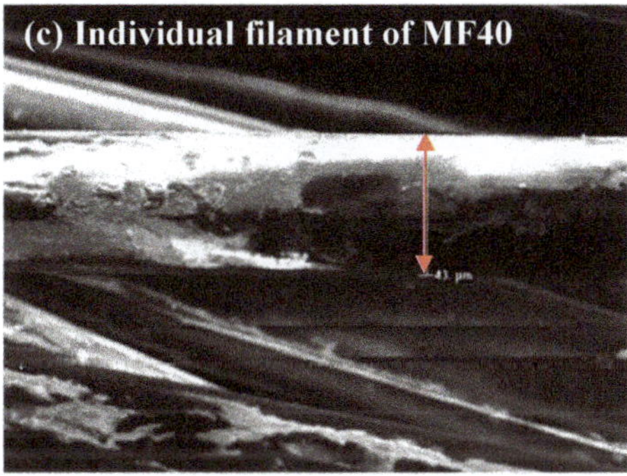

Figure 1. (**a**) Macro-synthetic MAC fiber, (**b**) multifilament fibrillated microfiber, and (**c**) diameter of an individual MF40 fiber.

Unidirectional composites were produced using the filament winding method shown in Figure 2a,b. Composites with continuous fibers allowed for measurement of reinforcement potential that is independent of the fiber length, delamination, or orientation effects. The experimental plan for mechanical tests is presented in Table 1 and includes tension tests on individual yarns and composite uniaxial tension. Testing variables included the fiber structure and content to study their affect on the tensile stress–strain response and damage parameters such as crack spacing and crack width [14,19].

2.1. Sample Preparation Using Filament Winding

A filament winding system was configured to fabricate continuous cement fiber laminates with aligned fiber yarns [13,17]. A computer-controlled system used stepper motors to pull the yarns and wind the sample on a mandrel. System components included the feed section, guidance assembly, and take-up mandrel. Labor-intensive tasks in production and panel making were reduced through this automated system. Servo-drives were programmed for automation of three sections of fiber feed, guide (fiber impregnation), and the take-up (molding) sections, as shown in Figure 2a.

The various sections included the stepper motors, positioning encoders, limit switches for safe interlocking, and a computer with control software interface, as shown in Figure 2b. The feed section used a spool of fibers that would unwind and was immersed into a wetting tank prior to immersion in an impregnation chamber. The sample was then wound on a rotating mold. Using a LabView© (2014, National Instruments, Austin, TX, USA) interface, a closed-loop system controls two stepper motors to feed and slide the yarns through and rotate the mandrel. The stepper motors in the take-up section controlled the winding, pulling, and transverse sliding of the composites.

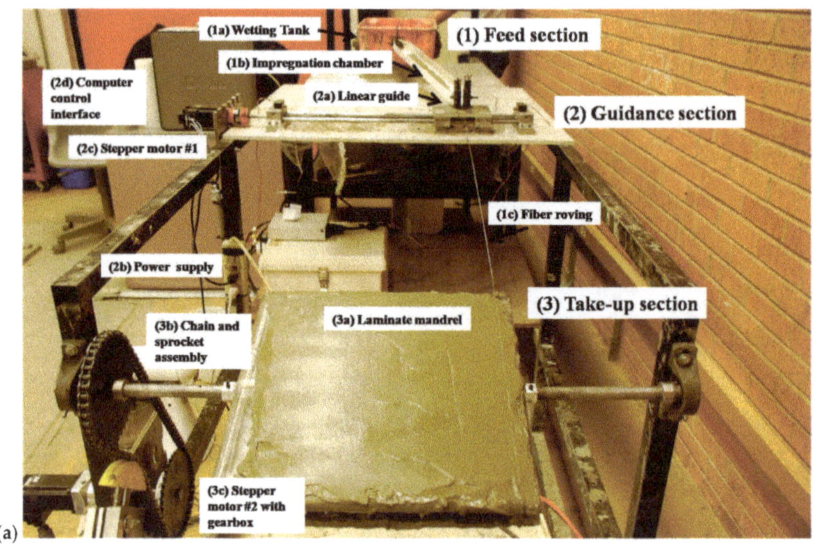

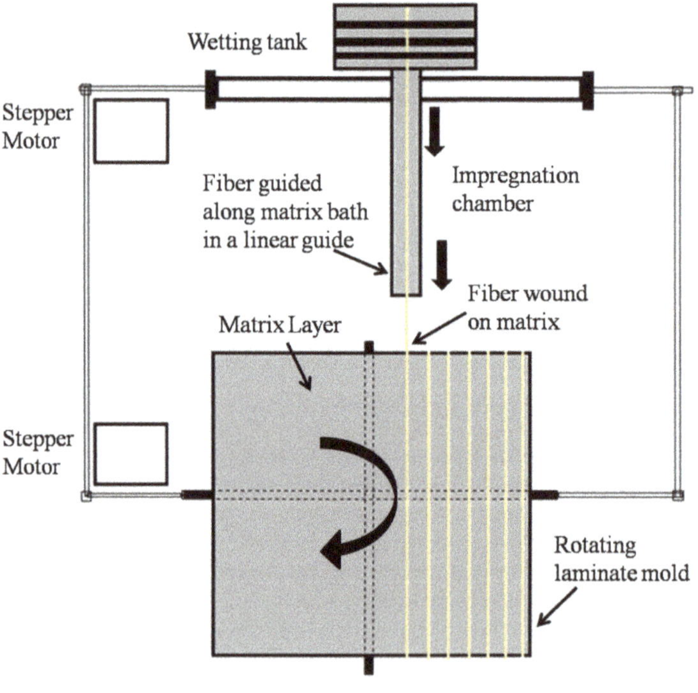

Figure 2. (**a**) Filament winding setup with the impregnation chamber, fiber guide section, and rotating mandrel; and (**b**) schematics of the steps section.

2.2. Mix Design

The control mix consisted of Portland cement, fly ash, and fine silica sand was used as the basic formulation of the composite design listed in Table 1. The control mortar mix design used a blend of 48% Portland cement type I/II, and 7% by weight of class F fly ash, a sand to cementitious solid ratio of 45% by weight, and water to binder ratio of 0.35. A naphthalene base high range water reducer manufactured by BASF was used at a dosage of 0.03% by weight of cement. Samples were made with the two polymeric fibers introduced earlier, namely macrofibre MAC and multifilament MF fibers, using 1%, 2.5%, and 4% volume fractions. Direct tension tests were conducted on a minimum of four replicate samples for each mix design.

Table 1. Summary of Tension specimens with continuous fibers.

Test Type	Yarn Type	Sample Variables	Curing	Yarn V_f%
Fiber Tension	MAC	150, 200, and 250 mm	N/A	
Fiber Tension	MF40	150, 200, and 250 mm	N/A	
Composite Tension	MAC	Volume fraction	28 days	1.0, 2.5, 4.0
Composite Tension	MF40	Volume fraction	7, 28 days	1.0, 2.5, 4.0

3. Testing Program

3.1. Tensile Response of Fibers

Fiber tension tests were conducted under displacement control mode to measure elastic modulus, strain capacity, ultimate strength, toughness, and mode of failure. The setup is shown in Figure 3a with a specimen under the applied load. An actuator displacement rate of 0.4 mm/min was used. Preliminary tests were conducted using fiber lengths of 150, 200, and 250 mm to address the length effect. Follow up studies used a sample length of 150 mm and a minimum of five replicate samples per series. The load was measured using a load cell rated at a capacity of 1300 N, while the elongation was recorded by an extensometer with a 50 mm gage length. A close-up view of the failed specimens with the extensometer attached is shown in Figure 3b,c.

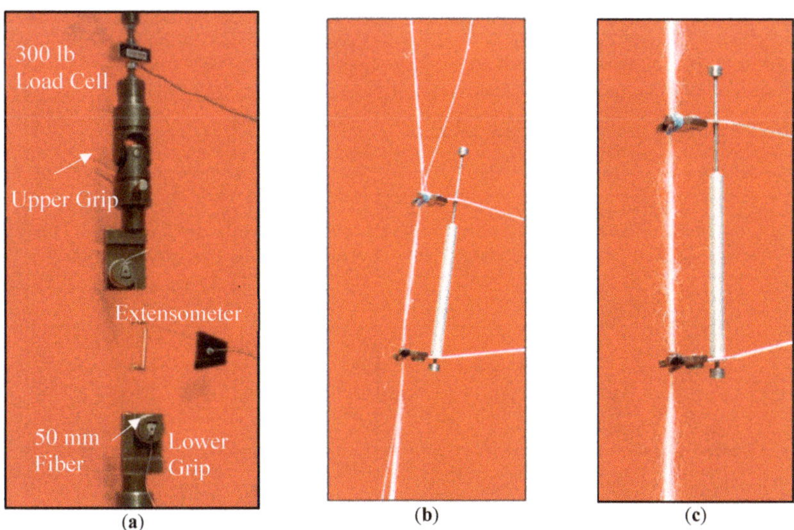

Figure 3. (**a**) Test setup for yarn tension tests, (**b**) failed MAC fiber, and (**c**) failed MF40 fiber.

Stress–strain behavior for the two types of MAC and MF fiber yarns are presented in Figure 4 and a summary of test results is given in Table 2. The initial stress–strain curve started with a stiff response up to a stress level of about 5–7 MPa.Beyond that level, the stiffness decreased due to fiber yielding. The general behavior was linear for the monofilament samples up to the failure; however, significant nonlinearity was observed for the microfiber yarn. Gradual transition of the stress–strain response of microfiber yarns to a nonlinear behavior started from 50% ultimate strain capacity without a clearly marked yield point. Beyond this level, the stiffness reduced gradually until failure.

Table 2 summarizes the single fiber tensile test results for both fiber types representing values of initial elastic modulus, E_1, and a post yield modulus, E_2. The macro-synthetic fiber, MAC, hadcomparatively higher initial and post-yield modulus, and showed a sudden failure compared to a progressive failure of the individual filaments of MF40 yarn. The ultimate strength was reached in a gradual manner for MF40 as opposed to a sharp end for MAC. Figure 3b,c show the failed MAC and MF40 specimens, respectively. With a strain capacity in the range of 12%, microfibers deformation was almost twice as much as the monofilament fibers, as shown in Figure 4. Compared to MAC, the MF40 microfiber exhibited significant crazing. This response was more pronounced when the strain wasmeasured using the actuator signal, as shown in Figure 4b, which also included the relative slipping of the individual filaments past each another, resulting in an apparent tensile strain as high as 75% for the MF series. These slip mechanisms lead to a higher strain capacity of the MF compared to MAC fibers.

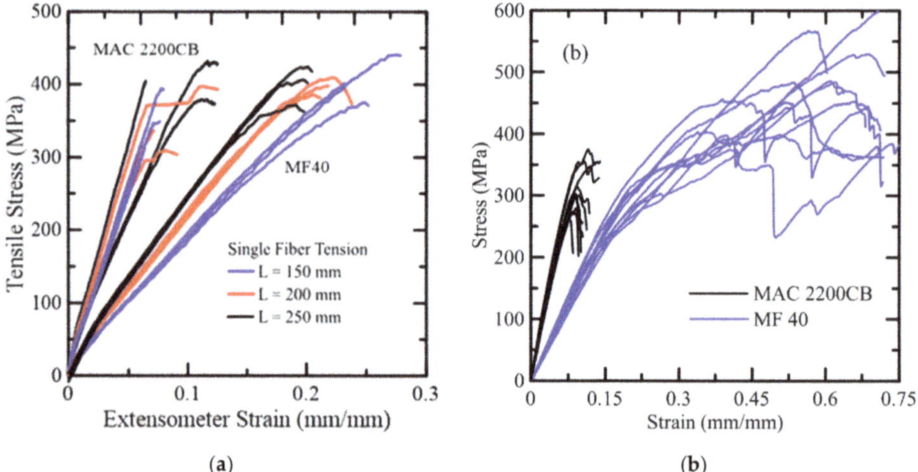

Figure 4. (**a**) Effect of sample length on the initial response of the MAC and MF40 fibers. (**b**) Tensile stress versus actuator strain comparing MAC and MF40 failure for fiber length of 150 mm.

Table 2. Single fiber tests result for MAC and MF40 fibers, gauge length 150 mm.

Fiber		Max Load	Max Elongation	Tensile Strength	Elastic Modulus, E_1	Post-Yield Modulus, E_2	Work to Fracture
		N	mm	MPa	MPa	MPa	J
MAC	Avg.	245.3	4.4	394	9239	4566	0.70
	Std Dev.	20.0	0.9	32.8	1813	918.3	0.25
MF	Average	293.8	6.3	405	4985	3058	1.59
	Std Dev.	33.0	0.7	45.5	1112	479.3	0.21

The difference in strain capacity between the two fiber compositions resulted in the toughness of MF being twice that of MAC and is attributed to the structure of multifilament yarns, which by distributing the damage among multiple fibers, promoted a progressive failure mechanism. This led to a 43% higher strain capacity than the macro MAC fiber, which was an inherently stiffer system (MAC = 9.2 GPa, MF40 = 5 GPa), as shown in Table 2. The post-yield reduced modulus for MAC was 4.6 GPa, which was 50% higher than post-yield modulus of MF40, which was at 3 GPa. Due to their strain capacity, finer MF40 fibrils required as much as 220% higher work to fracture. Multi-filament yarns uniformly distributed the load within the filaments, which failed sequentially over the failure strain range.

3.2. Tension Tests on Continuous Fiber Composites

A closed-loop servo-hydraulic test system, as shown in Figure 5, was used in actuator displacement control mode to conduct direct tension tests on the SHCC composites. Test coupons had nominal dimensions of 300 × 62 × 13 mm. The specimen was held using hydraulic grips with the pressure maintained between 1.7 and 2 MPa. Elongation was measured along a gage length of 90 mm using two linear variable differential transformers (LVDT) of 6 mm range and their average response was recorded along with the applied load and actuator displacement.

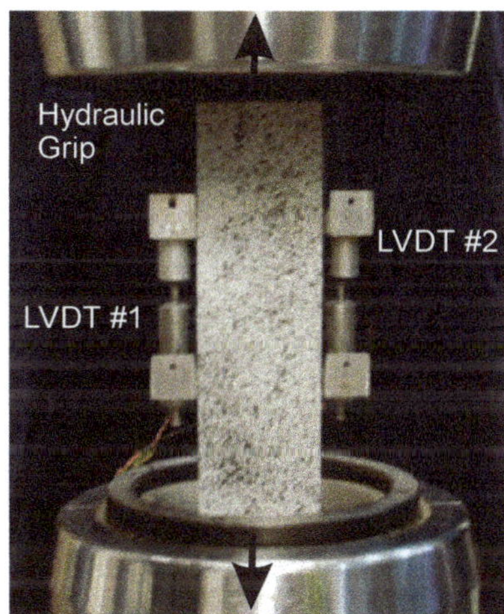

Figure 5. Tensile testing setup used to measure the characteristic response shown on the typical stress–strain response.

The characteristic stress–strain, crack spacing, and crack widening responses are summarized in Figure 6a–c. The observed stages of damage zones have been identified schematically in this figure and used in the discussion of results. The typical stress–strain response is predominantly linear up to point A, which is represented as the bend over point (BOP), this is referred to as Stage I. This is followed by the formation of the first crack in the specimen and initiation of Stage II. Between points B and C, there isthe formation of multiple distributed cracks and the initiation of fiber–matrix debonding. The bond exhibited by fibers resulted in crack bridging as the key toughening mechanism, which prevented the localization of individual cracks and promoted additional cracking. When a sufficient number of cracks had formed, stage III was initiated wherein crack saturation and widening of existing cracks occurred, leading to localized damage between points C and D. Crack saturation occurred due to limitations of the bond when stress in the matrix was insufficient to cause further cracks. Finally, in stage IV, tensile failure, fiber debonding, and slip occurred, and they were irreversible [7]. Beyond point D, the specimen significantly lost its load-carrying capacity and ultimately underwent complete failure.The experiments addressed the composite performance of the laminates using the correlation between the fiber–matrix bond, multiple cracking, crack widening, and crack saturation density.

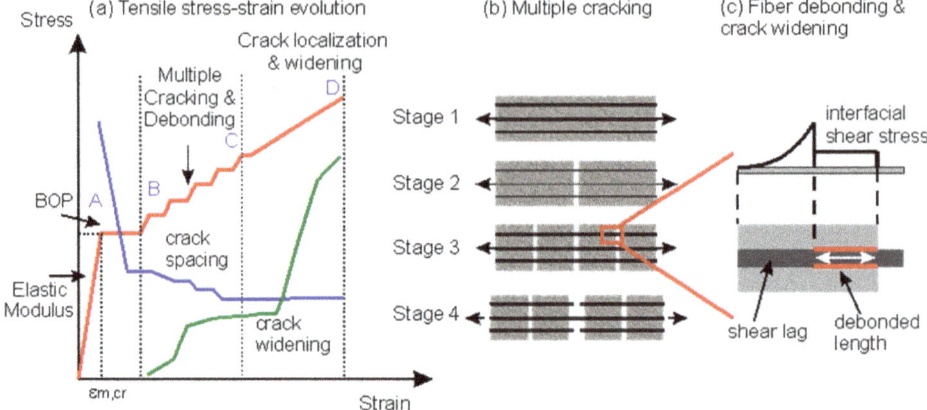

Figure 6. (**a**) Tensile testing results and regions of characteristic response shown on the typical stress–strain response, (**b**) four stages of linear elastic, cracking, multiple cracking, localization, and pullout, and (**c**) the interface debonding and pullout which contribute to crack widening.

Figure 7 shows the tensile response of the MF 40 fiber composite at two different fiber contents of 1% and 2.5% for curing durations of 7 and 28 days. The stress–strain response can be classified using the four stages as defined in Figure 6. In stage I, due to the linear behavior of matrix and fiber layers, the average strain in the longitudinal direction was uniform for the composite, fiber, and matrix. An increasing load initiated matrix cracking and stress was transferred to the fiber. Depending on the fiber content and bond, the first cracking was initiated in the form of a micro-crack and propagated along the width of the specimen at the bend over point (BOP) stress level, which is associated with the tensile strength of the matrix. Figure 7 indicates that BOP was directly correlated to the fiber content and curing age, and characterized by the elastic modulus, first crack strength, and strain.

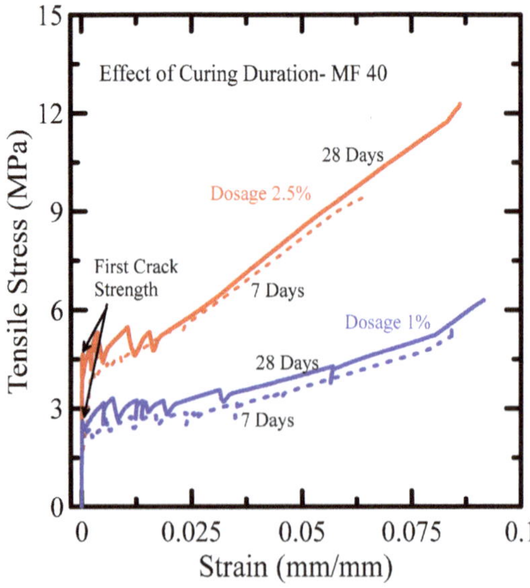

Figure 7. Effect of curing duration on the tensile response of microfilament based composites.

After initiation, a micro-crack may propagate in a stable manner due to fiber bridging, leading to a gradual reduction of matrix stiffness. The overall aspect of fiber contents is addressed in Figure 8a,b. The tensile stress–strain response of MAC and MF40 composites with a control matrix at various fiber dosages are presented at distinct stages of cracking for replicate test coupons. Higher fiber content increased the first crack strength [15] since a higher energy demand for the matrix crack propagation is imposed, which increased the apparent first crack strength. Formation of multiple parallel cracks was designated as stage II, as shown in Figure 6a. The dominant strain hardening behavior initiated after the first crack with additional parallel cracks occurring sequentially since the stiffness of the fiber phase allows for it to carry the load released by the matrix failure. The parallel cracks formed until the minimum crack spacing was reached, which correlated with the overall stiffness of the fibers, as shown in Figure 8c,d. The crack width and spacing were affected by the bond parameters and fiber content. This stage ended with fiber debonding as new crack formation seized and the existing cracks widened, matrix stress reduced, and fibers began to either get pulled out from the matrix or underwent fracture. In samples with 2.5% fiber, a 20% increase in BOP stress, and 35% increase in ultimate strain, which led to a 30% increase in UTS and 75% increase in toughness from 7 to 28 days of curing, was observed.

Summary results are presented in Table 3. The effect of fiber content on the first crack strength was more pronounced for MF40 composites in comparison to the MAC fibers. This was because of the bond surface area and distribution of thefibers throughout the matrix reducing the fiber to fiber specific spacing and reducing the minimum flaw size. The fiber bridging effect on the growing cracks was enhanced due to their distribution. The switch over from Stage I to II depended on the dosage of fibers available for bridging. The first cracking stress was higher for laminates with a higher fiber volume fraction for both MAC and MF40. At the same time for each category, the increasing volume fraction increasedthe first crack strength. MAC fibers showed an average 1.4 MPa stress level for 1% and 2.5% fiber dosages, while at 4% dosage, the stress at first crack increasedto 2.6 MPa. The MF40 fibers, on the other hand, had a lower first cracking stress of 2 MPa for 1% fiber dosage and an average of 3.8MPa and 4.4 MPa for the 2.5% and 4% replicates, as shown in Table 3. Regardless of the fibers used, stiffness, strength, and ductility increased significantly as the dosages increased. With increase in fiber content from 1%, 2.5%, and 4%; the pre-crack stiffness increased from 14.3 ± 7.4, to 21 ± 3.3, and 24 ± 14 GPa, respectively. The post-crack stiffness changes with the increase in the fiber content from 35 ± 9, 62 ± 32, to 177 ± 28 MPa for the MF fiber, which is much lower than the initial stiffness however much extends for a much larger strain range. The post-crack stiffnesses were significantly different in the two fiber systems, as shown by the reported values in Table 3. The crack spacing of composites at the crack saturation stage is shown in Figure 8c,d and point to the efficiency of the MF system.

The strain capacity of both macro- and micro-fiber systems, even at a 1% dosage level, exceeded 5%, which is an impressive level of deformation with significant energy absorption. The ultimate tensile strength for MAC composite replicates varied from 7.45 to 13.2 MPa. This tensile strength was significantly high and appropriate for the structural application of PP-based cementitious composites. Post-cracking stiffness increasedfrom 81 to 197 MPa over the entire strain range and depended on the various fiber contents. The overall toughness increasedfrom 0.79–0.83 MPa. MAC fibers at 1% and 2.5% fiber dosage showed similar stress–strain behavior, whereas at 4%, the overall composite stiffness and mechanical properties showed significant improvement. In all these systems, distributed cracking was the dominant mechanism, resulting in an increased overall toughness that was primarily due to a large strain range. The post-cracking behavior of the MF fibers were much improved compared to the MAC fibers, showing a stiffer post-crack response with distinct distributed cracking for 1% and 2.5% fiber dosages. However, for 4% MF40 dosage, several fine cracks close to each other resulted in a high crack density and toughness, as evident in Figure 8d. The first cracking stress for 2.5% and 4% fiber dosage of MF40 specimens were within 3.8 to 4.4 MPa;ultimate stress was 12.5 to 17.5 MPa; and toughness was 1.1 MPa to 1.37 MPa, which was 65% higher than MAC fibers at 4% dosage. The comprehensive result in these discussions can be found in Table 3.

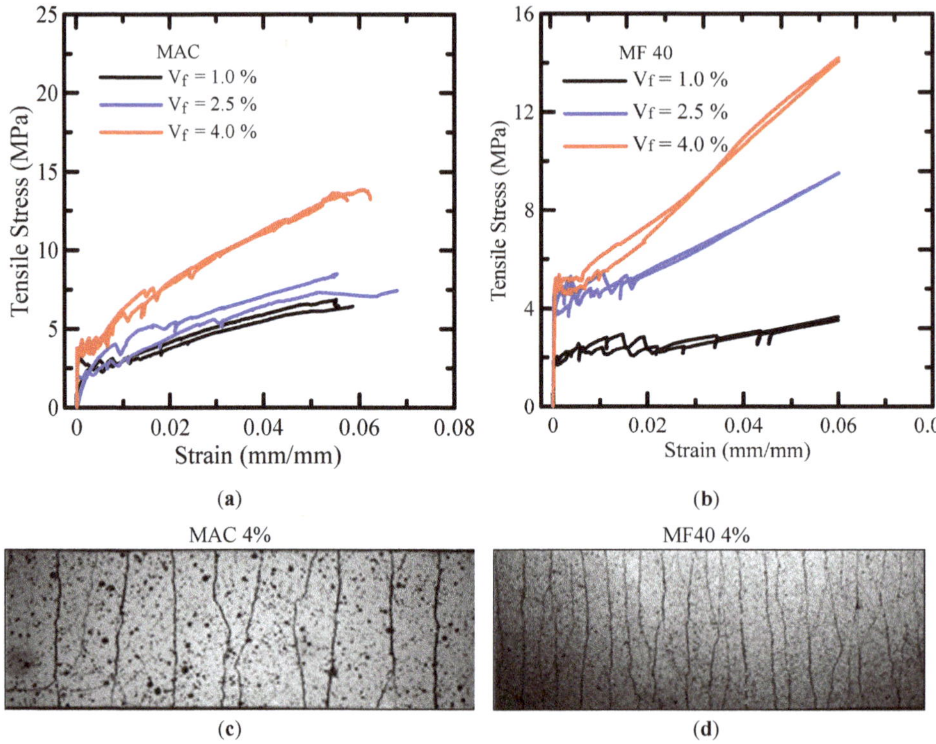

Figure 8. Effect of fiber content on the tensile stress–strain response: (**a**) MAC fiber and (**b**) MF40 fiber. Spacing distribution at crack saturation stage: (**c**) MAC composites and (**d**) MF40 composites.

Table 3. Stress Strain response parameters for control mix laminates with different fiber dosages.

Fiber Type, Vf	Replicate ID	Stress at First Crack MPa	Strain at First Crack mm/mm	UTS MPa	Strain at UTS mm/mm	Ultimate Strain mm/mm	Young's Modulus (LVDT) GPa	Post-Crack Modulus MPa	Work to Fracture (Stroke) N·ms	Toughness at 5% Strain MPa	Toughness at 10% Strain MPa
					MAC						
MAC 1%	#1	1.6	1.1×10^{-3}	7.4	0.12	0.16	1.4	94	90	0.16	0.43
	#2	1.5	1.0×10^{-3}	7.3	0.13	0.16	1.5	70	89	0.14	0.41
	#6	1.5	1.1×10^{-4}	8.3	0.13	0.15	13.6	80	65	0.09	0.37
	Avg	1.5	7.4×10^{-4}	7.4	0.13	0.16	2.8	81	89	0.15	0.42
MAC 2.5%	#2c	1.7	1.4×10^{-3}	8.6	0.13	0.17	1.2	88	143	0.16	0.48
	#4	1.7	2.7×10^{-4}	5.0	0.07	0.12	6.4	110	57	0.10	0.32
	#5	2.0	2.2×10^{-4}	8.8	0.12	0.16	9.1	180	131	0.17	0.51
	Avg	1.7	6.3×10^{-4}	7.5	0.11	0.15	7.6	126	110	0.15	0.44
MAC 4%	#5	2.8	8.8×10^{-5}	12.7	0.08	0.09	32.0	170	146	0.31	0.78
	#6	3.6	1.1×10^{-4}	12.9	0.08	0.11	33.6	170	205	0.35	0.89
	#7	3.0	7.1×10^{-5}	14.1	0.10	0.10	42.2	250	160	0.31	0.79
	Avg	3.1	9.0×10^{-5}	13.2	0.08	0.10	16.9	197	170	0.32	0.82
					MF40						
MF40 1%	#2	2.1	1.8×10^{-4}	5.3	0.13	0.14	11.7	36	96	0.13	0.33
	#3	2.7	1.6×10^{-4}	5.5	0.13	0.14	16.4	38	102	0.14	0.34
	Avg	2.4	1.7×10^{-4}	5.4	0.13	0.14	3.4	37	99	0.14	0.34
MF40 2.5%	#3	3.1	1.3×10^{-4}	8.6	0.13	0.17	24.5	120	170	0.23	0.66
	#5	4.5	1.1×10^{-4}	15.1	0.12	0.13	42.0	140	211	0.29	0.82
	#8	5.2	1.2×10^{-4}	9.9	0.08	0.12	43.6	206	168	0.25	0.69
	Avg	4.3	1.2×10^{-4}	12.5	0.10	0.13	27.7	156	190	0.27	0.76
MF40 4%	#3	4.7	1.5×10^{-4}	19.1	0.11	0.12	30.6	190	259	0.33	1.04
	#5	5.1	1.2×10^{-4}	16.4	0.10	0.12	41.7	200	220	0.32	0.99
	#7	5.2	1.1×10^{-4}	16.9	0.13	0.14	47.6	200	271	0.27	0.89
	Avg	5.0	1.3×10^{-4}	17.5	0.11	0.13	23.5	197	250	0.31	0.97

3.3. Digital Image Correlation

Digital image correlation (DIC) is a full-field displacement measuring approach that tracks the physical points of a speckle pattern on the surface of a specimen under deformation. Developed by Sutton et al. [20] and Bruck et al. [21], it is widely applied to experimental stress analysis [22–24]. For each subset region of a sample, the corresponding deformed position is found by searching in the vicinity that renders the correlation coefficient with the maximum likelihood or minimum cross-correlation function [23,24]. Commercial software VIC 3D-7 was used for measurement of crack density, spacing, and damage evolution [4,21,25,26].

Formation of a network of cracks and local strain fields are shown in Figures 9 and 10. The relative displacements of two points, as well as crack width and spacing parametersmeasured using the DIC, was compared with the LVDTs in Figure 9a,b. The DIC absolute and relative displacements along two horizontal segments were obtained at 10 s intervals and compared with the mean LVDT responses. The correlationwas close, as shown in Figure 9b, which validates the non-contacting DIC method since it provides a full-range response.

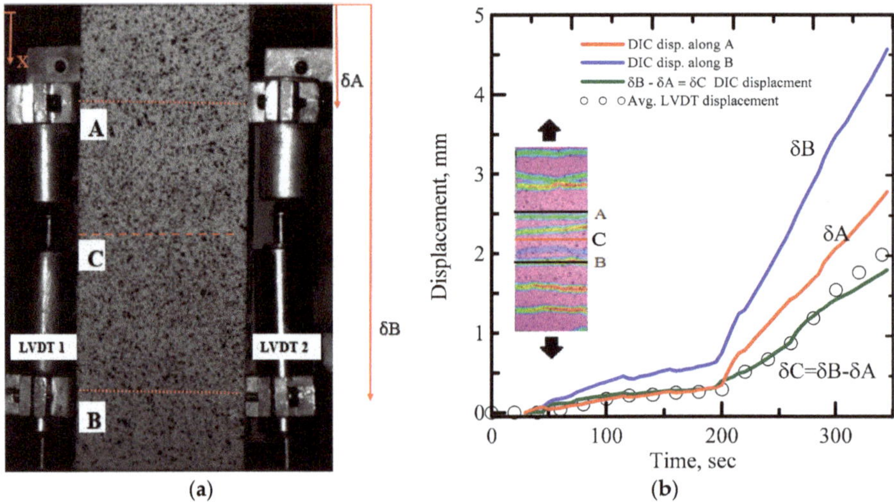

Figure 9. (a) Sample with LVDT mounted on sides. (b) DIC vs. LVDT displacement correlation.

The width of each crack was tracked from initiation to development to saturation stages and represented in Figure 10 showing the sequential formation of nine distributed cracks propagating throughout the width and observed as a function of time of a representative specimen. This data was post-processed to generate the crack width and spacing response up to the failure, as shown in Figure 10a,b representing the contour of longitudinal V displacement versus Y location for a MAC 4% (replicate 1) when all the cracks were formed. Each crack was marked as a discontinuity in displacement field, V(x), along with the Y location of the sample. The displacement discontinuity was measured as the crack width, as shown, and the crack spacing was marked as the distance along coordinate Y between any two cracks, as shown in Figure 10c.

Experimental stress versus time was compared with the crack formation, propagation, and widening, as shown in Figure 11a, using the individual crack openings measured using DIC post-processing. Results indicated that not all cracks were active at any given time during the loading history and the definition of strain may be significantly dependent on the gage length and the specific region of the specimen. Note that some of the cracks formed and then remained dormant before they opened further during subsequent loading stages.

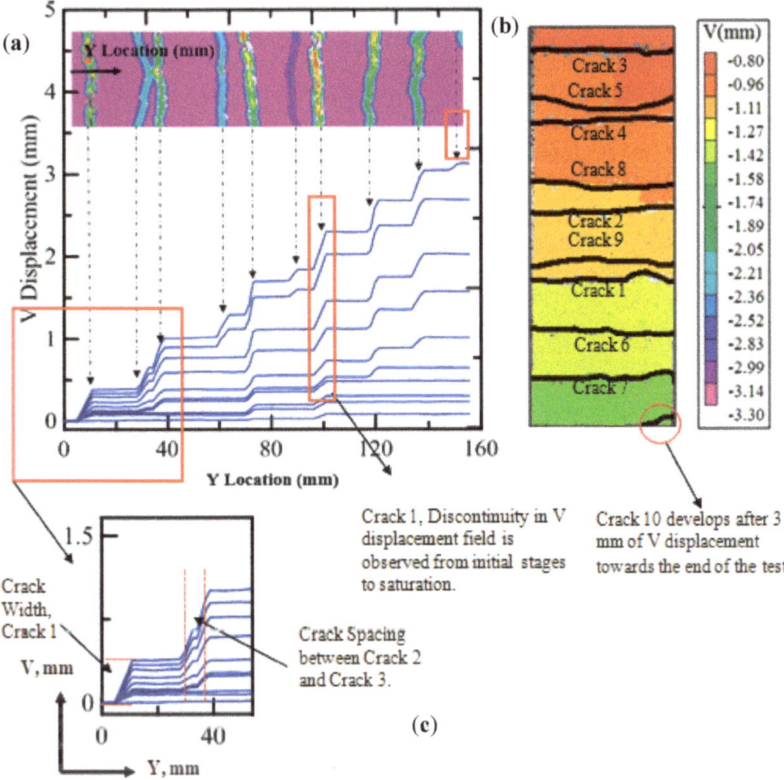

Figure 10. (a) Distribution of longitudinal strain is reported at distinct time steps, (b) DIC V displacement contour showing multiple crack formation at saturation stage, and (c) crack width and crack spacing estimation.

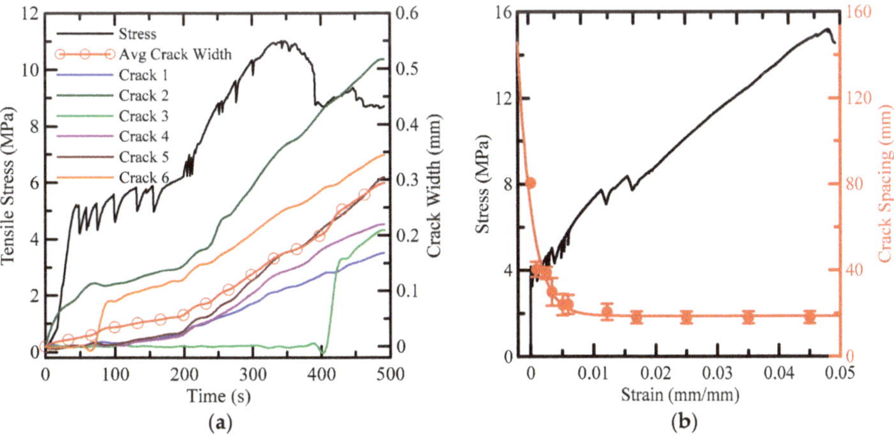

Figure 11. (a) Sequence of formation and individual stress–crack width response for MAC 4%, and (b) tensile stress–strain and crack-spacing response.

As shown in Figure 11a, Cracks 1, 5, and 3 developed early on within the first 100secs. of the test and widened within the time range of 250 to 400 secs. toreach a maximum 0.4 mm opening at the ultimate strength. However, Cracks 6 and 9 openedwhile the sample had reached its maximum stress capacity and the remaining cracks developed at a saturation crack width. The development of new cracks when the sample approachedmaximum stress indicates that multiple cracking stages of the overall composite was ending and the sample response was approaching the saturation stage. After the saturation stage, the majority of the cracks opened uniformly, indicating that the fiber phase was the primary load-carrying component. A stable crack spacing at this point and increased strains resulted in crack widening during the last stage offailure by fiber pullout. The crack spacing was measured as the distance between two cracks, as marked on the contour. At every strain, the number of cracks and their individual spacing was measured. The measure crack was plotted as function of applied strain and compared to tensile stress in Figure 11b. The mean crack spacing from these values indicatedthe damage induced at that point. An increasing strain reducedthe average crack spacing up to the saturation point. Figure 11b shows a saturation crack spacing of 20 mm at 0.015 strain. These data were further processed and shown as the relationship between crack spacing and applied strain.

3.4. Correlation of Fiber Size and Type on Crack Width and Spacing

The correlation of the representative stress–strain response with the distributed cracking on the two continuous fiber composites is shown in Figure 12. The filament structure of MF fibers developed the bond with the cementitious matrix. The interstitial spaces between the multiple filaments were used for penetration of the matrix phase, resulting in a superior mechanical bond and anchorage. The tensile strength exceeded 10 MPa, which was much higher than the mono-filament MAC fibers, which shows limited improvement in performance. A summary of the results of all the MAC and MF at different fiber contents are shown in Figure 13a,b, showing the correlation between the fiber content and crack saturation spacing measured from representative tests. With a tensile strength of about 8 MPa, the first cracking strength of MAC composites was quite similar to the plain matrix. The crack spacing–strain response implies that there were denser cracks with smaller individual crack spacing and lower saturated cracking MF fibers as compared to MAC fibers.

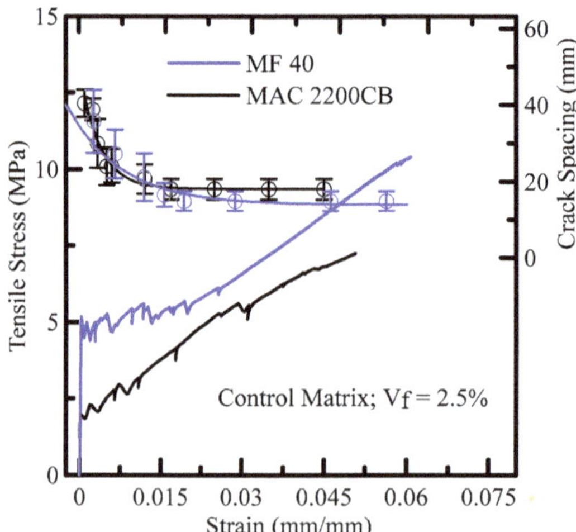

Figure 12. The tensile response of composites with MF40 fiber versus those with MAC fiber.

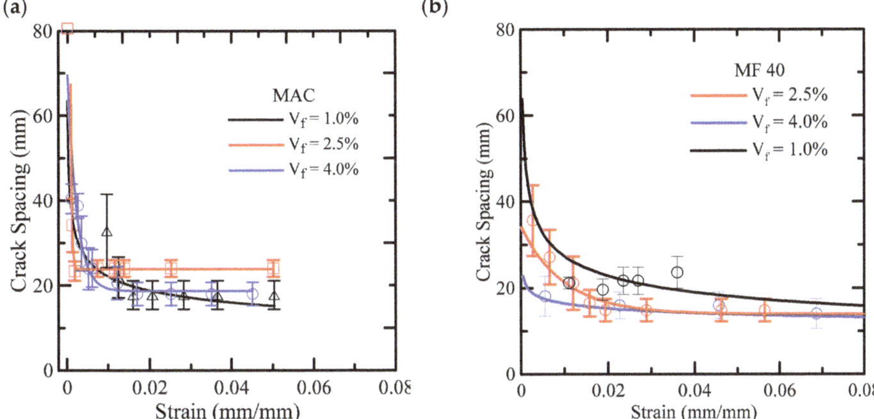

Figure 13. Tensile cracking behavior of representative fiber composites: (**a**,**b**) effect of tensile strain on the crack spacing formation for MAC and MF40 fiber composites, respectively.

The intensity of crack formation increased at higher dosages. The saturation crack spacing reduced from 25 to 15 mm with increasing fiber content, as shown in Figure 13a,b. At a dosage of 1%, the crack saturation was at 3% strain, while at dosages of 2.5% and 4%, new cracks continued to develop at higher strain levels of 6–7%, suggesting localized failure with lower dosages due to fewer cracks, and a dominance of crack-widening mechanisms.

3.5. Optical Microscopy

Toughening mechanisms were also observed by means of optical microscopy. The fiber reinforcement improved the ductility through several mechanisms that includedparallel cracking, crack bridging and deflection, fiber pullout, and fracture. The failure at the fiber–matrix interface was due to the transfer of shear stresses between the two phases, which exceeded the interfacial shear strength.

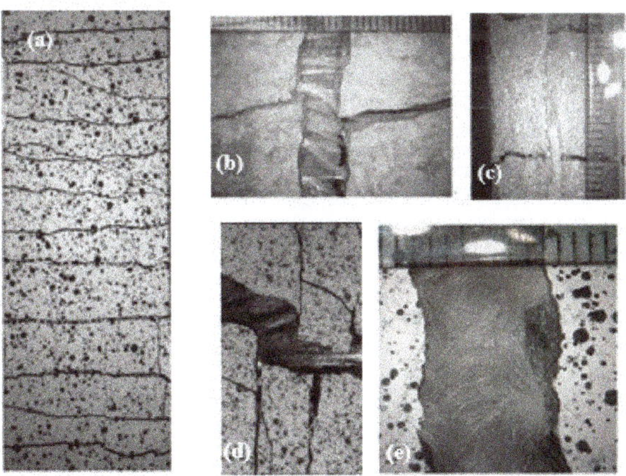

Figure 14. (**a**) Distributed saturation cracking in MAC 4% composites under tension, fiber bridging the crack (**b**) along the width and (**c**) along the thickness, (**d**) MF40 filaments debonding and pull out, and (**e**) MF40 filaments buckling after unloading (scale markers correspond to 1 mm).

The micrographs of multiple cracking are shown in Figure 14a–c for a representative MAC 4% specimen. Crack bridging is a key toughening mechanism within continuous fiber composites, which is shown along the width and thickness in Figure 14c. The fibers prevent significant strains and relaxation of the composite by bridging the distributed cracks and thereby slowing crack propagation, which allows for higher toughness. Figure 14d shows the delamination and pullout of MF40 fibers along with the transverse cracking with respect to the fiber direction. The superior bond exhibited by MF40 ledto fiber fractures accompanied by pullout. Figure 14e shows the crack bridging provided by MF40 fibrillated fibers, which began to buckle due to unloading as the matrix unloadedand compressedthe fibers. Fiber pullout was irreversible, and the final stage of tensile failurewas associated with unloading of the cracks and buckling of bridging fibers.

4. Conclusions

The effect of the macrosynthetic and bundled multifilament polypropylene fiber types on the distributed cracking and tensile stress–strain response of strain hardening cement composites were studied at different volume contents. Results indicate that tensile properties increased considerably with increasing fiber content. While the first cracking and ultimate tensile strength increased by about 200%, the post-crack modulus increased by over 400% as the volume fraction of microfilament micro MF fibers increased from 1–4%. Composites with monofilament macro MAC fibers with the increase in fiber content from 1–4% showed a more gradual increase of 100%, 78%, and 140% for first cracking, ultimate strength, and post-crack modulus, respectively. Comparing the two fiber types at 4% dosage, bundled microfibers exhibited higher first crack strength, ultimate strength, and toughness, which was 51%, 30%, and 65% higher than the mono-filament macro fiber systems. The tensile strength of the two systems compared at an average of 17.3 MPa versus 13.2 MPa for micro and macro fibers respectively. At the low fiber dosages, the performance of the macrofiber was slightly better. The nature of the open space between the multifilament structure of MF fibers allowed for penetration of the matrix and mechanical anchorage of the filaments, thus improving the interface bonding.

Four stages of composite stress–strain response consisting of the linear elastic stage upto the bend over point, the distributed cracking, and the crack widening zones were discussed in detail. The reduction of tensile stiffness during the distributed cracking provided for significant toughening and ductility of the composites. The general decrease in the crack spacing until saturation crack spacing was a key component of the material behavior. Evolution of crack spacing corresponding to the load was measured using quantitative DIC and correlated with the stress–strain response. At low dosages, crack formation was limited, and toughening through crack widening was more dominant. However at higher fiber dosages, especially with multifilament fiber yarns, denser crack distribution capacity and lower saturation crack spacing were observed. This enables better composite action with the cementitious matrix than the macrosynthetic fibers. The proposed structurally efficient, resilient, and durable sections promise to compete with several conventional building materials, such as timber and light gage steel, based sections for lightweight construction and panel applications.

Author Contributions: Conceptualization, B.M. and S.S.; Methodology, V.D. and J.B.; Software, V.D.; Validation, J.B. and H.M. Formal Analysis, B.M.; Investigation, V.D. and J.B.; Writing—Original Draft Preparation, VD.; Writing—Review & Editing, J.B.; Visualization, H.M.; Supervision, B.M. and S.S.; Project Administration, B.M.; Funding Acquisition, B.M.

Funding: This research was partially funded by the BASF corporation.

References

1.　　Peled, A.; Mob asher, B.; Bentur, A. Modern concrete technology. In *Textile Reinforced Concrete*; Taylor and Francis: London, UK, 2017; p. 473.

2. Mechtcherine, V.; Silva, F.A.; Butler, M.; Zhu, D.; Mobasher, B.; Gao, S.L.; Mäder, E. Behaviour of strain-hardening cement-based composites under high strain rates. *J. Adv. Concr. Technol.* **2011**, *9*, 51–62. [CrossRef]

3. Dey, V.; Zani, G.; Colombo, M.; di Prisco, M.; Mobasher, B. Flexural impact response of textile-reinforced aerated concrete sandwich panels. *Mater. Design* **2015**, *86*, 187–197. [CrossRef]

4. Kim, S.W.; Yun, H. Crack-damage mitigation and flexural behavior of flexure-dominant reinforced concrete beams repaired with strain-hardening cement-based composite. *Compos. Part B Eng.* **2011**, *42*, 645–656. [CrossRef]

5. Kim, S.W.; Park, W.S.; Jang, Y.I.; Feo, L.; Yun, H.D. Crack damage mitigation and shear behavior of shear-dominant reinforced concrete beams repaired with strain-hardening cement-based composite. *Compos. Part B Eng.* **2015**, *79*, 6–19. [CrossRef]

6. Esmaeeli, E.; Barros, J.A. Flexural strengthening of RC beams using Hybrid Composite Plate (HCP): Experimental and analytical study. *Compos. Part B Eng.* **2015**, *79*, 604–620. [CrossRef]

7. Yao, Y.; Silva, F.A.; Butler, M.; Mechtcherine, V.; Mobasher, B. Tension stiffening in textile-reinforced concrete under high speed tensile loads. *Cem. Concr. Compos.* **2015**, *64*, 49–61. [CrossRef]

8. Ghasemi, S.; Zohrevand, P.; Mirmiran, A.; Xiao, Y.; Mackie, K. A super lightweight UHPC-HSS deck panel for movable bridges. *Eng. Struct.* **2016**, *113*, 186–193. [CrossRef]

9. Wille, K.; Naaman, A.E.; Parra-Montesinos, G.J. Ultra-high performance concrete with compressive strength exceeding 150 MPa (22 ksi): A simpler way. *ACI Mater. J.* **2011**, *108*, 46–54.

10. Xu, M.; Wille, K. Fracture energy of UHP-FRC under direct tensile loading applied at low strain rates. *Compos. Part B Eng.* **2015**, *80*, 116–125. [CrossRef]

11. Kim, Y.Y.; Lee, B.Y.; Bang, J.W.; Han, B.C.; Feo, L.; Cho, C.G. Flexural performance of reinforced concrete beams strengthened with strain-hardening cementitious composite and high strength reinforcing steel bar. *Compos. Part B Eng.* **2014**, *56*, 512–519. [CrossRef]

12. Wille, K.; Xu, M.; El-Tawil, S.; Naaman, A.E. Dynamic impact factors of strain hardening UHP-FRC under direct tensile loading at low strain rates. *Mater. Struct.* **2016**, *49*, 1351–1365. [CrossRef]

13. Mobasher, B.; Pivacek, A. A filament winding technique for cement based cross-ply laminates. *Cement Concr. Compos.* **1998**, *20*, 405–415. [CrossRef]

14. Mobasher, B.; Peled, A.; Pahilajani, J. Distributed cracking and stiffness degradation in fabric-cement composites. *Mater. Struct.* **2006**, *39*, 317–331. [CrossRef]

15. Mobasher, B. *Mechanics of Fiber and Textile Reinforced Concrete*; CRC Press: Boca Raton, FL, USA, 2011; p. 473, ISBN 9781439806609.

16. Li, C.Y. Mechanical Behavior of Cementitious Composites Reinforced with High Volume Content Of Fibers. Ph.D. Thesis, Arizona State University, Tempe, AZ, USA, May 1995.

17. Pivacek, A.; Haupt, G.J.; Mobasher, B. Cement based cross-ply laminates. *Adv. Cement Based Mater.* **1997**, *6*, 144–152. [CrossRef]

18. BASF–Master Builders Solutions, Technical Document, MasterFiber MAC 2200CB Synthetic Macrofiber with Chemical Bond for Low Deflection Applications. Available online: https://assets.master-builders-solutions.basf.com/en-us/basf-masterfiber-mac-2200-cb-tds.pdf (accessed on 5 May 2019).

19. Peled, A.; Mobasher, B. Tensile behavior of fabric cement-based composites: Pultruded and cast, ASCE. *J. Mater. Civil Eng.* **2007**, *19*, 340–348. [CrossRef]

20. Sutton, M.A.; Wolters, W.J.; Peters, W.H.; Ranson, W.F.; McNeil, S.R. Determination of displacements using an improved digital correlation method. *Image Vision Comput.* **1983**, *1*, 133–139. [CrossRef]

21. Bruck, H.A.; McNeil, S.R.; Sutton, M.A.; Peters, W.H. Digital image correlation using newton-raphson method of partial differential correction. *Exp.Mech.* **1989**, *29*, 261–267. [CrossRef]

22. Destrebecq, J.-F.; Toussaint, E.; Ferrier, E. Analysis of cracks and deformations in a full scale reinforced concrete beam using a digital image correlation technique. *Exp.Mech.* **2011**, *51*, 879–890. [CrossRef]

23. Koerber, H.; Xavier, J.; Camanho, P.P. High strain rate characterisation of unidirectional carbon-epoxy IM7-8552 in transverse compression and in-plane shear using digital image correlation. *Mech. Mater.* **2010**, *42*, 1004–1019. [CrossRef]

24. Gao, G.; Huang, S.; Xia, K.; Li, Z. Application of digital image correlation (DIC) in dynamic notched semi-circular bend (NSCB) tests. *Exp. Mech.* **2015**, *55*, 95–104. [CrossRef]

25. VIC-3D 8 Manual, Correlated Solutions. Available online: http://www.correlatedsolutions.com/supportconte nt/VIC-3D-8-Manual.pdf (accessed on 24 May 2019).
26. Das, S.; Aguayo, M.; Dey, V.; Kachala, R.; Mobasher, B.; Sant, G.; Neithalath, N. The fracture response of blended formulations containing limestone powder: Evaluations using two-parameter fracture model and digital image correlation. *Cem. Concr. Compos.* **2014**, *53*, 316–326. [CrossRef]

 *applied sciences*

Article

Mechanical Behaviour of TRC Composites: Experimental and Analytical Approaches

Marco Carlo Rampini, Giulio Zani *, Matteo Colombo and Marco di Prisco

Department of Civil and Environmental Engineering, Politecnico di Milano, 20133 Milan, Italy; marcocarlo.rampini@polimi.it (M.C.R.); matteo.colombo@polimi.it (M.C.); marco.diprisco@polimi.it (M.d.P.)
* Correspondence: giulio.zani@polimi.it; Tel.: +39-02-2399-8772

Received: 15 March 2019; Accepted: 3 April 2019; Published: 10 April 2019

Featured Application: Optimisation of textile reinforced concrete (TRC) composites for new constructions and for the seismic retrofitting of existing buildings.

Abstract: Textile reinforced concrete (TRC) is a promising high-performance material that has been employed with success in new constructions, as well as a strengthening layer of existing structural components. In this work, we document the optimisation procedure of textile-based composites for new construction and for the seismic retrofitting of under-reinforced concrete elements and masonry buildings. The study, aimed at maximising the material performances avoiding waste of economic resources, was addressed by means of a series of uniaxial tensile tests conducted on a wide set of alkali-resistant (AR) glass fabrics and TRCs. The samples differed in terms of cement-based matrices, embedded textiles and addition of dispersed microfibers. The results highlight the effects of fabric characteristics and introduction of short fibres on the mechanical behaviour, proposing novel comparison parameters based upon the load bearing capacity and the deformation response of the composites. The application of simplified analytical models borrowed from the literature finally revealed the limitations of the available predictive approaches, suggesting future lines of investigation.

Keywords: textile reinforced concrete; TRC; fabric reinforced cementitious mortar; FRCM; glass fabric; high performance concrete; retrofitting; ACK model; stochastic cracking model

1. Introduction

The growing interest in cost-effective solutions for the structural upgrading of existing buildings and infrastructures has gradually oriented research towards the optimisation of high-performance cement-based composites originally conceived for new lightweight constructions. Such materials, known as textile reinforced concretes (TRCs) [1,2] and fabric-reinforced cementitious mortars (FRCMs) in their recent developments [3–5], are generally employed in the form of thin layers and have proven able to significantly enhance the load-bearing and deformation capacities of underperforming structures [6]. Among economy, ease of application, fire safety, durability and compatibility with the hosting substrates, one of the main advantages is the limited increase in the global mass and, hence, the containment of the inertial forces activated during seismic motions.

Considering the large surfaces targeted by the retrofitting interventions, a primary objective is to avoid material wastage; in this sense, there is a major need of guidelines and simplified predictive models that can effectively assist the identification of optimum design solutions. Being the tensile behaviour of TRC strongly influenced by the matrix composition [7,8], the geometrical and chemo-mechanical characteristic of the embedded fabrics [9–11] and the possible presence of short fibres [12], in this paper we aim at further clarifying the involved phenomena, introducing comparison parameters that may represent a starting point for future developments.

The investigation procedures were set in the context of recent legislative initiatives, which resulted in new Italian regulations: (1) the CNR DT-215 [13], National Research Council technical instructions for the design, execution and control of static consolidation interventions through the use of FRCMs; and (2) the mandatory "Guidelines for the identification, qualification and control of fibre-reinforced inorganic-matrix composites, referred to as FRCMs", recently issued by the Supreme Council for Public Works [14]. Against this background, experimental results pertaining to 70 fabric specimens and 72 composites tested in uniaxial tension are presented, discussed and modelled by means of a well-established analytical approach. In particular, the ACK model originally proposed by Aveston, Cooper and Kelly [15,16] and extended to E-glass fibre reinforced Inorganic Phosphate Cement (IPC) by Cuypers and Wastiels [17] was critically assessed, highlighting the merits, the predictive limitations and the room for improvement.

2. Mechanical Characterisation and Selection of Base Materials

FRCM samples were manufactured out of two alternative mix designs: (i) a flowable high-strength micro-concrete for applications comprising temporary formworks; and (ii) a commercial ready-mix thixotropic mortar developed for repairing. One layer of alkali-resistant (AR) glass fabric, alternately selected amongst seven products, was always placed at mid-thickness, by way of a traditional hand lay-up technique. The dispersion of short fibres into the matrix was also explored, with a view to improve energy absorption, control crack openings and guarantee greater structural performances at serviceability. Relevant material properties are described in the following.

2.1. Cement-Based Materials

The compositions of the two cement-based matrices are displayed in Tables 1 and 2. Matrix M1 was a fine-grained self-compacting very high performance concrete (VHPC), characterised by an average cubic compressive strength f_{cc} of 93.55 MPa and a flexural tensile strength f_{ctf} of 14.26 MPa, while matrix M2 was a shrinkage-compensated thixotropic mortar exhibiting an average cubic compressive strength f_{cc} of 58.94 MPa and a flexural tensile strength f_{ctf} of 7.02 MPa (Table 3).

Table 1. Matrix M1 composition.

Component	(kg/m^3)
Cement I 52.5	600
Sand 0–2 mm	976.46
Water	209
Superplasticiser	44
Blast furnace slag	500

Table 2. Matrix M2 composition.

Component	(kg/m^3)
Ready-mix admixture	1840
Water	276
Expansive agent	18.4

Table 3. Bending, tensile and compressive strengths of matrices M1 and M2: discrete values, average values, standard deviations and shape coefficients *m* of the two-parameter Weibull distributions.

Specimen	Matrix M1			Matrix M2		
	f_{ctf} (MPa)	f_{ct} (MPa)	f_{cc} (MPa)	f_{ctf} (MPa)	f_{ct} (MPa)	f_{cc} (MPa)
N1	16.49	7.30	100.42	8.26	3.66	69.08
N2	14.81	6.55	98.36	7.15	3.16	65.20
N3	14.01	6.20	93.86	8.46	3.74	59.94
N4	14.61	6.47	90.66	7.28	3.22	56.77
N5	12.74	5.63	88.97	5.11	2.26	49.52
N6	12.90	5.71	89.01	5.85	2.59	53.12
Average	14.26	6.31	93.55	7.02	3.10	58.94
(std)	(1.39)	-	(4.91)	(1.32)	-	(7.35)
(std%)	(10%)	-	(5%)	(19%)	-	(12%)
m (Weibull)	11.58	-	-	7.15	-	-

Bending and compressive tests were carried out on six nominally identical prismatic specimens ($40 \times 40 \times 160$ mm^3) according to UNI EN 196 [18] and the tensile strengths f_{ct} were deduced from the bending results via the formula proposed in the Model Code 2010 (MC2010) [19]:

$$f_{ct} = f_{ctf} \frac{\alpha \cdot h^{0.7}}{1 + \alpha \cdot h^{0.7}},\tag{1}$$

where *h* is the beam depth (40 mm) and α is a coefficient that decreases as the concrete brittleness increases; on a first approximation, α was taken equal to 0.06 (value referred to normal strength concrete) for both M1 and M2. Elastic moduli were estimated from the average compressive strength (42.9 GPa for M1 according to MC2010), or from data provided by the manufacturer (28 GPa for M2). Concerning the commercial mortar, it is important to underline the discrepancy between the mean flexural results of Table 3 (7.02 MPa) and the average values declared by the producer (10.1 MPa, corresponding to an f_{ctm} of 4.47 MPa), probably caused by a different casting procedure. Beam specimens were in fact manufactured without paying much attention to the material compaction and this resulted in a macroscopic porosity.

2.2. Alkali-Resistant Glass Fabrics

The seven AR-glass fabrics depicted in Figure 1 were selected from a broad set proposed by the manufacturer, following preliminary considerations aimed at ensuring the typical trilinear tensile behaviour of TRC and covering diverse structural interventions, where textile geometry and tensile capacity play a significant role.

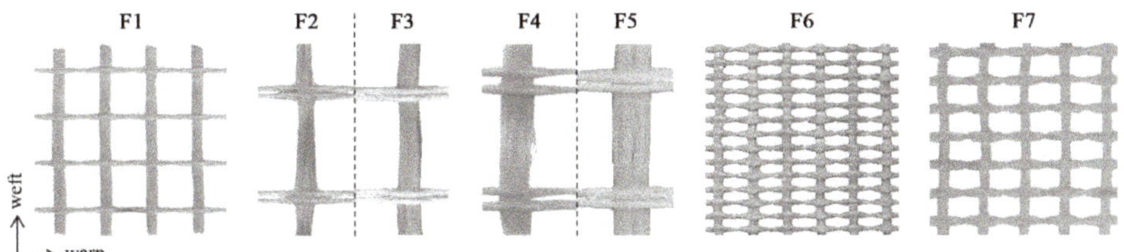

Figure 1. Overview of the investigated alkali-resistant glass fabrics (70×70 mm^2 samples).

The main characteristics of the fabrics are given in Table 4, where it is possible to observe that grid spacings between 5 and 38 mm were explored and both styrene-butadiene rubber (SBR) and epoxy coatings were considered. As displayed in Figure 1, Fabrics F2 and F3 as well as F4 and F5 were identical, with the only exception being the coating nature; on close observation it was possible to detect that slight geometrical alterations were present, since SBR-coated fabrics tended to be more squashed at the manufacturing stage. To the exclusion of Fabrics F6 and F7, the reinforcements were balanced in the warp and in the weft directions, as proved by the equivalent thicknesses t_{eq} collected in Table 4. This geometrical parameter was evaluated according to the following steps: (i) calculation of the equivalent wire section A_w as the ratio between the Tex (g/km) and the glass density (2680 kg/m^3); (ii) calculation of the number of wires n_w over 1 m, according to the spacing; and (iii) calculation of t_{eq} as the ratio between the global area (equivalent wire section times the number of wires $A_w \cdot n_w$) and 1 m width.

Table 4. Alkali-resistant glass fabrics characteristics.

		F1	F2	F3	F4	F5	F6	F7
Fabrication technique		Leano weave	Double leano weave	Double leano weave	Double leano weave	Double leano weave	Leano weave	Leano weave
Coating nature		SBR	Epoxy	SBR	Epoxy	SBR	SBR	SBR
Warp	Wire spacing (mm)	18	38	38	38	38	5	10
	Roving fineness (Tex)	2 × 1200	2 × 1200	2 × 1200	2 × 2400	2 × 2400	2 × 1200	2 × 2400
	Filament diameter (μm)	19	19	19	27	27	19	27
	Equivalent reinforcement thickness* t_{eq} (mm)	0.050	0.046	0.046	0.093	0.093	0.179	0.179
Weft	Wire spacing (mm)	18	38	38	38	38	12	14.3
	Roving fineness (Tex)	2400	2 × 2400	2 × 2400	4 × 2400	4 × 2400	2400	2 × 1200
	Filament diameter (μm)	27	27	27	27	27	27	19
	Equivalent reinforcement thickness* t_{eq} (mm)	0.050	0.046	0.046	0.093	0.093	0.071	0.062

* calculated over a width of 1 m.

For each direction of the reinforcements, five uniaxial tensile tests were performed according to the strip method [20] of Figure 2, obtaining the average peak loads $P_{max,avg}$ given in Table 5. The samples (70 × 400 mm^2 in size) were clamped to an electromechanical press (a clamping force of about 8 kN was applied) with a maximum load capacity of 30 kN and a constant machine crosshead displacement (stroke) rate of 100 mm/min was taken as the feedback parameter. Epoxy resin tabs were preliminary created at the ends of each specimen (see Figure 2), to prevent stress localisation and slippage phenomena within the clamping zones.

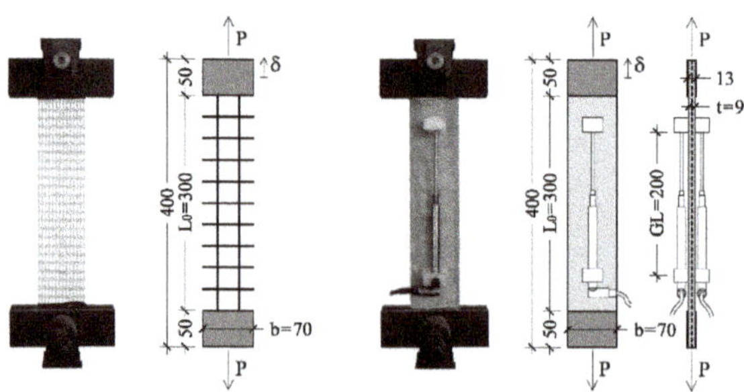

Figure 2. Fabric and textile reinforced concrete (TRC) specimens: uniaxial tensile test setups and nominal geometries (measures expressed in mm).

Referring to the standard deviations (std%) of Table 5, smaller pick distances (Fabrics F6 and F7) were generally associated to smaller dispersion of the mechanical results, in particular in the warp direction. The reason for this was the greater redundancy of the sample, which resulted less exposed to uneven stress distributions due to imprecise fabric geometry or clamping misalignments. The glass areas A_f given in Table 5 represent the cross-sections of the assembled rovings actually embedded within the 70 mm specimen widths, i.e. subjected to the tensile load P; as one should note, A_f does not correspond to the product between the equivalent thickness t_{eq} and 70 mm, since the strip width was generally not an exact multiple of the grid spacing. The fabric efficiency factor EF_f was instead calculated as follows:

$$EF_f = \frac{P_{f,max,avg}}{A_f \cdot \sigma_{fu}} ,$$ (2)

where σ_{fu} is the glass filament strength, assumed equal to 2000 MPa according to the manufacturer data. The parameter EF_f provides crucial information on the rate of utilisation of the reinforcing material and may hence be related to cost-efficiency considerations. As an example, despite Fabric F4 being characterised by a 43% smaller A_f than Fabric F7 (warp direction), the average peak loads were nearly the same, meaning that in Fabric F7 several glass filaments were basically ineffective. This was explained by the greater capability of epoxy resin to impregnate the glass filaments, limiting telescopic failure modes [21].

Table 5. Mechanical properties of the investigated fabrics: average maximum tensile loads and efficiency parameters (70 mm wide specimens).

		F1	F2	F3	F4	F5	F6	F7
	A_f (mm^2)	3.582	3.582	3.582	7.164	7.164	12.537	12.537
	$P_{f,max,avg}$ (kN)	5.72	6.41	6.21	12.50*	11.44*	15.98	12.20*
Warp	(std)	(0.27)	(0.29)	(0.31)	(0.67)	(0.63)	(0.15)	(0.20)
	(std%)	(4.80%)	(4.53%)	(4.92%)	(5.34%)	(5.48%)	(0.93%)	(1.63%)
	EF_f	0.80	0.89	0.87	0.87	0.80	0.64	0.49
	A_f (mm^2)	3.582	3.582	3.582	7.164	7.164	5.373	4.478
	$P_{f,max,avg}$ (kN)	5.36	4.81	5.51	11.70	10.09*	8.69	6.34
Weft	(std)	(0.27)	(0.33)	(0.76)	(0.75)	(0.82)	(0.71)	(0.41)
	(std%)	(5.13%)	(6.86%)	(13.78%)	(6.42%)	(8.09%)	(8.15%)	(6.40%)
	EF_f	0.75	0.67	0.77	0.82	0.70	0.81	0.71

* average of four samples.

Average load vs. stroke displacement (P-δ) and nominal fabric stress vs. normalised displacement (σ_f-δ/L_0) curves are plotted in Figures 3 and 4, respectively; the nominal fabric stress σ_f was calculated as $P_{f,max,avg}/A_f$, while the normalised displacement was obtained as the ratio between the stroke displacement and the free specimen length of Figure 2 (about 300 mm). As clearly shown in Figure 4a, heavy-duty textiles F6 and F7 markedly deviated from Fabrics F1–F5, confirming their lower efficiency. It is also worth noticing that F6 and, to a greater extent, F7 became stiffer as the applied tensile load increased, as a result of a strong fabric "Poisson" effect that is addressed below.

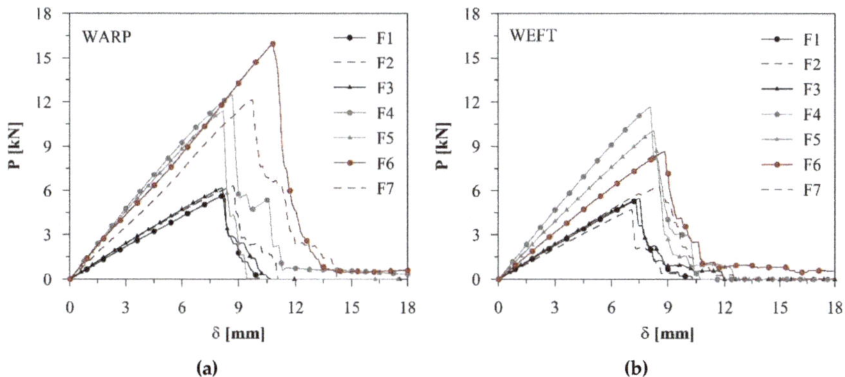

Figure 3. Fabric average tensile responses in terms of load vs. displacement: in the warp direction (**a**); and in the weft direction (**b**).

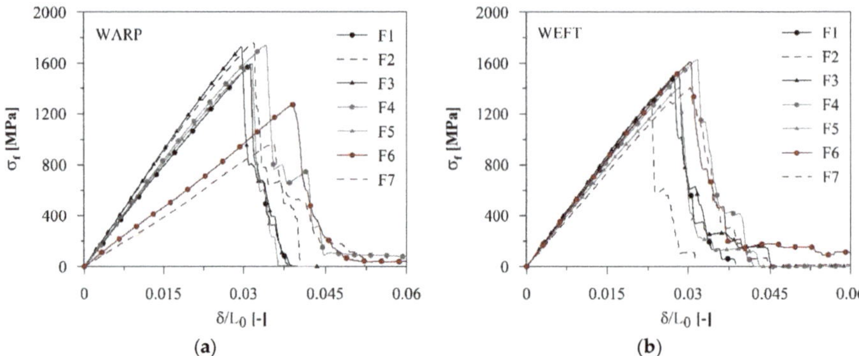

Figure 4. Fabric average tensile responses in terms of fabric stress vs. normalised displacement: in the warp direction (**a**); and in the weft direction (**b**).

2.3. Short Fibres

Hybrid reinforcing solutions were explored by including in the cement-based mortars the short fibres described in Table 6: (i) straight high-carbon steel microfibers (diameter 0.21 mm, length 13 mm, aspect ratio 62, tensile strength 2750 MPa), encoded as S; and (ii) high alkali resistance polyvinyl alcohol fibres (equivalent diameter 0.16–0.24 mm, length 18 mm, aspect ratio 90, yield strength 790–1160 MPa), encoded as PVA. Nominal volume fractions V_{sf} of 1% were considered, although it is fair to note that segregation of steel fibres in the fresh-state VHPC led, in some cases, to effective volume fractions of about 0.5%, as proven by the wash-out of the fibres segregated in the mixer bowl; however, as discussed in the following, no significant changes in the response were observed.

Table 6. Geometrical and mechanical properties of the short fibres (high-carbon steel S and polyvinyl alcohol PVA).

Characteristics	S fibres	PVA fibres
Material	High-carbon steel	Polyvinyl alcohol
Length l_f (mm)	13	18
Diameter d_f (mm)	0.21	0.16–0.24
Aspect ratio l_f/d_f	62	~90
Tensile strength (MPa)	2750	790–1160
Modulus of elasticity (GPa)	200	30

As shown in the literature [12], synergy effects can be achieved thanks to the mechanical stabilisation ensured by the textiles, potentially improving the material strength, the fracture behaviour and the bond between the yarns and the matrix.

3. Mechanical Characterisation of TRC Composites

TRC specimens ($70 \times 400 \times 9$ mm³ in size) were tested in uniaxial tension after at least 28 days of natural curing, according to the scheme in Figure 2. Four steel plates ($70 \times 50 \times 1$ mm³) were glued at the ends, to prevent stress localisations within the clamping regions, where a transverse force of about 12 kN was imposed. The tests were displacement controlled imposing a constant stroke rate of 0.02 mm/s and two linear variable differential transformer (LVDT) transducers measured integral crack opening displacements astride a gauge length (GL) of about 200 mm. Average results of each set of three nominally identical samples are depicted in Figures 5 and 6, in terms of nominal TRC stress ($\sigma_{TRC} = P/A_{TRC}$, with $A_{TRC} = t \cdot b$) vs. normalised displacement (δ/L_0) along the warp reinforcing direction; relevant geometrical and mechanical quantities are collected in Table 7. The curves reflect the typical trilinear behaviour of TRC, consisting of a first linear-elastic branch, a second multiple crack formation phase and a third region dominated by the fabric response [22]; please note that marked discontinuities of the average third-branch responses generally indicate the failure of one of the three nominal identical specimens belonging to the set, after which the results are averaged on the remaining samples. It is hence important to observe that the peaks of the curves are representative of the best performing sample, while the numbers collected in Table 7 are always the average of the three individual values.

Focussing on the final state—theoretically associated to the rupture of the AR-glass textile—it was possible to notice that some samples failed due to a loss of textile/matrix bond, mostly caused by an insufficient anchorage length. In fact, the length of the load introduction zones in Figure 2, albeit in the range of the minimum values prescribed by the Italian guidelines [14], is about half of the one suggested by recent recommendations for TRC tensile tests [23]. Fabric slippage most likely occurred on M2-based composites, since the greater porosity (noticed by visual inspection) of the thixotropic mortar penalised the bond between the rovings and the cementitious phase.

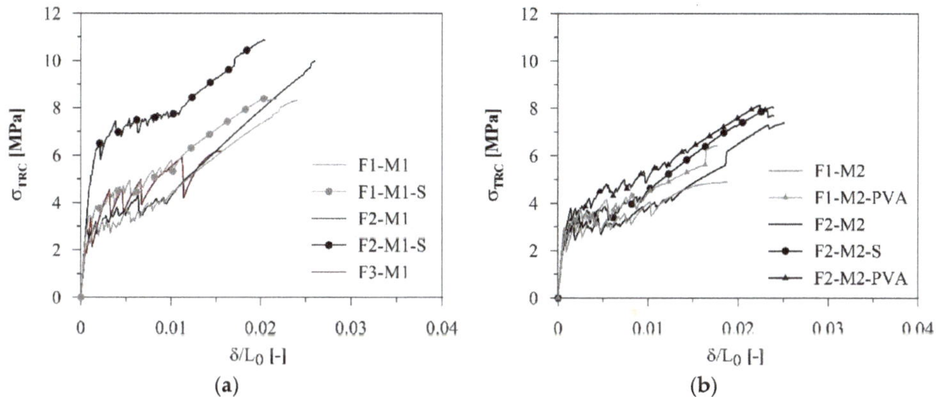

Figure 5. Average tensile response in terms of nominal stress vs. normalised displacement for M1-based (**a**) and M2-based (**b**) composites reinforced with Fabrics F1–F3 (warp direction).

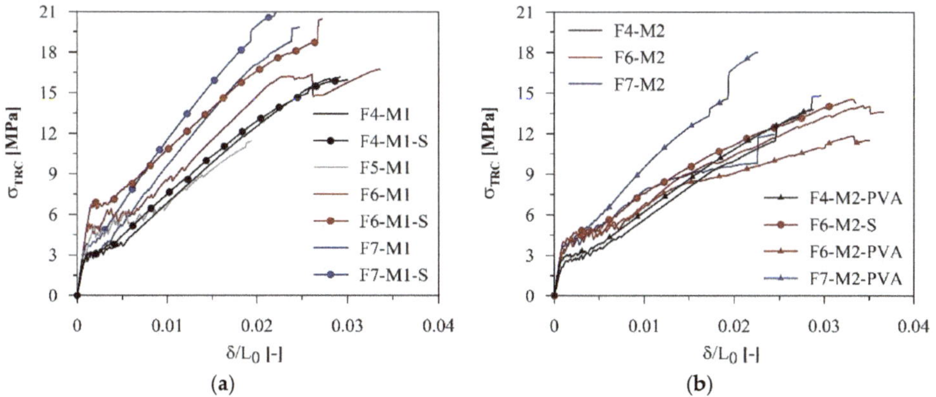

Figure 6. Average tensile response in terms of nominal stress vs. normalised displacement for M1-based (**a**) and M2-based (**b**) composites reinforced with Fabrics F4–F7 (warp direction).

Table 7. Tensile test results for TRC composites: average values (avg.) and standard deviations (std).

		t (mm)	b (mm)	L_0 (mm)	P_{max} (kN)	δ_u (mm)	$\sigma_{TRC,max}$ (MPa)	$\sigma_{TRCf,max}$ (MPa)	δ_u/L_0 (–)	σ_I (MPa)
F1-M1	avg.	8.34	70.66	295.67	4.60	6.66	7.82	1283.82	0.0225	3.35
	(std)	(0.49)	(1.24)	(0.58)	(0.13)	(0.44)	(0.45)	(36.83)	(0.0014)	(0.15)
F1-M1-S	avg.	8.41	70.71	295.00	5.05	6.24	8.49	1408.65	0.0211	2.88
	(std)	(0.17)	(0.76)	(1.00)	(0.16)	(0.16)	(0.30)	(45.14)	(0.0006)	(1.18)
F1-M2	avg.	10.40	70.59	293.00	3.55	5.01	4.84	991.96	0.0171	3.46
	(std)	(0.68)	(0.19)	(1.00)	(0.18)	(0.44)	(0.11)	(49.49)	(0.0015)	(0.49)
F1-M2-PVA	avg.	10.60	70.46	293.33	4.05	4.48	5.46	1131.77	0.0153	2.96
	(std)	(0.60)	(1.03)	(1.15)	(0.38)	(0.92)	(0.85)	(107.46)	(0.0032)	(0.15)
F2-M1	avg.	7.92	70.62	294.00	5.36	7.50	9.59	1497.15	0.0255	3.34
	(std)	(0.05)	(1.12)	(1.00)	(0.14)	(0.20)	(0.39)	(38.75)	(0.0006)	(0.78)
F2-M1-S	avg.	8.87	70.46	292.67	6.08	5.03	9.74	1698.58	0.0172	6.25
	(std)	(0.10)	(0.54)	(1.15)	(0.61)	(0.97)	(1.08)	(169.29)	(0.0033)	(1.53)
F2-M2	avg.	9.99	70.25	293.67	4.53	6.56	6.48	1265.55	0.0223	3.02
	(std)	(0.18)	(0.90)	(0.58)	(1.12)	(0.97)	(1.68)	(313.91)	(0.0033)	(0.39)
F2-M2-S	avg.	9.73	70.45	293.33	5.64	7.24	8.24	1575.79	0.0247	3.23
	(std)	(0.31)	(0.57)	(0.58)	(0.43)	(0.29)	(0.71)	(120.65)	(0.0010)	(0.22)
F2-M2-PVA	avg.	9.81	70.50	294.00	5.70	6.81	8.27	1591.94	0.0232	2.75
	(std)	(0.61)	(1.33)	(1.73)	(0.08)	(0.23)	(0.54)	(23.05)	(0.0008)	(0.46)
F3-M1	avg.	8.95	70.45	294.33	3.75	2.78	5.96	1047.76	0.0094	4.64
	(std)	(0.41)	(0.19)	(1.53)	(0.19)	(2.22)	(0.48)	(52.64)	(0.0075)	(1.93)
F4-M1	avg.	8.89	70.66	293.33	9.84	8.04	15.67	1373.17	0.0274	2.97
	(std)	(0.29)	(0.57)	(0.58)	(0.68)	(0.63)	(0.93)	(94.33)	(0.0022)	(0.43)
F4-M1-S	avg.	9.11	70.61	293.33	9.80	7.99	15.24	1368.26	0.0272	2.96
	(std)	(0.10)	(1.00)	(2.08)	(0.56)	(0.76)	(0.74)	(78.02)	(0.0024)	(0.17)
F4-M2	avg.	10.86	70.61	294.33	9.13	7.65	11.91	1274.61	0.0260	2.53
	(std)	(0.15)	(1.01)	(0.58)	(1.34)	(0.73)	(1.84)	(187.71)	(0.0024)	(0.36)
F4-M2-PVA	avg.	11.17	70.56	292.33	10.61	8.32	13.51	1481.42	0.0285	2.99
	(std)	(0.51)	(0.63)	(1.53)	(0.31)	(0.38)	(1.13)	(43.17)	(0.0012)	(0.26)
F5-M1	avg.	8.77	70.58	294.00	6.46	5.12	10.45	902.39	0.0174	3.86
	(std)	(0.12)	(0.89)	(1.00)	(0.82)	(0.88)	(1.15)	(114.31)	(0.0030)	(0.38)
F6-M1	avg.	9.01	69.70	295.00	10.78	8.25	17.15	859.64	0.0280	5.23
	(std)	(0.34)	(0.65)	(1.73)	(0.60)	(1.43)	(0.78)	(48.21)	(0.0050)	(1.33)
F6-M1-S	avg.	8.83	70.58	295.33	11.56	7.63	18.58	921.68	0.0258	6.07
	(std)	(0.44)	(1.05)	(0.58)	(0.77)	(0.60)	(1.65)	(61.05)	(0.0021)	(0.76)
F6-M2	avg.	9.51	70.29	293.67	9.50	10.39	14.22	757.92	0.0354	2.84
	(std)	(0.17)	(0.54)	(3.51)	(0.35)	(0.41)	(0.51)	(27.62)	(0.0012)	(0.36)
F6-M2-S	avg.	9.67	70.58	295.00	10.06	10.00	14.73	802.07	0.0339	4.15
	(std)	(0.10)	(0.76)	(1.00)	(0.40)	(0.25)	(0.41)	(32.27)	(0.0008)	(0.44)
F6-M2-PVA	avg.	10.99	70.46	294.67	8.87	9.67	11.46	707.14	0.0328	4.11
	(std)	(0.42)	(0.56)	(1.53)	(0.62)	(0.69)	(0.90)	(49.46)	(0.0025)	(0.19)
F7-M1	avg.	8.55	70.43	294.00	11.05	6.86	18.38	881.37	0.0233	2.54
	(std)	(0.68)	(0.33)	(0.00)	(0.96)	(0.53)	(1.28)	(76.22)	(0.0018)	(0.85)
F7-M1-S	avg.	8.54	70.53	295.33	11.76	6.26	19.54	938.03	0.0212	3.77
	(std)	(0.54)	(0.63)	(0.58)	(1.33)	(0.48)	(1.99)	(105.96)	(0.0016)	(1.20)
F7-M2	avg.	10.62	69.96	293.00	6.91	5.63	9.31	550.79	0.0192	3.83
	(std)	(0.20)	(0.44)	(1.00)	(1.62)	(2.17)	(2.28)	(128.93)	(0.0074)	(0.48)
F7-M2-PVA	avg.	10.32	70.13	294.00	10.32	5.79	14.38	823.55	0.0197	3.30
	(std)	(0.65)	(1.23)	(1.00)	(1.73)	(0.77)	(3.15)	(138.05)	(0.0027)	(0.42)

4. Discussion of the Results

4.1. Effect of Fabric Coating

As already introduced, the nature of the coating acts on both the efficiency of the AR-glass fabric and the strength of the composite. This is explained by the greater surface roughness observed in epoxy-impregnated textiles, which increases the adhesion with the surrounding cementitious matrix, and by the greater mechanical anchorage offered by the weft yarns when the specimen is loaded along the warp direction. This phenomenon is clearly correlated to the better ability of epoxy resins to penetrate and impregnate the glass filaments, stiffening the nodal connections between weaved strands.

The latter effect assumes greater importance in the case of wide-spaced fabrics because, contrary to narrow-spaced textiles, the number of nodal connections within the sample width is limited and, in some combinations, the TRC is unable to exhibit the intended trilinear response [22].

To better highlight the effect of the coating nature, M1-based composites alternately reinforced with Fabrics F2 and F3 as well as F4 and F5 Were considered. Such couples of textiles were characterised by the same geometry, being the only difference the impregnation technique: SBR for F3 and F5 and epoxy for F2 and F4. In Figures 7 and 8, the comparison between the two average responses in terms of nominal stress vs. normalised displacement corroborates the mentioned explanations, since a general increase of the mechanical capacity amid epoxy-impregnated textiles can be observed. The use of epoxy coating also entailed an increased number of cracks and a more stable multi-cracking phase (stress drops appear smoother). Moreover, it is worth underlining that the flatter geometry of SBR coated textiles, explained by the greater pressure exerted by the impregnation rollers, led to an increase of the nominal cracking stresses (given a constant composite thickness, the effective cross-section of the mortar is bigger).

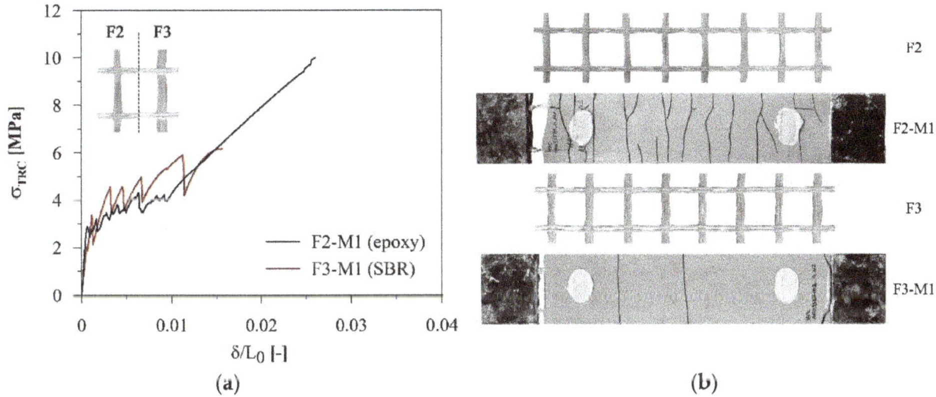

Figure 7. Effect of different coatings on the uniaxial tensile response (**a**); and on the cracking pattern (**b**) of M1-based composites reinforced with geometrically identical F2 and F3 fabrics.

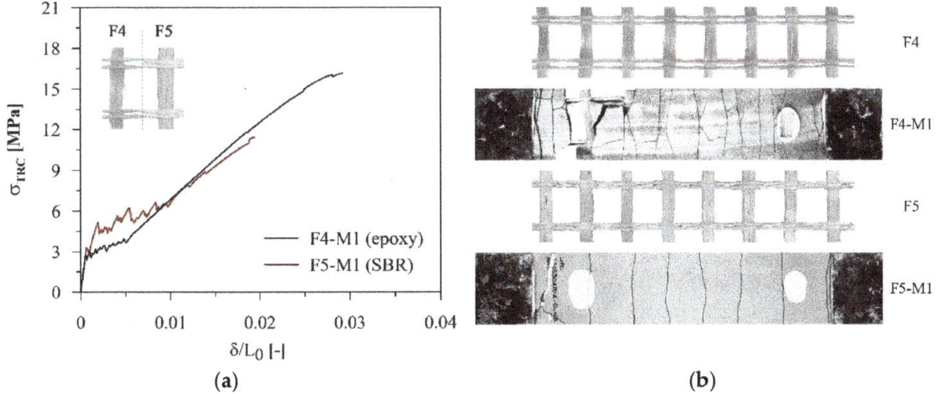

Figure 8. Effect of different coatings on the uniaxial tensile response (**a**); and on the cracking pattern (**b**) of M1-based composites reinforced with geometrically identical F4 and F5 fabrics.

4.2. Effect of Dispersed Short Fibres

The addition of short fibres impacts on the mechanical performances of TRC composites by increasing the first cracking stress, the nominal stress in the second and third branches—due to the improvement of the bond between the textile and the matrix—and the number of cracks, hence ensuring a better behaviour in terms of durability [12].

Average tensile responses of different composites reinforced with Fabrics F2 and F6, with or without the addition of short steel and PVA fibres, are displayed in Figure 9. It was possible to notice an overall improvement of the mechanical capacity for all types of matrices and added fibres, with the exception of F6-M2-PVA composites, where the benefits on the mechanical response were not visible, probably because of a combined effect of the narrow-spaced grid of the textile and the porosity of the matrix (in this case the addition of fibres may increase the number of defects, reducing bond and promoting early fabric slippage). The effect of hybrid reinforcing technologies is also assessed in Figure 10, where a densification of the cracking patterns can be generally observed.

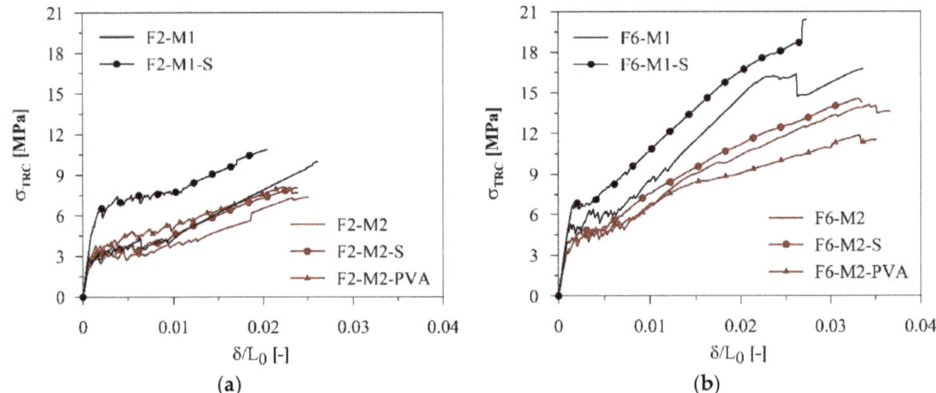

Figure 9. Effect of short fibres addition on the uniaxial tensile response of TRC composites reinforced with Fabrics: F2 (**a**); and F6 (**b**).

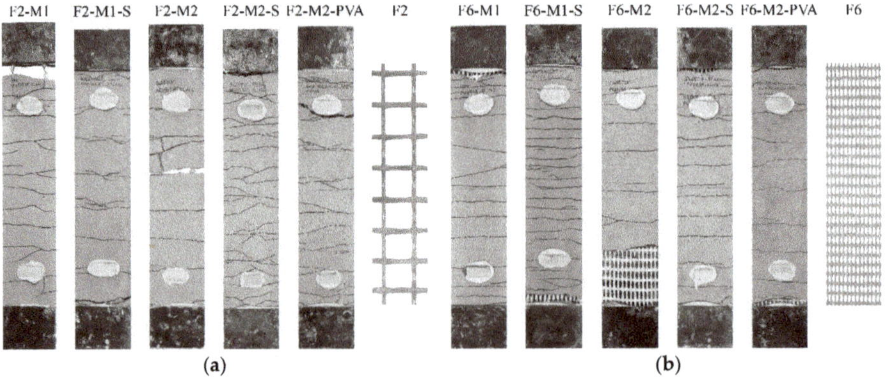

Figure 10. Cracking patterns with and without short fibres for TRC composites reinforced with Fabrics: F2 (**a**); and F6 (**b**).

4.3. Definition of Comparison Parameters

In light of the foregoing evidence on the influence of each component (fabric, coating, matrix and dispersed short fibres) on the TRC tensile behaviour, it was deemed necessary to establish a quantitative comparison approach between different responses to validate the observed trends in the perspective to promote the use of such materials among designers. The following considerations were made after excluding from the whole set the composites manufactured with Fabrics F3 and F5, on account of the observations drawn in Section 4.1.

Two different families of values were introduced: (i) efficiency factors (*EF*), which quantify the global load capacity of the systems and the exploitation of the AR-glass fibres; and (ii) ductility/energy absorption parameters, which describe the behaviour of a composite system in the multi-cracking phase (regarded as significant, since in seismic retrofitting applications hysteretic energy dissipation is expected to occur mainly in this region). Ductility coefficients may also help to better underline the effect of hybrid reinforcing solutions.

In addition to the standard efficiency factor, EF_{TRC}, defined as the ratio between the maximum capacity of the composite and the one of the plain fabric (see Equation (3) and Figure 11a), a second factor $EF_{TRC,f}$ was introduced (see Equation (4) and Figure 11a), obtained dividing the ultimate TRC stress referred to the fabric equivalent section, $\sigma_{TRC,f}$, by the glass filaments strength σ_{fu}:

$$EF_{TRC} = \frac{P_{TRC,max}}{P_{f,max}} = \frac{\sigma_{TRC,f,max}}{\sigma_{f,max}} , \tag{3}$$

$$EF_{TRC,f} = \frac{\sigma_{TRC,f,max}}{\sigma_{fu}} = \frac{\sigma_{TRC,f,max}}{2000 \text{ MPa}} . \tag{4}$$

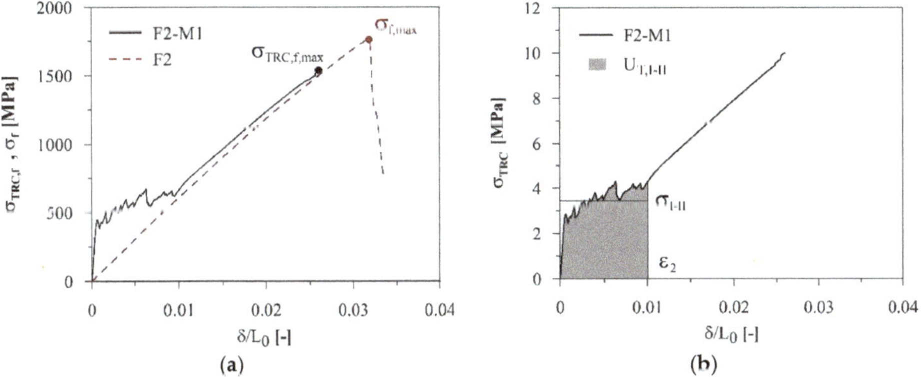

Figure 11. Identification of relevant mechanical parameters on the uniaxial tensile response: efficiency (a) and ductility/energy absorption (b) variables.

The ductility/energy absorption parameters depicted in Figure 11b are defined as: (i) the value of normalised displacement $\varepsilon_2 = \delta_2/L_0$ corresponding to the end of the multi-cracking phase; (ii) the total absorbed energy per unit volume $U_{T,I-II}$ within the equivalent strain range 0-ε_2; and (iii) the average stress σ_{I-II} (Equation (5)) obtained by imposing the equivalence of the area under the stress–strain curve in the first two stages of the response:

$$\sigma_{I-II} = \frac{U_{T,I-II}}{\varepsilon_2} . \tag{5}$$

The introduced parameters, averaged over each nominally identical set, are collected in Table 8.

Table 8. Relevant mechanical parameters for TRC composites under uniaxial tension: average values (avg.) and standard deviations (std).

		$\sigma_{f,max}$ (MPa)	$\sigma_{TRC,f,max}$ (MPa)	EF_{TRC} (-)	$EF_{TRC,f}$ (-)	$U_{T,I\text{-}II}$ (J/m^3)	ε_2 (-)	$\sigma_{I\text{-}II}$ (MPa)
F1-M1	avg.	1596.87	1283.82	0.80	0.64	0.0283	0.0089	3.13
	(std)	(76.62)	(36.83)	(0.02)	(0.02)	(0.0079)	(0.0015)	(0.39)
F1-M1-S	avg.	1596.87	1408.65	0.88	0.70	0.0421	0.0097	4.30
	(std)	(76.62)	(45.14)	(0.03)	(0.02)	(0.0114)	(0.0023)	(0.21)
F1-M2	avg.	1596.87	991.96	0.62	0.50	0.0357	0.0114	3.13
	(std)	(76.62)	(49.49)	(0.03)	(0.02)	(0.0057)	(0.0014)	(0.17)
F1-M2-PVA	avg.	1596.87	1131.77	0.71	0.57	0.0345	0.0098	3.49
	(std)	(76.62)	(107.46)	(0.07)	(0.05)	(0.0051)	(0.0005)	(0.36)
F2-M1	avg.	1789.50	1497.15	0.84	0.75	0.0358	0.0101	3.43
	(std)	(81.01)	(38.75)	(0.02)	(0.02)	(0.0139)	(0.0019)	(0.84)
F2-M1-S	avg.	1789.50	1698.58	0.95	0.85	0.0786	0.0116	6.79
	(std)	(81.01)	(169.29)	(0.09)	(0.08)	(0.0052)	(0.0004)	(0.31)
F2-M2	avg.	1789.50	1265.55	0.71	0.63	0.0302	0.0100	2.96
	(std)	(81.01)	(313.91)	(0.18)	(0.16)	(0.0126)	(0.0035)	(0.18)
F2-M2-S	avg.	1789.50	1575.79	0.88	0.79	0.0329	0.0096	3.43
	(std)	(81.01)	(120.65)	(0.07)	(0.06)	(0.0047)	(0.0011)	(0.08)
F2-M2-PVA	avg.	1789.50	1591.94	0.89	0.80	0.0466	0.0110	4.24
	(std)	(81.01)	(23.05)	(0.01)	(0.01)	(0.0088)	(0.0018)	(0.22)
F4-M1	avg.	1744.44	1373.17	0.79	0.69	0.0167	0.0054	3.09
	(std)	(93.24)	(94.33)	(0.05)	(0.05)	(0.0034)	(0.0007)	(0.25)
F4-M1-S	avg.	1744.44	1368.26	0.78	0.68	0.0124	0.0043	2.90
	(std)	(93.24)	(78.02)	(0.04)	(0.04)	(0.0030)	(0.0008)	(0.31)
F4-M2	avg.	1744.44	1274.61	0.73	0.64	0.0189	0.0066	2.84
	(std)	(93.24)	(187.71)	(0.11)	(0.09)	(0.0053)	(0.0012)	(0.27)
F4-M2-PVA	avg.	1744.44	1481.42	0.85	0.74	0.0166	0.0055	3.00
	(std)	(93.24)	(43.17)	(0.02)	(0.02)	(0.0030)	(0.0005)	(0.29)
F6-M1	avg.	1274.63	859.64	0.67	0.43	0.0666	0.0107	6.02
	(std)	(11.80)	(48.21)	(0.04)	(0.02)	(0.0300)	(0.0027)	(1.21)
F6-M1-S	avg.	1274.63	921.68	0.72	0.46	0.0394	0.0061	6.19
	(std)	(11.80)	(61.05)	(0.05)	(0.03)	(0.0231)	(0.0023)	(1.01)
F6-M2	avg.	1274.63	757.92	0.59	0.38	0.0433	0.0095	4.53
	(std)	(11.80)	(27.62)	(0.02)	(0.01)	(0.0107)	(0.0017)	(0.32)
F6-M2-S	avg.	1274.63	802.07	0.63	0.40	0.0304	0.0068	4.45
	(std)	(11.80)	(32.27)	(0.03)	(0.02)	(0.0036)	(0.0007)	(0.06)
F6-M2-PVA	avg.	1274.63	707.14	0.55	0.35	0.0356	0.0082	4.33
	(std)	(11.80)	(49.46)	(0.04)	(0.02)	(0.0073)	(0.0012)	(0.25)
F7-M1	avg.	972.32	881.37	0.91	0.44	0.0320	0.0073	4.27
	(std)	(15.90)	(76.22)	(0.08)	(0.04)	(0.0135)	(0.0019)	(0.92)
F7-M1-S	avg.	972.32	938.03	0.96	0.47	0.0109	0.0031	3.46
	(std)	(15.90)	(105.96)	(0.11)	(0.05)	(0.0035)	(0.0002)	(0.84)
F7-M2	avg.	972.32	550.79	0.57	0.28	0.0358	0.0079	4.47
	(std)	(15.90)	(128.93)	(0.13)	(0.06)	(0.0100)	(0.0014)	(0.49)
F7-M2-PVA	avg.	972.32	823.55	0.85	0.41	0.0251	0.0057	4.36
	(std)	(15.90)	(138.05)	(0.14)	(0.07)	(0.0046)	(0.0008)	(0.31)

Looking at the efficiency factors trends plotted in Figure 12, it was possible observe that:

- A more loosely packed matrix (i.e., M2) generally implied lower EF_{TRC} (solid black dots) in the composites, thus confirming the previously drawn conclusion about the internal fabric slippage in M2-based systems (see Section 3).
- The values of $EF_{TRC,f}$ (grey crosses) were lower in the composites reinforced with greater amounts of AR-glass (F6 and F7, compared with F1, F2 and F4) and this clearly indicated a material waste, while EF_{TRC} (solid black dots) was only representative of the interaction between matrix and

fabric. Focussing on F6-M1 and F7-M1 systems, it was possible to observe that close $EF_{TRC,f}$ values did not imply similar EF_{TRC} values, probably due to different internal slippage (see Figure 6a).

- The addition of short fibres implied, as expected, a general increase of the EF_{TRC} parameters. The only exception was represented by F6-M2-PVA samples, possibly because the addition of short fibres in a matrix embedding a narrow-grid textile introduced a defect at the fabric–matrix interface, favouring internal slippage. In addition, it is important to stress that the higher porosity of matrix M2 entailed a lower fibre pull-out strength; moreover, the increment of EF_{TRC} was higher in the cases with low-grammage textiles (F1 and F2) because, given the fixed volume fraction of short fibres, the percentage increase of the reinforcement ratio was higher.

- The composites manufactured with epoxy-impregnated fabrics (F2–F4) were more efficient, probably due to a greater chemo-mechanical bond (greater surface roughness combined with stiffer nodal connections between warp and weft yarns, as discussed in Section 4.1). Moreover, the two composite efficiency factors (EF_{TRC} and $EF_{TRC,f}$) were closer, due to the greater efficiency of the plain fabric (see the warp EF_f values of Table 5).

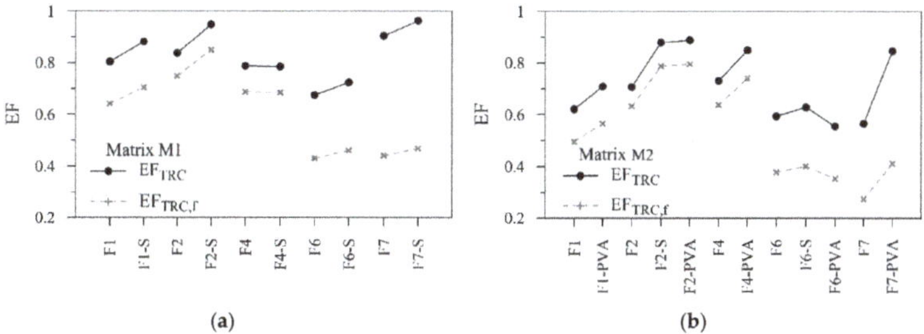

Figure 12. Efficiency factors in uniaxial tension for M1-based (**a**) and M2-based (**b**) TRC composites.

Figure 13 reveals the effect of short fibres addition on the ductility/energy absorption parameters. Normalising each average value with respect to the corresponding plain TRC one, was noticed that:

- In the case of low to medium grammage fabrics (F1 and F2), a general increase of the average stress σ_{I-II} was observed. Moreover, greater control over damage development was achieved. An overall increase of the total absorbed energy was observed, in particular in M1-based composites, confirming preliminary considerations.

- In the case of heavy-duty fabrics (F6 and F7, already characterised by a great reinforcement volume fraction), the responses became stiffer and showed a reduction of ε_2 and $U_{T,I-II}$; consequently, the improvement of the mechanical capacity could only be assessed through the efficiency factors.

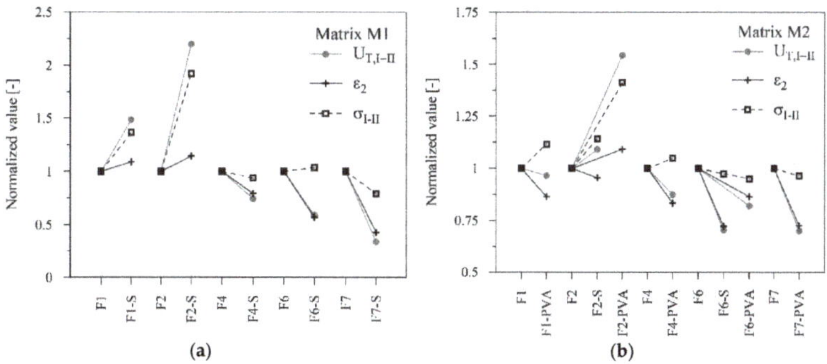

Figure 13. Ductility/energy absorption parameters in uniaxial tension for M1-based (**a**) and M2-based (**b**) TRC composites: effect of fibres addition.

5. Analytical Modelling

As proven by empirical results, variations in the tensile response of TRC and hybrid composites depend upon the characteristics of base materials (fabrics, coatings, matrices and dispersed fibres) and their positive or negative interactions when combined into a complex system. The evaluation of the tensile capacity of TRC is generally assessed experimentally, by performing uniaxial tensile tests that represent the basis of current engineering approaches [23] for the structural design of building interventions.

In this context, there is a strong need to implement available analytical tools into robust predictive models aimed at assisting the experimental optimisation procedures. The following subsections deal with the modelling of the macro-mechanical tensile responses by means of well-established simplified approaches, critically assessed in the perspective of highlighting future lines of investigation.

5.1. Description of the ACK and the Stochastic Cracking Models

The Aveston, Cooper and Kelly theory [15,16] for fibre-reinforced composites was originally developed to describe the tensile stress–strain curve of brittle matrix composites, reinforced with quasi-unidirectional fibres characterised by a volume fraction greater than the critical one. The response (see Figure 14) is described by means of a three stages curve: pre-cracking zone (Stage I), multiple cracking zone (Stage II) and post-cracking zone (Stage III). The basic assumptions of this model are:

- The fibres are considered capable of carrying loads only along their longitudinal axis.
- The matrix–fibre bond is assumed to be weak.
- Once the matrix and the fibres debond, a pure frictional shear stress τ_0 is considered. This shear stress is assumed constant along the debonded interface.
- Poisson effects (transverse contraction) of both the fibres and the matrix are neglected.
- Global load sharing is used for fibres.
- In a section orthogonal to the applied load, matrix normal stresses are considered uniform.

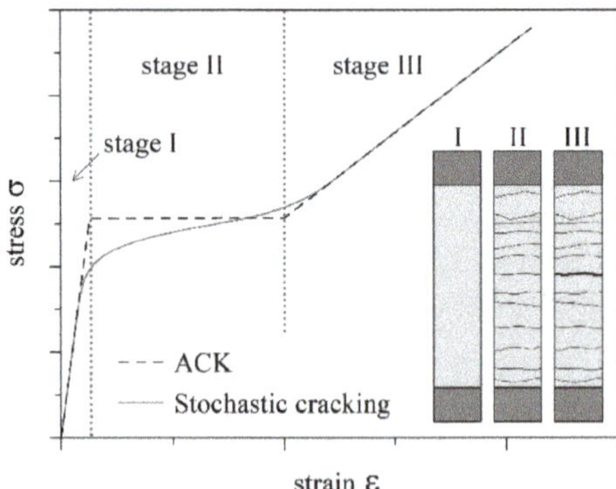

Figure 14. Typical stages of the tensile response in the ACK and Stochastic cracking models and illustrative cracking patterns.

The first branch of the response is linear elastic and the bond between the matrix and the fibres is assumed to be perfect. In this stage, the stiffness of the composite $E_{c,I}$ is function of the fibres and the matrix volume fractions (V_f, V_m) and their stiffness (E_f, E_m) and it is computed by means of the well-known rule of mixtures, via the following equation:

$$E_{c,I} = E_m V_m + E_f V_f .\tag{6}$$

The two volume fractions can be decreased by means of fibres and matrix efficiency factors (η_f, η_m) connected to imperfect matrix–fibre adhesion, warping or misalignment of the unidirectional fibres and eventual presence of inclusions or air voids in the matrix [24]. The modification of the ACK theory is described as follows:

$$E_{c,I} = E_m V_m^* + E_f V_f^* \tag{7}$$

$$V_f^* = V_f \cdot \eta_f \tag{8}$$

$$V_m^* = V_m \cdot \eta_m . \tag{9}$$

According to the ACK model, the matrix has a deterministic tensile failure stress σ_{mu}. When this value is reached, the composite shows multiple cracking (Stage II). The composite multiple cracking stress σ_{mc} is computed as:

$$\sigma_{mc} = \frac{E_{c1} \cdot \sigma_{mu}}{E_m} . \tag{10}$$

When the first crack appears, matrix and fibres debond and the pure constant frictional shear stress is considered, τ_0 provides the stress transfer between fibres and matrix. From the equilibrium in the longitudinal direction, it is possible to obtain the transmission length δ_0, equal to the distance from a crack face at which the matrix reaches again the stress σ_{mu}:

$$\delta_0 = \frac{V_m \, r \, \sigma_{mu}}{V_f \, 2 \, \tau_0} . \tag{11}$$

where r is the equivalent fibres radius. Since, according to the ACK theory, the matrix has a unique tensile failure stress, multiple parallel cracks are simultaneously introduced in the specimen, until saturation is reached. In Stage II, the internal stress leads to a final state where the distance among cracks is between δ_0 and $2\delta_0$ (with an average value equal to X = 1.337·δ_0 [25]).

In Stage III, when the multi-cracking phenomenon is over, only fibres contribute to the stiffness of the composite:

$$E_{c,III} = E_f \cdot V_f^* . \tag{12}$$

An extension of the original ACK theory to E-glass fibre reinforced cementitious composites was proposed by Cuypers and Wastiels under the name of stochastic cracking model [17]. They considered the stochastic nature of the tensile strength of the matrix, through the use of a two-parameter Weibull distribution function [26] and they assumed the same distribution for the crack spacing. Because of the non-deterministic nature of the matrix cracking stress, it is necessary to underline the difference between the mean saturated crack distance at the end of Stage II (X) and the mean actual crack spacing (x). While the X value can be evaluated experimentally, observing the crack pattern at the end of a tensile test, both the values of the transmission length δ_0 and the actual crack distance x are functions of the applied stress. Depending on the value of the stress σ_c applied to the composite, x may be smaller than, equal to or larger than $2\delta_0$. In particular, x tends to X according to the following formula:

$$x = X \left\{ 1 - \exp\left[-\left(\frac{\sigma_c}{\sigma_{Rc}} \right)^m \right] \right\}^{-1} , \tag{13}$$

where σ_{Rc} is the reference average cracking composite stress, computed according to Equation (10). The stress σ_{mu} is taken as the average tensile strength f_{ctm} of Table 3 and m is the width of the strength distribution (respectively, the first and the second parameter of the Weibull function, see Table 3). The analytical stress–strain curve of the composite (see Figure 14) is described by means of the following non-linear functions:

$$\varepsilon_c = \begin{cases} \frac{\sigma_c}{E_{c1}} \left(1 + \frac{\beta \cdot \delta^*}{x} \right), & x \geq 2\delta_0 \\ \sigma_c \left(\frac{1}{E_f V_f^*} - \frac{\beta \cdot x}{4\delta^* \cdot E_{c1}} \right), & x < 2\delta_0 \end{cases} \tag{14}$$

with

$$\beta = \frac{E_m V_m^*}{E_f V_f^*} , \tag{15}$$

and where δ^* is evaluated according to Equation (11), with the evolving matrix stress in place of σ_{mu}.

5.2. Implementation of the Stochastic Cracking Model

To apply the stochastic cracking model to the set of TRC composites and compare the obtained analytical responses with the experimental results along the warp direction, some assumptions regarding the base materials and the efficiency of the composites were made.

As regards the failure strength f_{ctm} of the two matrices, different approaches were considered, following the observations of Section 2.1: (i) average values of the f_{ct} tensile strengths, computed with Equation (1); (ii) average tensile strengths empirically derived, according to the MC2010, from the average compressive strengths; and (iii) data provided by the premixed mortar manufacturer (the obtained values are collected in Table 9).

Table 9. Average matrix failure strengths f_{ctm}.

Approach	Matrix M1	Matrix M2
(i) from bending tests	6.31 MPa	3.10 MPa
(ii) from compressive tests	4.59 MPa	3.46 MPa
(iii) manufacturer data	n.a.	4.47 MPa

Analyses "A" (see Section 5.3) considered the average strengths calculated from the three-point bending flexural strengths (f_{ctm} = 6.31 MPa for M1 and f_{ctm} = 3.10 MPa for M2), while Analyses "B" instead assumed f_{ctm} = 4.59 MPa for M1 and f_{ctm} = 4.47 MPa for M2; the former value is the one obtained from the average compressive strength, while the latter is the one provided by the thixotropic mortar manufacturer. Regarding M2, the selection of the highest tensile strength was consistent with a greater compaction of the mortar (as resulting in the composites, produced by means of a hand lay-up technique) and the stabilising effect offered by polyacrylonitrile-based microfibres originally introduced in the pre-mixed powder to minimise shrinkage effects.

For both analyses, the matrix elastic modulus E_m and the Weibull shape coefficient m were assumed equal to 42.9 GPa and 11.58 for the M1-based composites and 28.0 GPa and 7.15 for the M2-based ones, respectively (please note that the m Weibull parameter associated to the tensile strength distribution was hence assumed equal to the experimental flexural one). The efficiency of the cement-based matrix, as defined in the original theory, was set at η_m = 1. Regarding the textiles, an elastic modulus E_f of 70 GPa was assumed, as suggested by the glass filaments manufacturer. The fabric volume fraction V_f was calculated as the ratio between A_f (Table 5) and the average cross section of the samples belonging to each set (see Table 7), while V_m was equal to 1-V_f. The reinforcement efficiency η_f was taken equal to EF_f, where EF_f is the fabric efficiency factor introduced in Section 2.2 (see Equation (2) and Table 4). This latter assumption followed the hypothesis that the filaments that effectively contribute to the resisting mechanism ($A_f^* = A_f \cdot EF_f$) work in parallel and each of them is capable of reaching the maximum tensile material strength (σ_{fu} = 2000 MPa).

The hypothesis seems to be confirmed by Figure 15, showing a reworking of the plain fabric tensile responses of Figure 4 carried out by calculating the nominal fabric stress σ_f^* as the ratio between the load P_f and the effective fabric area A_f^*. As one should note, the scatter of the seven average curves in terms of stiffness was limited, proving that a fabric efficiency factor merely estimated on the failure loads (see Equation (2)) may be sufficiently representative of the global textile behaviour.

The value of the average saturated crack spacing X was obtained experimentally, dividing the LVDT gauge length (the nominal value of 200 mm was assumed) by the number of cracks detected at the end of each tensile test (see Figure 10).

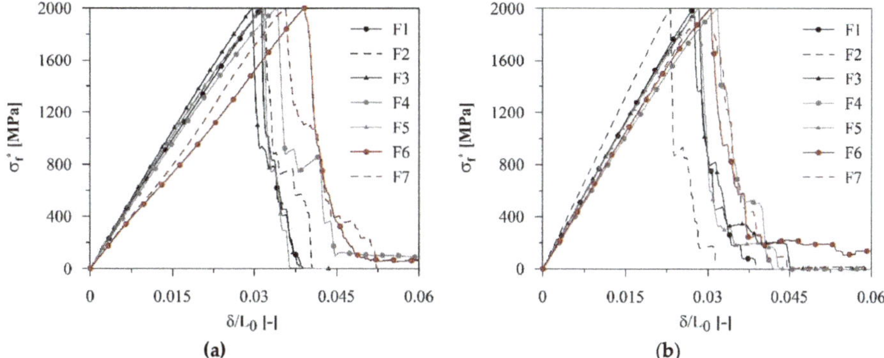

Figure 15. Fabric average tensile responses in terms of fabric stress on effective filaments σ_f^* vs. normalised displacement in the warp directions (**a**) and in the weft directions (**b**).

It is worth noticing that this simplified analysis could not simulate any of the contributions provided by dispersed short fibres, since the matrix was treated as an elasto-brittle material, with null post-cracking behaviour. Moreover, it is important to underline that the theoretical response was always trilinear-hardening, since volume fractions greater than the critical one were assumed a priori and fabric end-slippage phenomena—as the one occurring in the F3-M1 response of Figure 7a—could not be captured. Hence, in its present configuration, the algorithm cannot be regarded as sufficiently robust to represent a comprehensive predictive tool and, for this reason, its application in this research was limited to the assessment of ex-post modelling capabilities.

5.3. Assessment of Analytical Modelling Capabilities

The results obtained by applying the stochastic cracking model to the investigated sets of TRC systems are displayed in the following, with reference to M1- and M2-based composites reinforced with Fabrics F1 (Figure 16), F2 (Figure 17), F4 (Figure 18), F6 (Figure 19) and F7 (Figure 20). Average sample dimensions were considered and the analytical responses were cut at the maximum average stress value $\sigma_{TRC,max}$ (see Table 7). The analytical curves were compared with the experimental results in terms of composite stress σ_{TRC} vs. strain ε. Experimental strains were evaluated from the LVDT outputs, as the ratio between the average crack opening displacement COD and the measured gauge lengths GL. Please note that each curve in the figures is broken to the loss of the first transducer.

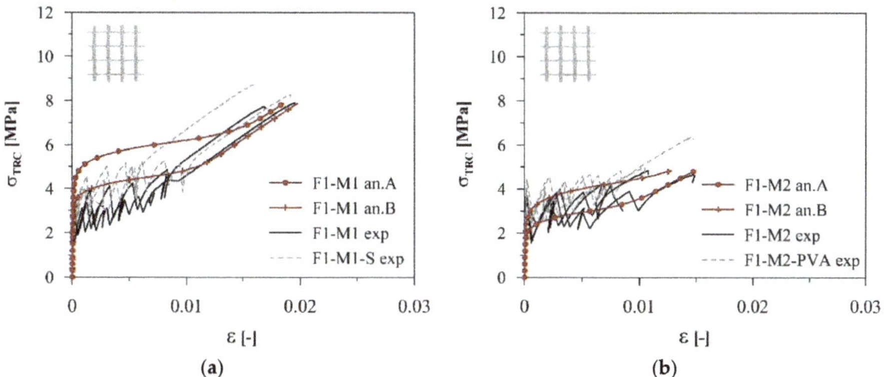

Figure 16. Comparison between the predicted analytical curves and the experimental responses in terms of nominal stress vs. strain for M1-based (**a**) and M2-based (**b**) composites reinforced with Fabric F1.

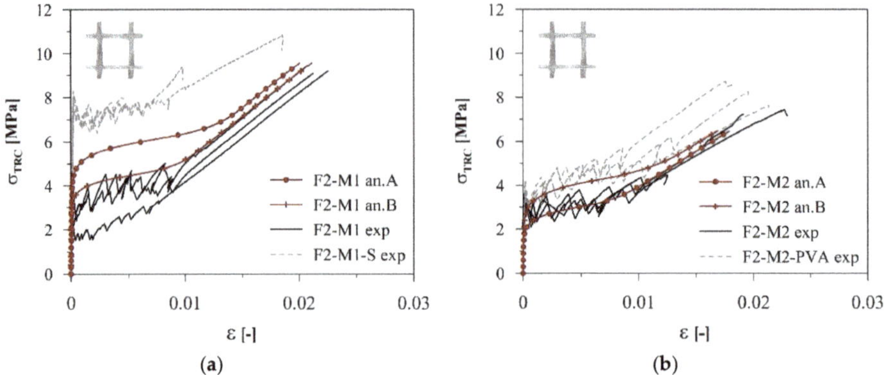

Figure 17. Comparison between the predicted analytical curves and the experimental responses in terms of nominal stress vs. strain for M1-based (**a**) and M2-based (**b**) composites reinforced with Fabric F2.

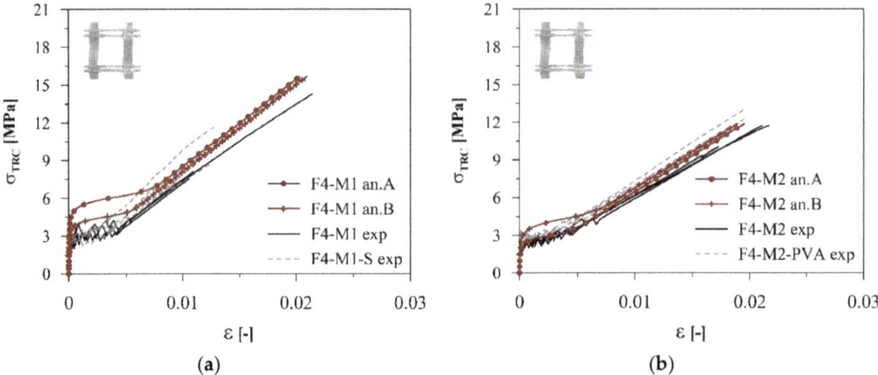

Figure 18. Comparison between the predicted analytical curves and the experimental responses in terms of nominal stress vs. strain for M1-based (**a**) and M2-based (**b**) composites reinforced with Fabric F4.

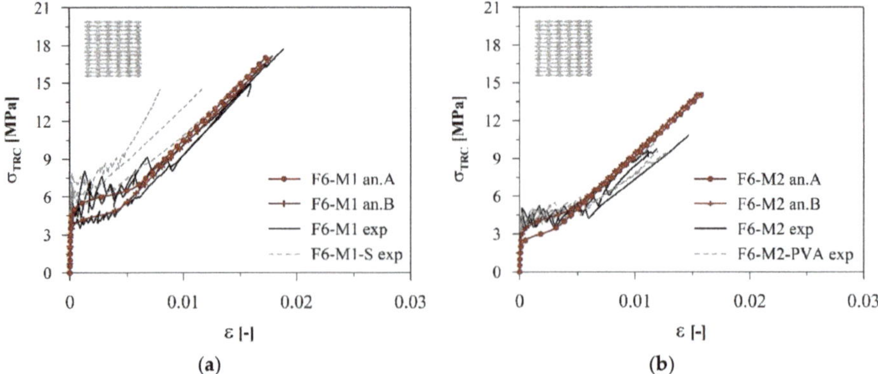

Figure 19. Comparison between the predicted analytical curves and the experimental responses in terms of nominal stress vs. strain for M1-based (**a**) and M2-based (**b**) composites reinforced with Fabric F6.

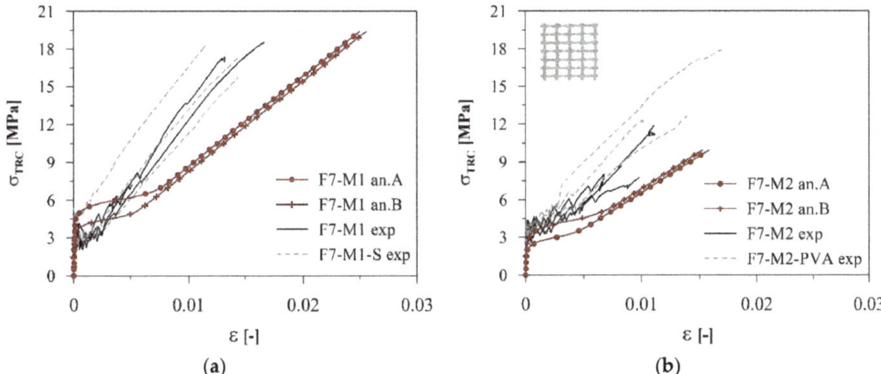

Figure 20. Comparison between the predicted analytical curves and the experimental responses in terms of nominal stress vs. strain for M1-based (**a**) and M2-based (**b**) composites reinforced with Fabric F7.

From the comparison, it was possible to notice that the analytical curves seem to properly fit the experimental results in almost any of the investigated TRCs. It was observed that:

- The stochastic model should fit the sequence of cracking stress values in the stable propagation phase. In all the cases investigated, the choice of a different tensile strength (type "B" analysis in place of type "A") allowed containing estimation errors. Moreover, it could be concluded that: (i) the two-parameter Weibull distribution function worked adequately; (ii) the m parameter, despite being obtained from flexural results, satisfactorily simulated the slope of direct tension curves; (iii) as expected, the effect of dispersed microfibers was not captured; and (iv) differences between the analytical and the experimental cracking stresses may be connected to mixed tensile-bending stress fields (unbalanced shrinkage could lead to a loss of planarity of the specimen).

- The transition point between the second and the third branch was satisfactorily caught by the analytical simulations. The ability of a simplified model to fit this response range was really important, bearing in mind that one of the main advantages of TRC—as an example in the case of retrofitting applications—is its energy dissipation capacity (Section 4.3). In these preliminary analyses, the use of mean saturated crack distances X obtained from experimental observation seemed an adequate compromise solution and again confirmed that the Weibull function can effectively describe discrete distributions of cracks. Future alternatives for the assessment of the X value might be based on the experimental evaluation of the frictional shear stress by means of pull-out tests and the use of refined approaches, such as the car parking problem solution.

- The hypothesis of evaluating the effective amount of AR-glass reinforcement multiplying the fibres volume fraction V_f by the fabric efficiency factor EF_f appeared to properly fit the third-branch slopes in almost all cases, with the only exception being F7-based composites. This was probably associated to the fabric–matrix interaction during the tensile loading: as confirmed by the experimental evidences graphically reconstructed in Figure 21b, during the loading phase along the warp direction, Fabric F7 exhibited a marked transverse deformation correlated to the "Poisson" effect in woven fabrics [27]. This deformability seemed to be negligible in the other cases, e.g., Fabric F2 (see Figure 22). In the composite system, the embedding matrix acted as a restraint to this transverse deformation, stiffening the third branch composite response with respect to the plain fabric behaviour. In Figure 21a, this effect is highlighted by the comparison between the composite and the fabric experimental curves. According to the literature, it can be stated that the "Poisson" effect is more significant in the case of unbalanced warp/weft fabrics [27], in terms of both equivalent diameter ratio and pick distances of the yarns; nevertheless, greater role is attributed to the warp crimp, found to have almost a linear correlation with the fabric transverse deformation [28]. Within the group of fabrics investigated in this work, F7 was clearly the more affected, due to the higher AR-glass grammage and the significant yarn diameter in the warp direction (this effect is greater also with respect to Fabric F6 where, even though the warp grammage is identical, the crimp is lower due to the smaller spacing of warp yarns).

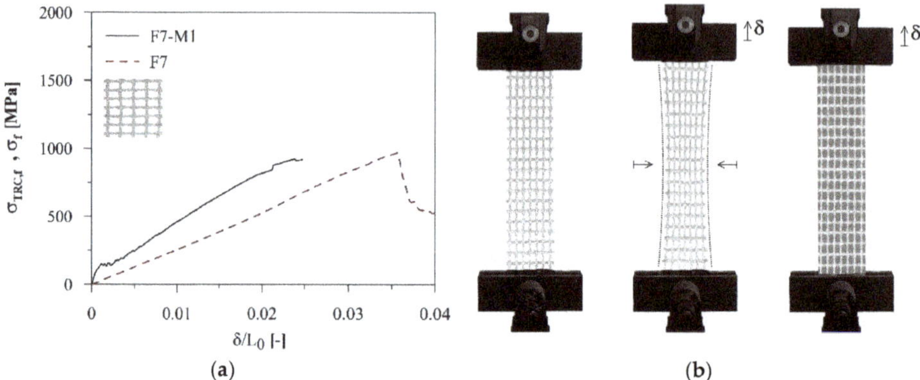

Figure 21. Comparison between F7-M1 composite and F7 fabric tensile responses (**a**); and explanation of the stiffening effect provided by the prevented transverse contraction of Fabric F7 (**b**).

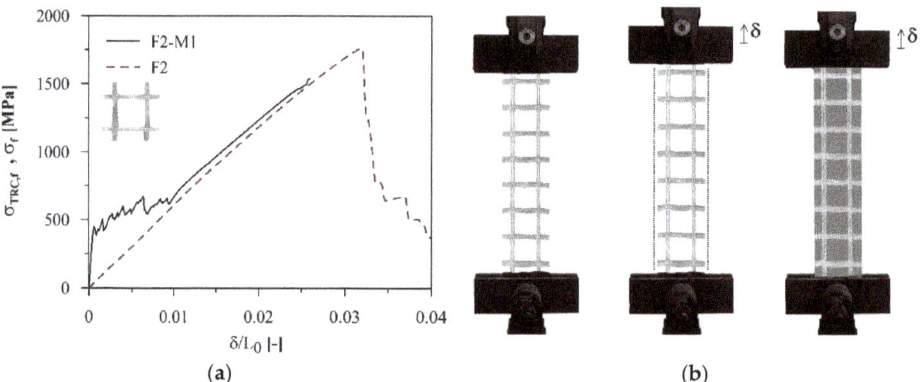

Figure 22. Comparison between F2-M1 composite and F2 fabric tensile responses (**a**); and negligible transverse contraction of Fabric F2 (**b**).

In view of the latter remarks, the stochastic model was again applied to the F7-based composites, adopting $\eta_f = EF_f$ only in the evaluation of the peak analytical stress, but imposing $\eta_f = 1$ in Equation (8) (in this sense, all filaments contributed to the resisting mechanism and $E_{c,III}$ was assumed equal to the elastic modulus E_f of the AR-glass filaments). This solution, in which all filaments (A_f) worked as parallel springs with $E_f = 70$ GPa, represents an upper bound of the experimental results in terms of third branch slope, as shown in Figure 23.

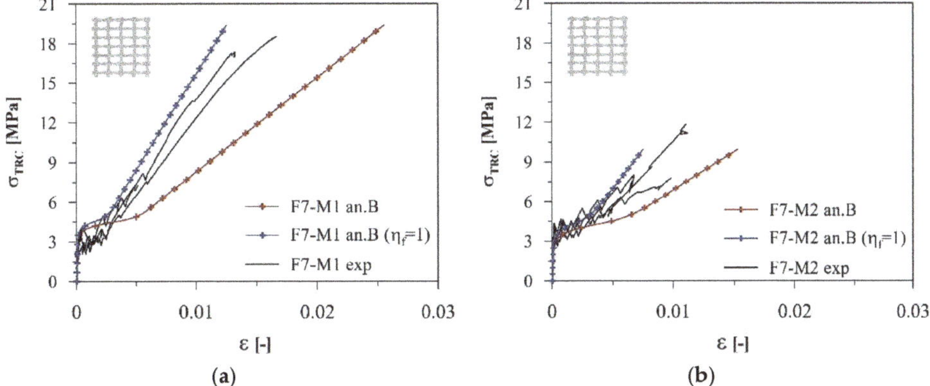

Figure 23. Effect of fabric efficiency on the predicted analytical response and comparison with the experimental data in terms of nominal stress vs. strain for M1-based (**a**) and M2-based (**b**) composites reinforced with F7 fabric.

6. Conclusions and Future Developments

The extensive experimental campaign presented in this work allowed better understanding the involved phenomena and the effect of individual components (matrix, textile and short fibres) on the tensile behaviour of TRC composites. In this sense, the use of the efficiency and the ductility/energy absorption parameters introduced in this work represent a systematic approach for the comparison of alternative TRC configurations and for the assessment of synergic contributions. From the interpretation of the experimental results, it can be concluded that:

- The coating nature and the fabric weaving had a significant influence on the global capacity of the composites, mainly in terms of global efficiency; both the reduction of the filaments that effectively participated in the mechanical response and the variation in the bond–slip behaviour at the fabric/matrix interface played a significant role.
- In general, the addition of short fibres was reflected in an increase of the mechanical capacities and a better behaviour in terms of durability. This effect seemed to be less visible in the case of narrow-spaced textiles, or when the matrix choice entailed a reduced fibre pull-out strength.
- The differences between matrices designed for different applications (new constructions vs. retrofitting) should be considered not only in terms of first cracking strength and initial elastic stiffness, but also in view of better understanding the global response of TRC composites. In particular, the internal slippage of the fabric—favoured by the use of less compact matrices—may significantly reduce the stiffness of the third branch of the curve, affecting the ultimate capacity of the composite.
- To obtain efficient tensile responses also when heavy-duty textiles are adopted, it might be necessary to slightly increase the thickness of the specimens, better controlling the internal slippage between the fabric and the surrounding matrix.

Cost-effective solutions were obtained by carefully combining the different components, in view of minimising the waste of material while improving the overall mechanical capacity. In this sense, the aims of the optimisation process were the maximisation of the efficiency (both the EF_{TRC} and the $EF_{TRC,f}$) and of the energy absorption capacity.

The application of the stochastic cracking model showed that it was possible to simulate the experimental tensile curves with a good correlation, starting from assumptions directly derived from standard tests results (Section 5.3). The fabric efficiency factor EF_f appeared to be directly related to the slope of the third branch in the stress–strain composite behaviour on condition that the transverse deformation of the fabric was limited. In the attempt to provide a predictive capacity to the model, it might be necessary: (i) to implement the effect of added fibres, so as to capture the second stage stress increase and the correct extent of the multi-cracking zone (this last directly correlated to the variation of the cracking pattern development due to fibres addition); (ii) to investigate the non-linear nature of the bond–slip behaviour between the fabric and the matrix by means of pull-out tests, even if this might overcomplicate the model; (iii) to better evaluate the cracking tensile stress of the employed matrix, for instance by means of direct tensile tests on more representative specimens (e.g., with the same nominal dimensions and the same mortar compaction of the TRC composites); and (iv) to develop a method to assess the fabric elastic behaviour, when transversely restrained by the mortar (furtherly investigating the effects of crimp, yarns diameter and grid spacing on the fabric "Poisson" effect).

Author Contributions: Conceptualisation, M.C.R., G.Z., M.C. and M.d.P.; methodology, M.C.R., G.Z. and M.C.; software, M.C.R. and G.Z.; validation, M.C.R. and G.Z.; formal analysis, M.C.R. and G.Z.; investigation, M.C.R. and G.Z.; data curation, M.C.R. and G.Z.; writing—original draft preparation, M.C.R. and G.Z.; writing—review and editing, M.C.R., G.Z., M.C. and M.d.P.; visualisation, M.C.R. and G.Z.; supervision, G.Z., M.C. and M.d.P.; and funding acquisition, M.d.P.

Funding: This research was funded by the ReLUIS interuniversity consortium (ReLUIS PR 5 2017).

Acknowledgments: The authors would like to acknowledge Gavazzi Tessuti Tecnici Spa, BASF Construction Chemicals Italia Spa, BEng. Nicola Borgioni, MEng. Giada Catalano and MEng. Laura Tiraboschi for their precious contribution to the research.

Conflicts of Interest: The authors declare no conflict of interest. The funders had no role in the design of the study; in the collection, analyses, or interpretation of data; in the writing of the manuscript, or in the decision to publish the results.

References

1. Peled, A.; Bentur, A.; Mobasher, B. *Textile Reinforced Concrete*, 1st ed.; CRC Press: Boca Raton, FL, USA, 2017; pp. 1–473.
2. Brameshuber, W. (Ed.) *Textile Reinforced Concrete—State-of-the-Art Report of RILEM TC 201-TRC*; RILEM Publications SARL: Bagneux, France, 2006.
3. De Felice, G.; De Santis, S.; Garmendia, L.; Ghiassi, B.; Larrinaga, P.; Lourenço, P.B.; Oliveira, D.V.; Paolacci, F.; Papanicolaou, C.G. Mortar-based systems for externally bonded strengthening of masonry. *Mater. Struct.* **2014**, *47*, 2021–2037. [CrossRef]
4. De Santis, S.; De Felice, G. Tensile behaviour of mortar-based composites for externally bonded reinforcement systems. *Compos. Part B Eng.* **2015**, *68*, 401–413. [CrossRef]
5. Rampini, M.C.; Zani, G.; Colombo, M.; Di Prisco, M. Textile reinforced concrete composites for existing structures: Performance optimization via mechanical characterization. In Proceedings of the 12th Fib International PhD Symposium in Civil Engineering, Prague, Czech Republic, 29–31 August 2018; pp. 907–914.
6. Koutas, L.N.; Tetta, Z.; Bournas, D.A.; Triantafillou, T.C. Strengthening of Concrete Structures with Textile Reinforced Mortars: State-of-the-Art Review. *J. Compos. Constr.* **2019**, *23*, 03118001. [CrossRef]
7. Mechtcherine, V.; Schneider, K.; Brameshuber, W. Mineral-Based Matrices for Textile-Reinforced Concrete. In *Textile Fibre Composites in Civil Engineering*; Triantafillou, T.C., Ed.; Elsevier Inc.: Amsterdam, The Netherlands, 2016; pp. 25–43.
8. Butler, M.; Mechtcherine, V.; Hempel, S. Durability of textile reinforced concrete made with AR glass fibre: Effect of the matrix composition. *Mater. Struct.* **2010**, *43*, 1351–1368. [CrossRef]
9. Colombo, I.G.; Magri, A.; Zani, G.; Colombo, M.; Di Prisco, M. Erratum: Textile reinforced concrete: Experimental investigation on design parameters. *Mater. Struct.* **2013**, *46*, 1953–1971. [CrossRef]
10. Peled, A.; Bentur, A. Mechanisms of fabric reinforcement of cement matrices: Effect of fabric geometry and yarn properties. *Beton- und Stahlbetonbau* **2004**, *99*, 456–459. [CrossRef]
11. Soranakom, C.; Mobasher, B. Geometrical and mechanical aspects of fabric bonding and pullout in cement composites. *Mater. Struct.* **2009**, *42*, 765–777. [CrossRef]
12. Barhum, R.; Mechtcherine, V. Influence of short dispersed and short integral glass fibres on the mechanical behaviour of textile-reinforced concrete. *Mater. Struct.* **2013**, *46*, 557–572. [CrossRef]

13. Consiglio Nazionale delle Ricerche. *CNR DT-215 Istruzioni per la Progettazione, l'Esecuzione ed il Controllo di Interventi di Consolidamento Statico Mediante L'utilizzo di Compositi Fibrorinforzati a Matrice Inorganica*; Consiglio Nazionale delle Ricerche: Rome, Italy, 2018. (In Italian)
14. Consiglio Superiore dei Lavori Pubblici, Servizio Tecnico Centrale. *Linea Guida per la Identificazione, la Qualificazione ed il Controllo di Accettazione di Compositi Fibrorinforzati a Matrice Inorganica (FRCM) da Utilizzarsi per il Consolidamento Strutturale di Costruzioni Esistenti*; Consiglio Superiore dei Lavori Pubblici, Servizio Tecnico Centrale: Rome, Italy, 2018. (In Italian)
15. Aveston, J.; Cooper, G.A.; Kelly, A. Single and multiple fracture. The Properties of Fibre Composites. In *Proc Conf National Physical Laboratories*; IPC Science & Technology Press Ltd.: London, UK, 1971; pp. 15–24.
16. Aveston, J.; Kelly, A. Theory of multiple fracture of fibrous composites. *J. Mater. Sci.* **1973**, *8*, 352–362. [CrossRef]
17. Cuypers, H.; Wastiels, J. Stochastic matrix-cracking model for textile reinforced cementitious composites under tensile loading. *Mater. Struct.* **2006**, *39*, 777–786. [CrossRef]
18. UNI EN 196. *Method of Testing Cement-Part 1: Determination of Strength*; UNI EN: Brussels, Belgium, 2005.
19. *fib Model Code 2010, Vol. 1, Bull. 65*; International Federation for Structural Concrete: Lausanne, Switzerland, 2012.
20. ISO 4606. *Textile Glass—Woven Fabric—Determination of Tensile Breaking Force and Elongation at Break by the Strip Method*; ISO: Geneva, Switzerland, 1995.
21. Cohen, Z.; Peled, A. Controlled telescopic reinforcement system of fabric-cement composites - Durability concerns. *Cem. Concr. Res.* **2010**, *40*, 1495–1506. [CrossRef]
22. Hegger, J.; Will, N.; Curbach, M.; Jesse, F. Tragverhalten von textilbewehrtem Beton. *Beton- und Stahlbetonbau* **2004**, *99*, 452–455. (In German) [CrossRef]
23. RILEM Technical Committee 232-TDT (Wolfgang Brameshuber). Recommendation of RILEM TC 232-TDT: Test Methods and Design of Textile Reinforced Concrete. *Mater. Struct.* **2016**, *49*, 4923–4927. [CrossRef]
24. Blom, J.; Cuypers, H.; Van Itterbeeck, P.; Wastiels, J. Determination of material parameters of a textile reinforced composite using an inverse method. In Proceedings of the ECCCM 13 Conference, Stockholm, Sweden, 2–5 June 2008.
25. Curtin, W.A. Stochastic Damage Evolution and Failure in Fibre-Reinforced Composites. *Adv. Appl. Mech.* **1999**, *36*, 163–253.
26. Weibull, W. A Statistical distribution function of wide applicability. *ASME J.* **1951**, *18*, 293–297.
27. Sun, H.; Pan, N.; Postle, R. On the Poisson's ratios of a woven fabric. *Compos. Struct.* **2005**, *68*, 505–510. [CrossRef]
28. Shahabi, N.E.; Saharkhiz, S.; Varkiyani, S.M.H. Effect of fabric structure and weft density on the Poisson's ratio of worsted fabric. *J. Eng. Fibers Fabr.* **2013**, *8*, 63–71. [CrossRef]

 *applied sciences*

Article

The Impact-Tensile Behavior of Cementitious Composites Reinforced with Carbon Textile and Short Polymer Fibers

Ting Gong *, Ali. A. Heravi, Ghaith Alsous, Iurie Curosu and Viktor Mechtcherine

Institute of Construction Materials, Technische Universität Dresden, 01187 Dresden, Germany
* Correspondence: ting.gong@tu-dresden.de; Tel.: +49-351-463-42852

Received: 19 August 2019; Accepted: 24 September 2019; Published: 27 September 2019

Abstract: The paper at hand focuses on the tensile behavior of ductile cementitious composites reinforced with short, randomly distributed, polymer fibers and a continuous carbon textile under quasi-static and impact loading. Strain-hardening cement-based composites (SHCCs) made of high strength fine-grained matrix with the addition of a 2% volume fraction of 6 mm-long ultra-high molecular weight polyethylene (UHMWPE) fibers and as-spun poly(p-phenylene-2,6-benzobisoxazole) (PBO-AS) fibers, respectively, were reinforced with one layer of carbon textile, which corresponds to a 0.68% volume fraction. The same fine-grained matrix reinforced with carbon textile only served as the reference material. The synergetic action of the two reinforcement types was investigated in uniaxial tension tests on composite specimens, as well as by means of single-yarn pullout tests at displacement rates of 0.05 mm/s in a hydraulic testing machine, and 8 m/s in a tensile split Hopkinson bar. The specimen's deformations, the formation of cracks, and the fracture processes were monitored optically and subsequently evaluated using digital image correlation (DIC).

Keywords: strain-hardening cement-based composites; textile reinforcement; short-fiber reinforcement; hybrid reinforcement; tension; impact loading; single yarn pullout

---

1. Introduction

Textile reinforced concrete (TRC) describes high-performance, cementitious composites containing two or three-dimensional fabrics made of carbon or alkali-resistant glass [1–3]. Their quasi-static tensile behavior is marked by an extensive strain-hardening phase, during which multiple controlled cracking develops in the fine-grained concrete matrix. TRC's high tensile ductility, strength, and stiffness enable their applications as thin retrofit layers on damaged structures and for strengthening existing structures that may deal with highly dynamic loading scenarios. However, the relatively coarse mesh size of the textile reinforcement does not allow for a sufficient in-plane and out-of-plane confinement of the surrounding mortar under high-speed loading, which can lead to a pronounced spalling/scabbing of the cementitious cover and a considerable degradation of the functionality of the strengthening layer. To eliminate this drawback, one can reinforce the cementitious matrix additionally, with short fibers. In particular, the use of ductile fiber-reinforced composites as matrix material promises to be highly instrumental for this purpose.

Strain-hardening cement-based composites (SHCCs) consist of fine-grained cementitious matrices and short, randomly distributed micro-fibers in volume fractions of up to 2%. They provide a suitable solution in respect to the desired increase in impact resistance of the strengthening layers, since SHCCs are characterized by a high inelastic deformability as a result of the successive formation of multiple, fine cracks under increasing tensile loading [4–6]. Their deformation behavior is expected to be well compatible with that of the textile reinforcement. In strengthening and retrofitting applications against dynamic loading scenarios, such as impact or blast, the textile reinforcement should offer a secure confinement of the strengthened reinforced concrete core (substrate) and ensure a favorable stress distribution, while the ductile SHCC's matrix should yield a better crack control, along with higher energy dissipation and damage tolerance.

The typical load-displacement behavior of carbon textile and textile reinforced cementitious composites is shown in Figure 1. The deformation span in which multiple cracking occurs is generally considerably smaller than the elongation at which the textile yarns fail. The addition of short fibers can enhance the overall composite response by increasing matrix cracking stress and possibly the tensile strength of the composite also [7–9]. Furthermore, an extension of the cracking deformation span is expected, maybe even to yarn failure with better stress distribution and crack control [10]. For fully exploiting the positive synergy of the short and continuous fiber reinforcements, appropriate material design principles must be followed. Silva et al. [11] and Barhum et al. [12] reported a decrease in composite strain capacity due to the restriction of crack opening by the short glass fibers, while Hinzen et al. [7] found that the strain capacity of the composite can be increased by adding a combination of short glass and Aramid fibers.

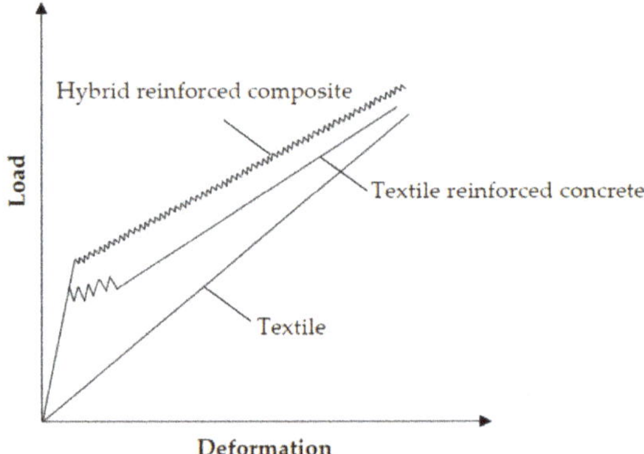

Figure 1. Schematic presentation of tensile load-deformation behavior of textile reinforcement, textile reinforced concrete, and hybrid reinforced composite.

For impact tensile loading, the strain rate's effect on the tensile behavior of hybrid reinforced composite largely depends on the material composition [11,13], which can be attributed to three mechanisms; namely, the strain rate's effects on (1) the cracking behavior of plain matrix; (2) the performance of fiber reinforcement, thus continuous carbon yarns and short polymer fibers; and (3) the interfacial characteristics of fiber reinforcement with the surrounding matrix [14–17]. The increasing loading rates influence, not only the tensile strength of the matrix, but the crack bridging behavior of continuous short fibers. Shim et al. [18] observed an increase in the tensile strength and modulus of Aramid textile under impact tensile tests, but a decrease in the failure strain. Zhu et al. [19] found that both the tensile strength and strain capacity increased under higher loading rate in the case of Kevlar 49 single yarns. The bonding properties between the continuous carbon yarn and the surrounding matrix depend not only on the loading rate, but on the type of cementitious matrix and the presence of short fibers. Yang and Li [20] and Ranade et al. [17] emphasized the rate sensitivities of the chemical bond properties of short fibers, and the corresponding negative effect on the strain capacity of an SHCC at higher strain rates. Curosu et al. [21] found that the increasing strain rate leads to an increase in both the tensile strength and stiffness of short ultra-high molecular weight polyethylene (UHMWPE) fibers but a decrease in their elongation at failure. Furthermore, a pronounced increase in the frictional bond between short polymer fibers and matrix was observed at higher displacement rates [22].

The results reported in the paper at hand are part of a more extensive study, in which material design concepts for hybrid fiber-reinforced composites are developed based on multi-scale experimental and analytical investigations at low and high loading rates. In this paper, only the influence of loading rate on the composite tensile behavior was analyzed, focusing on the effect of short-fiber reinforcement and short-fiber type. The reference materials in the current work are two types of strain-hardening cementitious composites, reinforced with ultra-high molecular weight polyethylene (UHMWPE) and as-spun poly(p-phenylene-2,6-benzobisoxazole) (PBO-AS) fibers, respectively [22,23]. Given the different crack-bridging properties of the reinforcing fibers, these two SHCCs yield different pre-peak strain capacities, which is interesting concerning the cracking behavior and deformation compatibility short fiber-reinforced matrix with the carbon textile. Both types of SHCC, as well as their constitutive cementitious matrix, were reinforced with one layer of carbon textile with a longitudinal reinforcing ratio of 0.68%. Besides the experiments at the composite level, single-yarn pullout tests from plain and fiber-reinforced matrices were performed with the same materials. The composite and pullout specimens were tested by means of a universal testing machine at a displacement rate of 0.05 mm/s and in a gravitational split-Hopkinson tension bar at displacement rates of up to 8 m/s. Besides the quantitative evaluation of the material tensile behavior in terms of stress–strain curves, the digital image correlation (DIC) facilitated a detailed description of the cracking processes under loading and a better interpretation of the material response measured.

2. Materials under Investigation

2.1. Cementitious Matrix

The fine-grained cementitious matrix was specifically designed for high-strength SHCC, being made with short UHMWPE fibers (hereafter called PE in this paper) produced by DSM, the Netherlands, under the brand name Dyneema®. This SHCC was previously investigated by the authors under quasi-static and impact tensile loading, in combination with Aramid and PBO fibers [22,23]. The matrix has a high content of cement, and has silica fume as the additional binder; see Table 1.

Table 1. Mixture composition of the high-strength, fine-grained cementitious matrix.

Components	kg/m^3
CEM I 52.5R-SR3/NA	1460
Silica fume	292
Quartz sand 0.06-0.2 mm	145
Superplasticizer	45
Water	315

The low water-to-binder ratio of 0.18 contributes to the high strength and density of the matrix, which was necessary for ensuring the proper anchorage of the hydrophobic PE micro-fibers. Only a small portion of very fine sand was used, with the maximum aggregate size of 0.2 mm, since the nature and geometry of the polymer micro-fibers and the necessity for a uniform fiber distribution in the matrix imposed limitations regarding the content and size of aggregates. Furthermore, this choice was dictated by the micromechanical conditions for tensile strain-hardening and multiple cracking in SHCC, which required a low fracture toughness of the matrix [24].

2.2. Short Micro-Fibers

Two types of short micro-fibers were investigated in this research, including the PE fibers and as-spun poly (p-phenylene-2,6-benzobisoxazole) (PBO-AS) fibers. The fibers possess high tensile strength and high moduli of elasticity. Table 2 presents their physical, geometrical, and mechanical properties. The short polymer fibers had a length of 6 mm and nominal diameters were 20 μm for PE and 13 μm for PBO-AS, respectively. The choice of relatively small length of fibers was based on the consideration of fresh SHCC workability and the dimensions of the textile mesh, as presented in the next section. Comparing them to the highly hydrophobic nature of PE fibers, PBO fibers exhibit a weak hydrophilic behavior and a subsequently higher bonding strength with the surrounding matrix. These two types of fibers present different levels of tensile strength, elasticity moduli, and bonding properties, which contribute to a better comparison of composite behaviors based on the additions of different short fibers.

Table 2. Properties of polymer fibers as provided by the producers [25,26].

Fiber Type	UHMWPE	PBO-AS
Producer	DSM	Toyobo
Brand	Dyneema®	Zylon®
Nominal diameter [μm]	20	13
Length [mm]	6	6
Density [kg/m^3]	970	1540
Tensile strength [MPa]	2500	5800
Modulus of elasticity [GPa]	80	180
Elongation at break [%]	3.5	3.5

2.3. Carbon Textile Reinforcement

TUDALIT-BZT2 produced by V.FRAAS, Germany, was used as textile reinforcement. The spacings between warp yarns (parallel to loading direction) and weft yarns were 12.7 mm and 16.0 mm, respectively; see Figure 2. Knitted filaments connected the warp and weft yarns to form a stable structure without a rigid connection. The physical and mechanical properties of the textile yarns are given in Table 3.

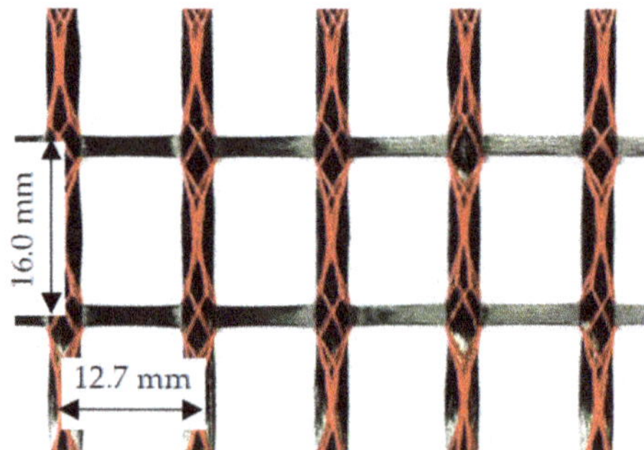

Figure 2. Geometry of the textile reinforcement under investigation.

Table 3. Properties of carbon textile TUDALIT-BZT2-V.FRAAS [27].

	Warp Yarn	Weft Yarn
Average yarn count [tex]	3300	800
Effective yarn cross-section [mm^2]	1.800	0.451
Average tensile strength [MPa]	1700	1700
Average modulus of elasticity [GPa]	170	152

In the current paper, the combination between TUDALIT-BZT2-V.FRAAS textile and the high-strength matrix presented in Section 2.1, but without short fiber, will be named TRC-M. This composite was additionally tested in order to better understand the role of short fiber reinforcement in the case of hybrid reinforcement. The combinations of textile and two SHCC compositions will be named TRC-PE and TRC-PBO, respectively.

3. Experimental Program

3.1. Specimens

Plates with dimensions of 260 mm × 90 mm × 20 mm were cast in a specially fabricated mold, which enabled us to fix the position of the textile in the middle of the plate's thickness, as shown in Figure 3a. The plates were produced using the lamination technique. The first layer of plain matrix or SHCC was cast in the mold before the placement of textile reinforcement. Subsequently, the textile mesh was gently pressed into the matrix/SHCC so that the latter could penetrate through the textile mesh. The ends of the textile yarns were clamped by the mold to ensure a fixed position in the middle. The second layer of the matrix was then cast on top, followed by leveling and smoothening. Note that under consideration of the typical anchorage issues related to carbon textiles, the ends of the textile yarns were protruding outside of the mold in order to enable their stronger anchorage by gluing them at the specimens' ends in the adapters; see Figure 3a. The specimens were demolded at the age of 24 hours, sealed in plastic sheets and subsequently cured for 27 days in a climatic chamber with a constant temperature of 20 °C and relative humidity of 65%.

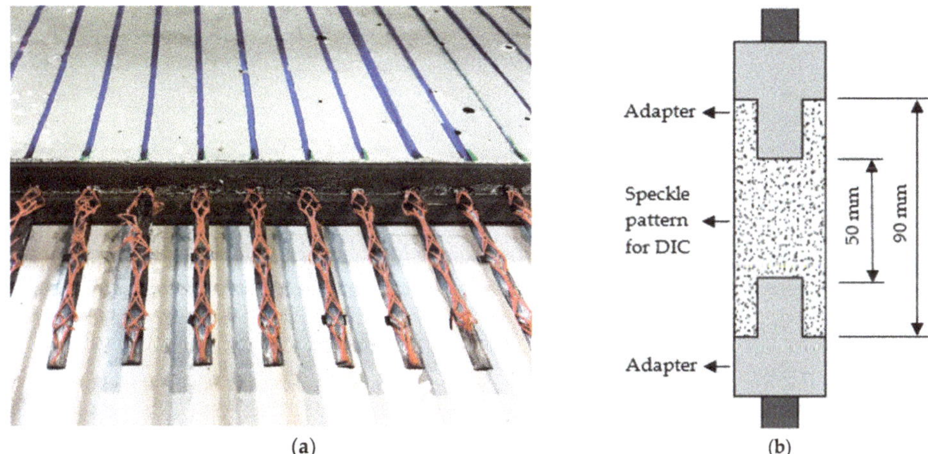

(a)   (b)

Figure 3. Specimen production for tension tests: (**a**) plate after demolding; (**b**) final specimen dimensions and fixities.

Prior to testing, the plates were cut into smaller specimens with dimensions of 90 mm × 40 mm × 20 mm, containing three warp yarns in the loading, i.e., longitudinal direction. The length of the middle gauge was 50 mm, and it covered four weft yarns. All specimens were reinforced with only one layer of textile, hence a longitudinal reinforcement ratio of 0.68% calculated based on the effective cross-sectional area of 1.8 mm^2 for one warp yarn. It should be noted that usually the TRC tensile specimens have a relatively large length in order to ensure a proper textile anchorage at the specimen ends and attain yarn rupture instead of yarn pullout. However, in dynamic tension experiments, such as in the split Hopkinson bars, the length of the specimens is limited by the condition of dynamic stress equilibrium. For this reason, a length of 90 mm was adopted for the composite specimens in this investigation. To avoid premature yarn pullout after the initial cracking, the longitudinal textile yarns had protruding ends at both ends of the specimens, which were bent over the specimens' ends and glued within the adapters by bi-component epoxy resin; see Figure 3b. The adapters were made of steel for quasi-static tension tests and aluminum for impact tension tests. In the last step, the middle, gauge portions of the specimens were sprayed to create a random black and white pattern needed for digital image correlation (DIC). All the tests were performed on the 28th day after casting.

As for single-yarn pullout tests, the initial plates were cut into smaller specimens, as shown in Figure 4. The 30 mm-long specimens contained two transversal weft yarns and one longitudinal warp yarn. The 30 mm-long protruding ends of the warp yarns were glued inside aluminum cylinders with a bi-component epoxy resin. Identical specimens were tested quasi-statically and dynamically.

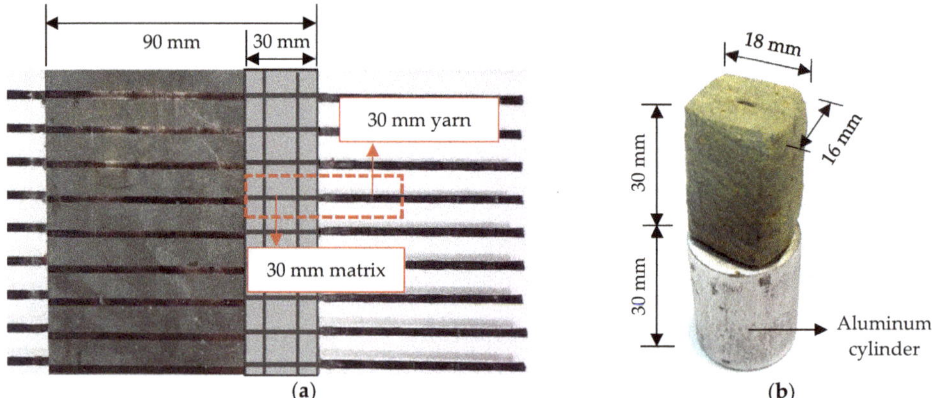

Figure 4. Specimen production for single-yarn pullout tests: (**a**) cutting configuration; (**b**) final specimen dimensions and fixities.

3.2. Setups for Quasi-Static Tension and Single-Yarn Pullout Tests

The specimens were first glued to steel adapters and subsequently clamped rigidly in the machine with the help of steel rods; see Figure 5a. The quasi-static uniaxial tension tests were performed in an Instron 8501 hydraulic testing machine under a controlled displacement rate of 0.05 mm/s. The deformations of the gauge portion were measured by two linear variable differential transducers (LVDTs) connected to the adapters on both sides of the specimens. Additionally, the deformations, formation of cracks, and fracture processes were monitored optically, and subsequently evaluated using digital image correlation (DIC). Images with a resolution of 3456 × 5184 pixels were taken with a Canon E05 700D camera at intervals of 5 seconds. The DIC evaluation was performed using the Aramis software by GOM GmbH.

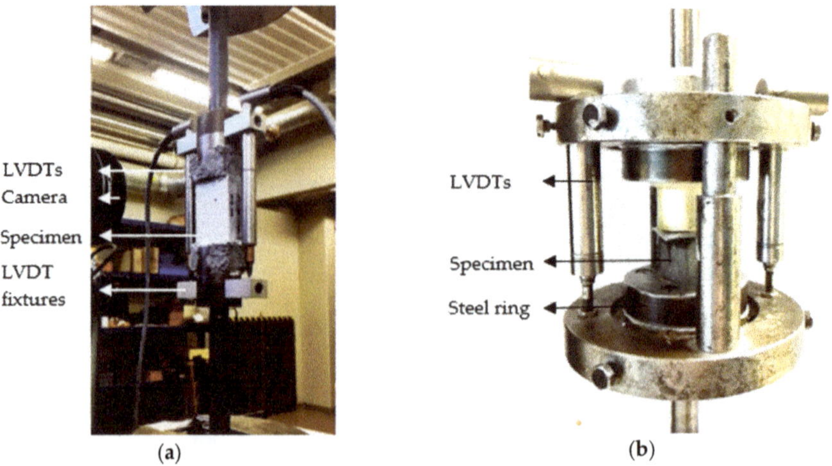

Figure 5. Test setups for: (**a**) quasi-static tension tests and (**b**) quasi-static single-yarn pullout tests.

The quasi-static single-yarn pullout tests were performed in a Zwick Roell 1445 testing machine at a displacement rate of 0.05 mm/s. The specimens were glued at their ends inside two aluminum rings, which were fixed in the machine, as shown in Figure 5b. LVDTs were fixed at both sides of the rings to capture the slip of yarns.

3.3. Setups for the Impact Tension Test and Single-Yarn Pullout Test

As shown in Figure 6, a gravitational split Hopkinson tension bar (SHTB) was used for both impact tension tests and single-yarn pullout tests [28]. The peak displacement rate in the tests was 8 m/s, which was reached by dropping a 30 kg striker from a height of 1 m. Both the input and output bars were made of brass in the case of impact tension tests and were aluminum for single-yarn pullout tests. The reason for these choices was to match the impedance of bars with adapters used in the two types of test to minimize the wave distortion by adapters.

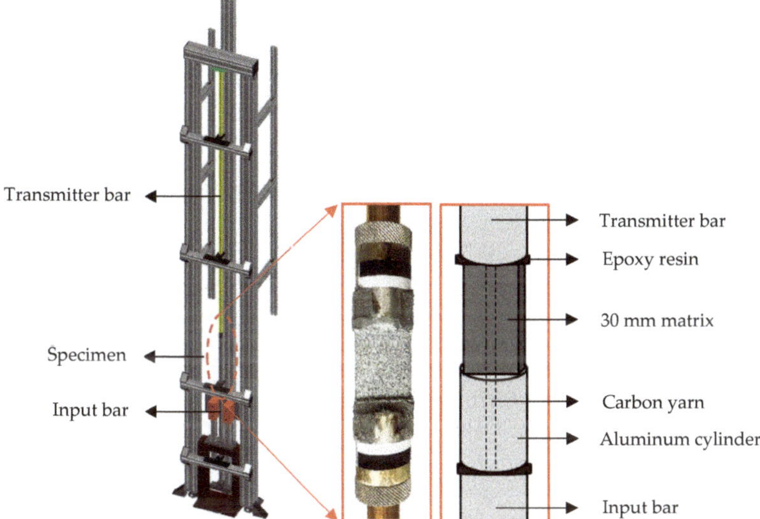

Figure 6. Testing configuration of the gravitational split Hopkinson tension bar (SHTB) for impact tension and single-yarn pullout test.

In the impact tension experiments, the dimensions of the specimens are imposed by the requirement of uniform stress along the sample; i.e., dynamic stress equilibrium. In this study, the length of the specimens was adopted based on preliminary tests on 50 mm-long SHCC specimens of cylindrical geometry, which were directly glued to the input and transmitter bars. However, the composites reinforced with fabric impose a plate-like geometry. The rectangular cross-section had a width larger than the diameter of the input and output bars, making necessary the usage of adapters between specimen and bars. The shape of the adapters was designed to target the reduction of any adverse effects on wave propagation due to impedance mismatch. After a series of calibration tests, it was found that the transmitted wave represents, reliably, the stress history in the sample. Therefore, the reaction forces were calculated based on the waves measured on the transmitter bar using three strain gauges glued axis-symmetrically around the bar. To ensure a higher accuracy of the results, the deformation of the samples was measured by an optical extensometer. A high-speed stereo camera system was used to monitor the crack formation in the loaded samples with a sampling rate of 50,000 frames per second.

In the single-yarn pullout tests, the pullout force was calculated based on the measurements on the transmitter bar. The slip was calculated based on the relative displacement of the two bars according to one-dimensional wave analysis and optically with the help of DIC. An aluminum cylinder was used to connect the carbon yarn with the aluminum input bar to ensure identical impedance with the bar. In this way, the speed at the endpoint of the input bar could be regarded as the real pullout speed of the carbon yarn. SIRIUS®HS-STG+ systems produced by DEWEsoft®were used for data acquisition with a sampling rate of 1 million per second, and a filter of 40 kHz was adopted to reduce the electrical noise while avoiding possible phase shift in the signal.

4. Results and Discussion

4.1. Results of Quasi-Static Tests

4.1.1. Uniaxial Tension Tests on Plate-Like Composite Specimens

The small specimen length and the insufficient yarn anchorage resulted in undesirable failure modes of the plate-like composite specimens. As mentioned in the previous sections, the carbon yarns were longer than the specimens and the protruding ends were bent over the specimens' edges and glued between the specimens and the adapters. Due to the relatively weak bond strength between the carbon yarns and the surrounding matrix, and because of the small specimen length, the failure of the yarns occurred in the bent-over segments in the adapters. Such a failure was mostly facilitated by the poor transversal properties of the carbon fibers and the resulting damage induced during bending. Thus, in this configuration, the strength of the carbon textile reinforcement cannot be fully exploited and the specimens fail under considerably lower loads compared to long specimens with a proper textile anchorage [2,8,11]. Nevertheless, despite these limitations, the comparative study offers an insight into the influence of the short fiber reinforcement on the composite behavior and addresses further improvements in terms of composite design and testing configuration.

The uniaxial quasi-static tensile stress–strain curves are plotted in Figure 7. The comparison of the representative curves indicates the contributions of SHCCs to the total response of the composite in the yarn pullout stage.

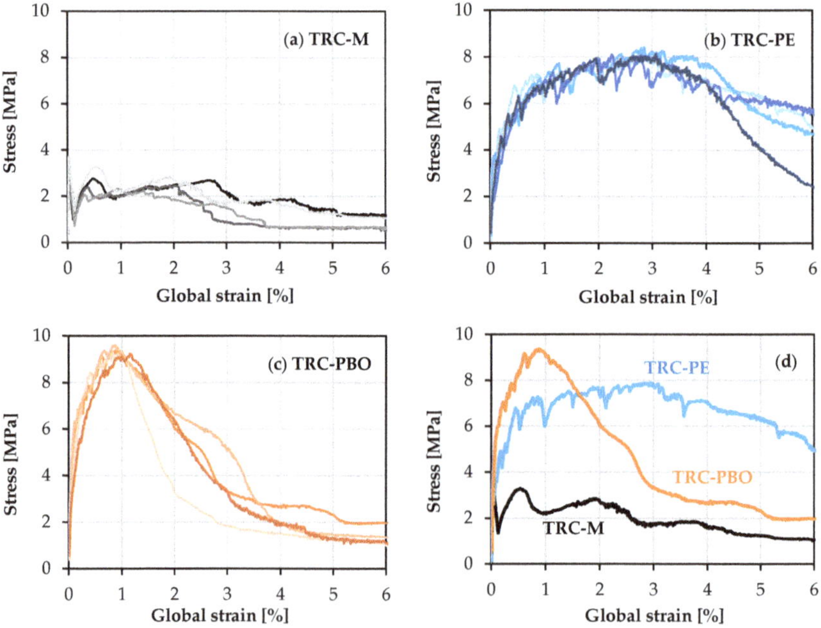

Figure 7. (a–c) Tensile stress–strain curves of the composites under quasi-static loading and (d) representative stress–strain relationships for the composites under investigation.

It should be noted that the specimens had already exhibited a few minor cracks prior to loading due to the forced mechanical clamping, which makes it challenging to define the accurate value of the first crack stress. However, the results of different materials are still comparable under the same testing conditions. Hence the results of the first crack stress, peak tensile stress, ultimate global strain, and the work-to-fracture of the tested specimens under uniaxial quasi-static tension tests are given in Table 4. The tensile strength of the composites was obtained by dividing the peak load by the composite cross-section. The global strains were calculated as specimen deformation over their gauge length; the ultimate value (strain capacity) corresponds to the peak load. Note that the global strain is a measure of the material deformability, and it is not associated with a uniform strain field in the specimens. It is rather representative of the extent of multiple cracking in the samples before peak loading. The post-peak behavior is associated with crack localization and widening accompanied by yarn pullout.

Table 4. Results of the uniaxial quasi-static tests on the investigated composites (average values; standard deviations are given in parentheses).

	First Crack Stress [MPa]	Peak Tensile Stress [MPa]	Ultimate Strain [%]	Work-to-Fracture [kJ/m^3]
TRC-M	3.4 (0.3)	3.4 (0.3)	–	–
TRC-PE	2.5 (0.5)	8.1 (0.2)	2.8 (0.4)	185.7 (31.3)
TRC-PBO	4.6 (1.1)	9.4 (0.2)	0.9 (0.2)	67.4 (12.1)

In the case of TRC-M without any discrete fiber reinforcement, the yarn failure and subsequent pullout are accompanied by the widening of the localization crack with no multiple cracking, yielding an average composite tensile strength of only 3.4 MPa, which is also the first crack stress; see Figure 7a. On the contrary, the materials reinforced additionally with short micro-fibers, i.e., the ones with SHCC matrix TRC-PE or TRC-PBO, exhibit strain-hardening behavior. The tensile stress increases after the first crack accompanied by the formation of multiple cracks; see Figure 7b,c.

The relatively low first crack stress of TRC-PE (2.5 MPa) can be traced back to the relatively high porosity of the matrix, as well as to the highly hydrophobic nature of short PE fibers. Due to the purely frictional bond, the fibers are only activated after crack formation, while prior to that they act as micro-defects [23], leading to even lower first crack stress than that measured for TRC-M (3.4 MPa). The short PBO-AS fibers, in contrast, possess a weak hydrophilic character. The smaller diameter results in higher aspect ratio and larger amount of fibers in the case of the same volume fraction of 2%. The lower diameter, higher stiffness, and strength of the PBO-AS fibers, as well as their adequate bonding to cementitious matrix ensures efficient confinement of the matrix already, prior to cracking [23]. Furthermore, PBO-AS fibers enable better control of micro-cracks, hence enhancing the first crack stress of the composite. In addition to their weak hydrophilicity, the high Young's modulus of PBO-AS fiber ensures narrow crack widths in TRC-PBO in comparison to those in TRC-PE. This influences both the strain at peak stress and the work-to-fracture of the corresponding composites. Work-to-fracture is the area under the stress–strain curves up to the peak load.

The potential of the material to develop multiple cracks and exhibit strain-hardening behavior can be characterized by the strain-hardening modulus; i.e., the ratio of tensile strength to first crack stress [17]. It can be observed that while TRC-PBO prevails in both first crack stress (4.6 MPa) and tensile strength (9.4 MPa), TRC-PE exhibits a higher strain-hardening modulus, with a first crack stress of 2.5 MPa and tensile strength of 8.1 MPa. Taking into consideration the strain capacity of 2.8% in the case of TRC-PE and 0.9% for TRC-PBO, it is not surprising that the TRC-PE yields a considerably higher work-to-fracture of 185.7 kJ/m^3 when compared to 67.4 kJ/m^3 in the case of TRC-PBO.

The ultimate strain capacity of the investigated composites is decided by both the average crack width and the average crack spacing, which is defined as the ratio of gauge length to the average number of cracks within the gauge length of the specimen. Figure 8 shows tensile stress–strain curves and corresponding average crack widths for representative specimens of TRC-PE and TRC-PBO. Due to the 5 second interval between individual image recordings, only a limited number of photos were taken during the tests. This explains the relatively low number of crack width measurement points in the case of TRC-PBO, as shown in Figure 8b.

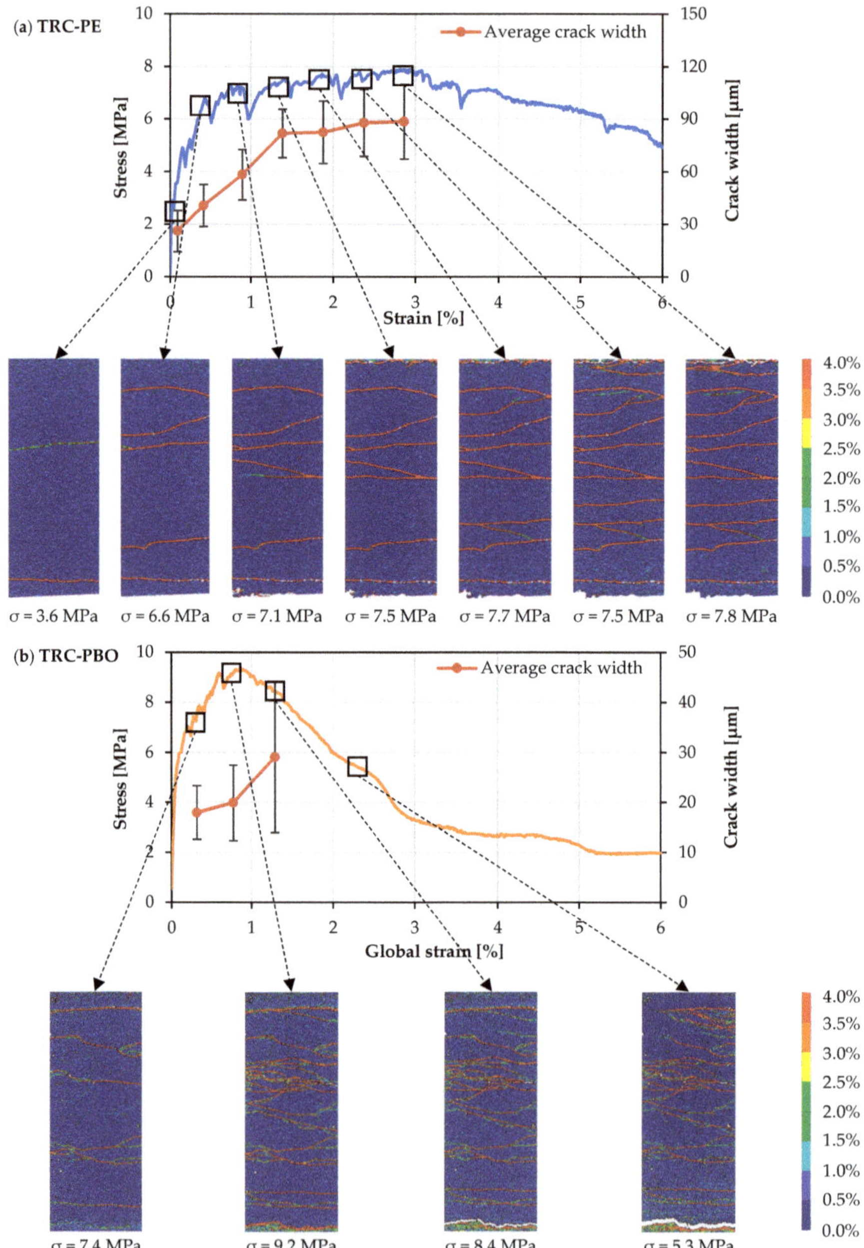

Figure 8. Tensile stress–strain curve and corresponding average crack widths of a representative specimen of both (**a**) TRC-PE and (**b**) TRC-PBO with corresponding composite stresses and crack patterns under quasi-static loading.

Despite both materials showing multiple cracking, the crack patterns are very different. At the same strain level, TRC-PBO exhibits finer and denser cracks along the specimen. The average number of cracks of TRC-PE and TRC-PBO at peak stress are 12 and 16, resulting in the average crack spacings of 4 mm and 3 mm, respectively; see Table 5. Though TRC-PBO exhibits more cracks, the low average crack width of 29 μm at peak load leads to an overall smaller strain capacity when compared to TRC-PE, for which the average crack width is 88 μm. The relationships between crack density, crack width, and applied stress have been investigated in order to achieve a better understanding of the crack pattern [13,14,16,29]. It can be observed that, at a similar stress level, TRC-PBO shows better crack control.

Table 5. Average crack width and spacing at the ultimate strain level for the representative specimens subjected to uniaxial quasi-static tension tests (localization crack excluded).

	TRC-M	TRC-HDPE	TRC-PBO
Average crack width at the ultimate strain level [μm]	–	88	29
Average crack spacing at the ultimate strain level [mm]	–	4	3

Note that SHCC specimens were also investigated without textile reinforcement. The reference SHCC matrices show inferior mechanical properties compared to the hybrid fiber-reinforced composites presented in this section. Thus, despite the undesirable failure mode of the textile yarns, their contribution is still significant. At the material level, the proper composite tensile behavior of SHCC and textile can be only highlighted with the help of large specimens loaded quasi-statically. This is, however, the matter of a different study by the authors [10].

For a more comprehensive analysis of the textile contribution in the pre-peak and failure localization phases presented above, single-yarn pullout tests were performed, using the same plain matrix and SHCCs, both under quasi-static and impact loading. The results presented in the next section demonstrate the effects of the addition of short polymer fibers on the anchorages of the carbon yarns coated with styrene-butadiene.

4.1.2. Single-Yarn Pullout Tests

The force-slip curves of quasi-static single-yarn pullout from plain and SHCC matrices are plotted in Figure 9. The slip of the carbon yarns was recorded by LVDTs attached directly to the specimens. The force-slip curve can be divided into three stages, as shown in Figure 9d. According to the pullout load-slip model presented in [30,31], stage I corresponds to the linear elastic stage, and is followed by the gradual debonding stage II, which terminates at the end of the debonding process. Stage III represents the pullout process influenced mainly by the yarn–matrix interfacial friction.

The relatively small embedment length of the yarns in combination with the weak affinity of both the carbon filaments and the styrene-butadiene coating to the cementitious matrix led to a complete pullout of the yarns. The peak forces for carbon yarn pulled out from plain matrix and PE-SHCC are nearly identical, with average forces of 441 N and 440 N, respectively. In contrast, the addition of PBO fibers led to a considerably higher bond strength, with an average peak pullout force of 530 N; see Table 6. This could be traced back to the mitigation of shrinkage-induced micro-cracking as ensured by PBO fibers [32]. However, a more detailed analytical investigation of the yarn-matrix interface should bring more clarity to this phenomenon. Note that in previous studies, such as [12], the addition of short alkali-resistant glass fibers and carbon fibers had a positive effect on the yarn-matrix bond strength. It could be that the poor wettability of the PE fibers, as well as their lower stiffness could be the reason for a lower bond strength in comparison to PBO-SHCC.

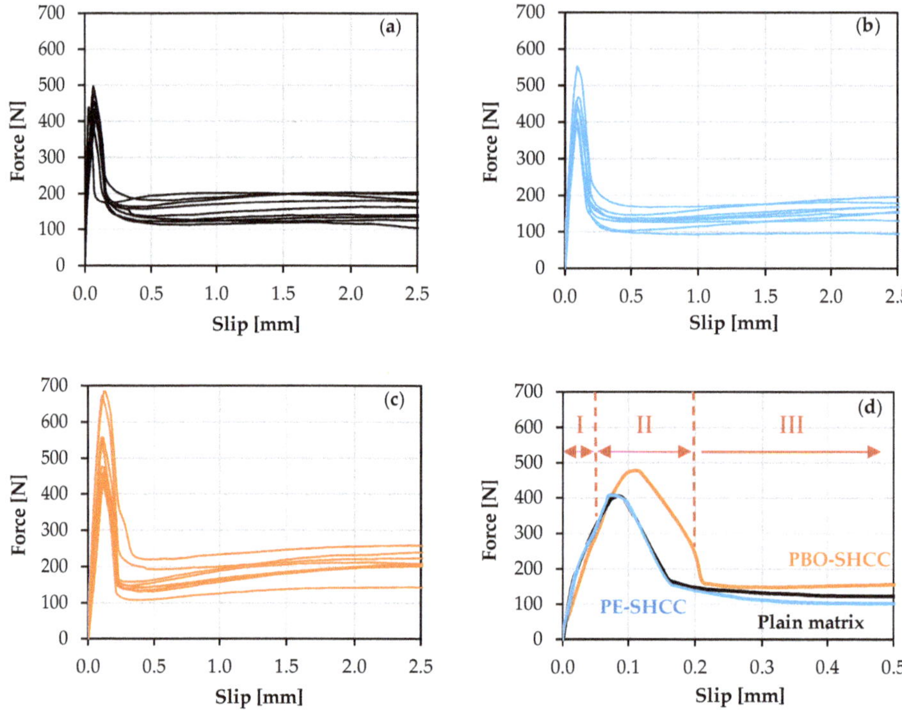

Figure 9. Quasi-static force-slip relationships of single carbon yarns pulled out from: (**a**) plain matrix; (**b**) PE-SHCC; (**c**) PBO-SHCC; and (**d**) representative force-slip relationships for different composites up to a slip extent of 0.5 mm.

Table 6. Peak debonding forces in quasi-static pullout tests with three different matrices (average values; standard deviations are given in parentheses).

	Single-Yarn Pullout from Plain Matrix	Single-Yarn Pullout from PE-SHCC	Single-Yarn Pullout from PBO-SHCC
Peak debonding force [N]	441 (34)	440 (53)	530 (94)

4.2. Results of the Impact Tension Tests

4.2.1. Dynamic Uniaxial Tension Tests

All three types of composites were tested under impact tensile loading with a peak displacement rate of 8 m/s, corresponding to an average strain rate of 160 s^{-1}; see dashed curves in Figure 10. The first crack stress of the composites is defined here as the first peak of the stress–strain curve, also detected with the help of DIC. The tensile strength is defined as the stress value at the peak of the ascending branch of the curve, while the ultimate global strain corresponds to the strain value at the peak stress prior to failure localization. The results of tensile strength, ultimate strain, and the work-to-fracture of the tested specimens are given in Table 7. The dynamic increase factor (DIF) of each parameter was calculated to demonstrate the rate effect on the tensile behavior of the composites.

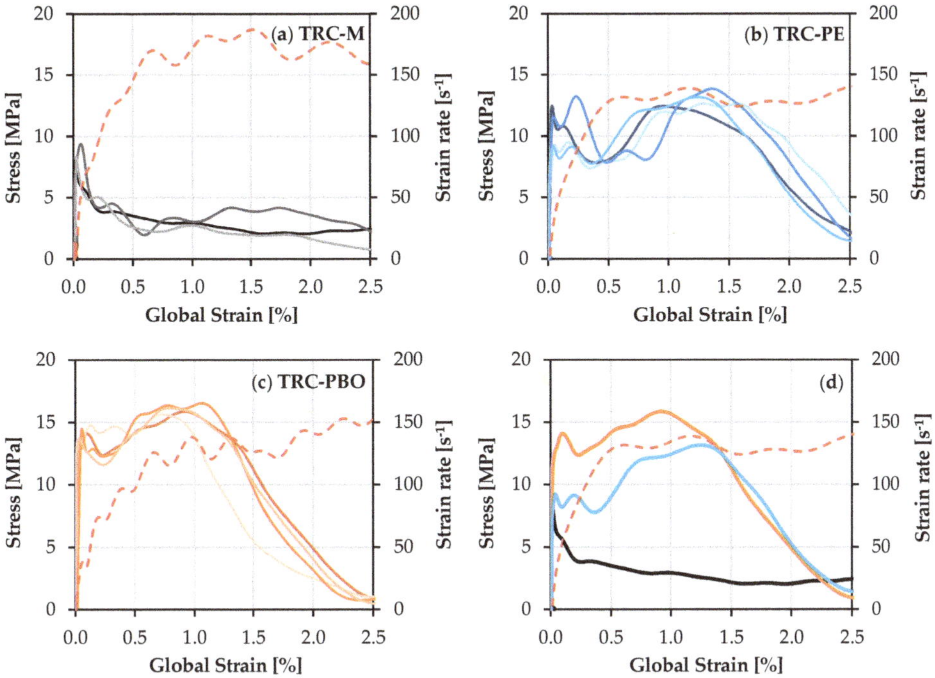

Figure 10. (a–c) Tensile stress–strain curves of the composites under impact loading (the global strain rate curves are represented by dashed lines) and (**d**) representative stress–strain relationships for the composites under investigation.

Table 7. Results of the uniaxial impact tension tests on the composites under investigation (average values; standard deviations are given in parentheses).

	First Crack Stress [MPa]	Peak Tensile Stress [MPa]	Ultimate Strain [%]	Work-to-Fracture [kJ/m^3]
TRC-M	8.6 (0.7)	4.5 (0.6)	0.3 (0.1)	14.7 (4.2)
DIF	2.5	1.3	–	–
TRC-PE	10.6 (1.6)	13.0 (0.6)	1.2 (0.2)	122.4 (21.9)
DIF	4.2	1.6	0.4	0.7
TRC-PBO	13.9 (0.5)	16.0 (0.4)	0.9 (0.1)	122.2 (22.6)
DIF	3.0	1.7	1.0	1.8

It is noteworthy that all the composites exhibit multiple cracks under impact tensile loading. TRC-PBO bears the highest average maximum tensile stress of 16 MPa, followed by TRC-PE, with an average tensile strength of 13 MPa. Both the first crack stress and the tensile strength of the composites are increased pronouncedly in comparison to the corresponding values measured in the quasi-static regime. Despite the multiple cracking occurring in TRC-M, the material exhibited a strain-softening behavior with a very short plateau immediately after the initial stress peak between 0.2% and 0.3% strain. This plateau can be attributed to the pullout behavior of the textile yarns, as demonstrated in the next section. Tensile stress–strain curves and corresponding average crack widths for representative specimens of TRC-M, TRC-PE, and TRC-PBO are plotted in Figure 11. The crack formation and fracture processes were captured by high-speed cameras and then evaluated with DIC. It can be observed that micro-cracking already occurred before the formation of the first crack. The first peak of the curve corresponds to the formation of the first macro-crack, which propagates through the entire specimen's cross-section, leading to a large decrease of the composite stiffness. The momentary loss of the load-carrying capacity is revealed by the drop in the stress–strain curves after the first peak, as shown in Figure 10d. The strain-hardening behavior lasted until the composite with SHCC matrices reached their tensile strength, and afterwards, no new cracks developed while only the localization crack continued to open.

It can be observed that for all types of composites, the DIF of the first crack stress is considerably higher than that of the tensile strength. The first crack of material occurs during the initial loading stage associated with increasing displacement rates; i.e., acceleration. Due to this, the structural inertia has a significant contribution to the apparent first crack stress; see also, [33]. The first crack occurrence attenuates the effect of strain rates in the rate sensitive matrix considerably.

Average crack widths and crack spacings at critical strain levels of 0.1%, 0.3%, and 0.4% for representative specimens subjected to uniaxial impact tension (except localization crack) are given in Table 8, according to the nearest frame recorded. Though the average crack widths keep increasing, along with the deformation for all three materials, TRC-PE and TRC-PBO possess superior crack control behavior with a steady and moderate growth in average crack width.

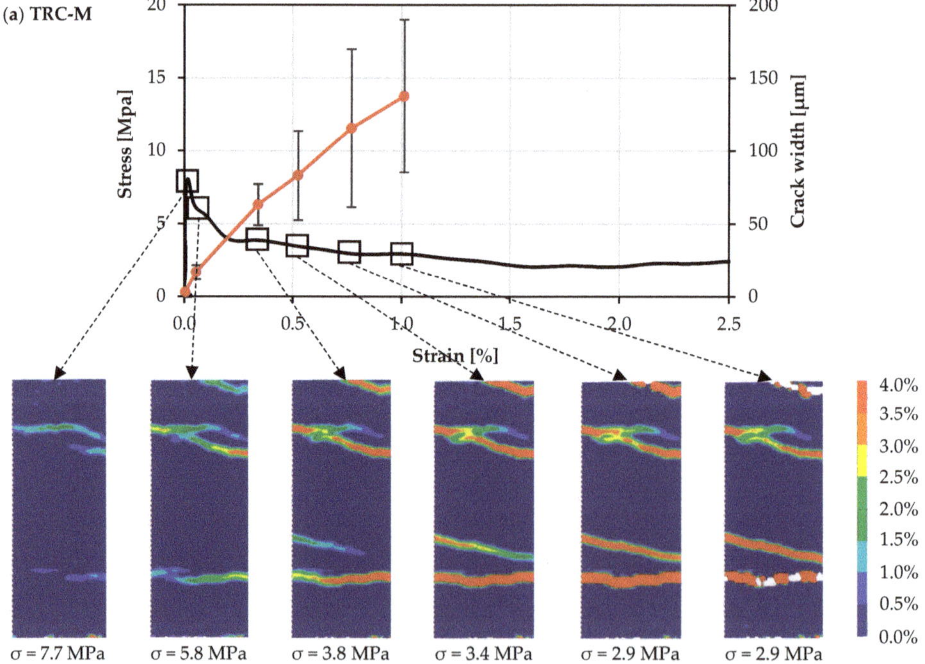

Figure 11. *Cont.*

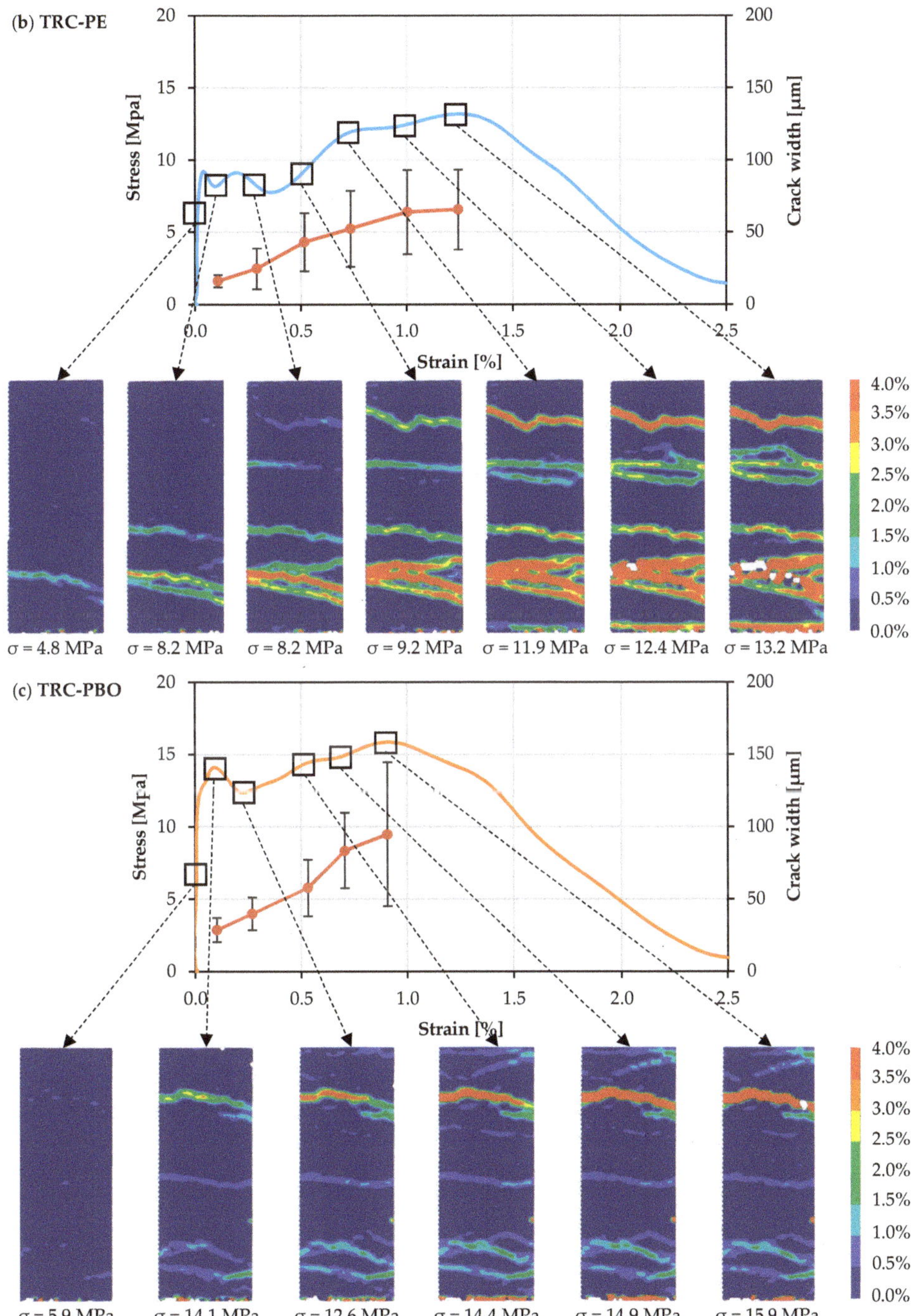

Figure 11. Tensile stress–strain curve and corresponding average crack widths of a representative specimens of (**a**) TRC-M, (**b**) TRC-PE, and (**c**) TRC-PBO, with corresponding composite stresses and crack patterns under impact loading.

Table 8. Average crack width and crack spacing at the strain levels of 0.1%, 0.3%, and 0.4% for the representative specimens subjected to uniaxial impact tension tests (localization crack excluded).

	TRC-M	TRC-PE	TRC-PBO
Average crack width at 0.1% strain level [μm]	17	16	29
Average crack spacing at 0.1% strain level [mm]	25	17	25
Applied stress at 0.1% strain level [MPa]	5.8	8.2	14.1
Average crack width at 0.3% strain level [μm]	63	25	40
Average crack spacing at 0.3% strain level [mm]	17	10	13
Applied stress at 0.3% strain level [MPa]	3.8	8.2	12.6
Average crack width at 0.4% strain level [μm]	83	43	58
Average crack spacing at 0.4% strain level [mm]	13	10	13
Applied stress at 0.4% strain level [MPa]	3.4	9.2	14.4
Average crack width at the ultimate strain level [μm]	138	72	107
Average crack spacing at the ultimate strain level [mm]	13	7	13

Compared to the quasi-static tension tests, at the same strain level, the average crack width of TRC-PBO is higher than that of TRC-PE, which is the smallest among the three composites. Furthermore, the ultimate average crack spacing of TRC-PBO is larger. Even though TRC-PE still possesses a higher ultimate strain capacity of 1.2% comparing to 0.9% of TRC-PBO, its corresponding DIF of 0.4 reveals a pronounced loss in pre-peak strain capacity, leading to a slight decrease in the work-to-fracture; see Table 7. TRC-PBO, on the contrary, maintained strain capacity at the same level as under quasi-static loading, which led, along with the considerably higher tensile strength, to a significant increase in work-to-fracture, with the DIF being 1.8. The average crack-width–strain curve of TRC-M exhibits a considerably steeper slope after the formation of the first crack, indicating a more rapid degradation in the composite stiffness, as there is no contribution by short fibers. However, the improvement in the strain capacity (multiple cracking) indicates a better energy absorption behavior of this material under impact loading in comparison to its performance under quasi-static loading. The reason behind this improvement is the enhancement of the yarn-matrix bond under dynamic loading, which will be discussed in the next section.

The stresses applied and the average crack width growth at the critical strain levels are plotted in Figure 12. Taking into consideration both the load-carrying capacity and the average crack width developed, TRC-PBO exhibits the highest stress levels and a favorable crack control behavior at each strain level. When comparing to TRC-M, a sudden loss in composite stiffness is avoided, which means that TRC-PBO is able to carry a higher impact load with a more steady development of cracks.

4.2.2. Dynamic Single-Yarn Pullout Tests

The dynamic pullout curves are plotted in Figure 13. They show the same pattern as those obtained from the quasi-static tests. Note that the curves exhibit oscillations in the pullout stage, which are due to the intrinsic digital noise of the strain gauges, significant for such low pullout forces. Similar to the quasi-static tests, the peak bond strength between carbon yarn and the matrix with the addition of short PBO fibers, is the highest among the three composites, with an average of 1455 N; see Table 9. Note that the SHTB could ensure only a limited displacement during one wave passage, which was 2.5 mm.

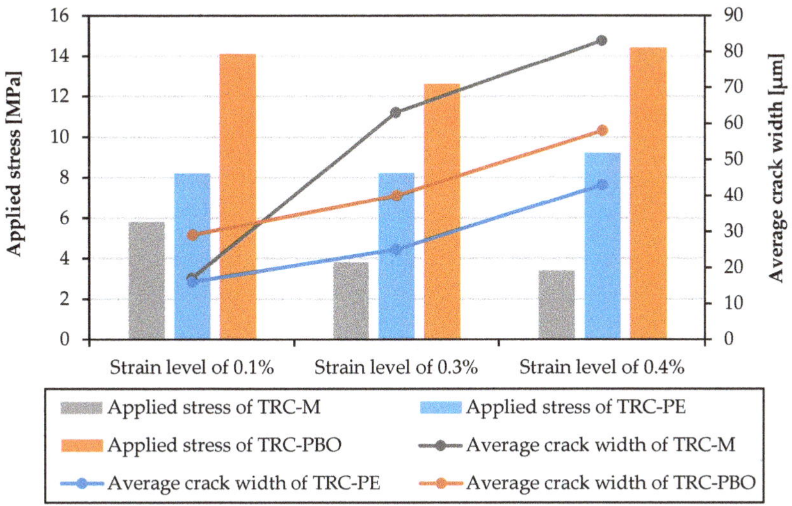

Figure 12. Average crack width and applied stress for representative specimens under impact tensile loading.

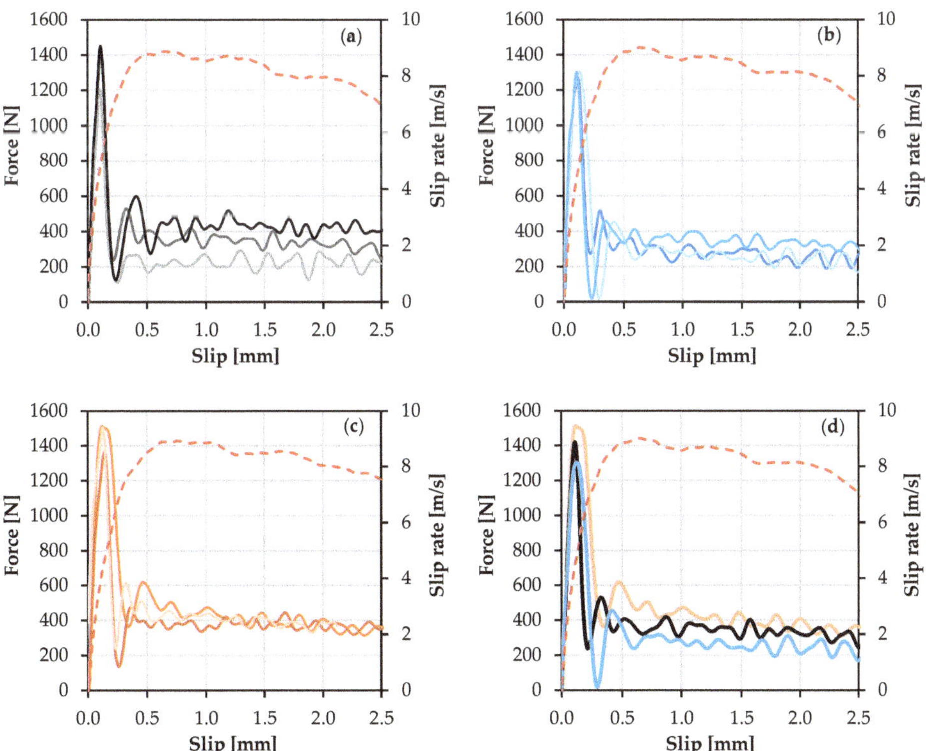

Figure 13. Dynamic force-slip relationships of single carbon yarns pulled out from: (**a**) plain matrix; (**b**) PE-SHCC; (**c**) PBO-SHCC; and (**d**) representative force-slip relationships for different composites up to a slip extent of 2.5 mm (the global strain rate curves are represented by dashed lines).

Table 9. Peak debonding force obtained from high-speed pullout tests with three different matrices (average values; standard deviations are given in parentheses).

	Single-Yarn Pullout from Plain Matrix	Single-Yarn Pullout from PE-SHCC	Single-Yarn Pullout from PBO-SHCC
Peak debonding force [N]	1355 (131)	1289 (22)	1455 (82)
DIF	3.1	2.9	2.7

It is obvious that the increasing displacement rate led to an increase in the peak pullout force with a dynamic increase factor (DIF) of around three for all three parameter combinations. The increased bonding between carbon yarn and matrix contributes to a higher tensile strength of the composites. Moreover, taking into consideration the increased tensile strength of the uncracked regions under a higher strain rate, higher stresses are needed to generate new cracks. In the previous section, it was shown that the number of cracks in TRC-M increased in impact tests, but it decreased in TRC-PE and TRC-PBO. In a future study, tension tests of single carbon yarn and short PE and PBO-AS fibers, and the pullout tests of the above-mentioned short fibers, need to be performed in order to attain a comprehensive understanding of the strain rate sensitivities of tensile properties and the bonding properties of the reinforcements to various matrices. What is more, as presented in [34], the bond strength could be enhanced by decreasing the spacing between weft yarns, upon the premise of a sufficient mesh spacing for short fibers to penetrate, which indicates the approach to further improve the bond properties of textiles by optimizing the mesh to an extent.

5. Conclusions

The tensile behavior of three composites reinforced by a carbon textile was investigated under quasi-static and impact tensile loading. Two of them contained additional short polymer fibers (SHCC matrices). The crack distributions were correlated to the strain of composites by means of digital image correlation. In this study, due to specific specimen geometry, only composites with SHCC matrices exhibited strain hardening behavior and multiple cracking in the quasi-static tension regime. Partly, this is related to material properties, and partly to the testing configuration, which did not allow for a sufficient anchorage of the textile reinforcement.

The dynamic loading leads to a pronounced increase in the tensile strength for all three materials, but the effects on the strain capacity are different. For composites containing both carbon textile and short PE fibers (TRC-PE), strain capacity decreases, accompanied by a smaller average crack width and larger average crack spacing. The average crack spacing in the composite containing carbon textile and short PBO fibers (TRC-PBO) is also considerably larger under impact tensile loading. However, in combination with the wider openings of the cracks, an ultimate strain similar to that in the quasi-static tests is obtained.

Additionally, the results of quasi-static and dynamic single-yarn pullout tests were presented. The high loading rate leads to a considerable increase in the yarn-matrix bonding strength for all three matrices under investigation. The bond is influenced by the type of short fiber reinforcement in the SHCC matrices; the addition of PBO fibers results in the highest bonding strength with the carbon yarn in both quasi-static and dynamic pullout tests.

Finally, it should be stressed that in the presented investigation, the premature rupturing of the textile reinforcements due to the specifics of the setup used, was observed, followed by the yarn's pullout. This phenomenon surely had an effect on the behavior of the composites that was recorded. This effect, which likely depends on the loading rate, needs further investigation. A corresponding experimental program is being planned by the authors. New ideas for preventing premature textile failure will be implemented. This program will involve other types of textile reinforcement as well, such as one made of UHMWPE fiber.

Author Contributions: Methodology, T.G., A.A.H. and I.C.; formal analysis, T.G.; investigation, T.G. and A.A.H.; data curation, T.G., A.A.H. and G.A.; writing—original draft preparation, T.G.; writing—review and editing, A.A.H., I.C. and V.M.; supervision, V.M. and I.C.; project administration, V.M.; funding acquisition, V.M.

Funding: This research was funded by German Research Foundation (Deutsche Forschungsgemeinschaft-DFG) within the framework of the Research Training Group GRK 2250.

Acknowledgments: The authors express their great gratitude to the German Research Foundation (Deutsche Forschungsgemeinschaft-DFG) for the financial support provided within the framework of the Research Training Group GRK 2250. Credit is also given to Mr. Kai Uwe Mehlisch for his valuable contribution in performing the experiments. The authors also express their great gratitude to Open Access Funding by the Publication Fund of the TU Dresden.

References

1. Mechtcherine, V. Novel cement-based composites for the strengthening and repair of concrete structures. *Constr. Build. Mater.* **2013**, *41*, 365–373. [CrossRef]
2. Butler, M.; Mechtcherine, V.; Hempel, S. Experimental investigations on the durability of fibre-matrix interfaces in textile-reinforced concrete. *Cem. Concr. Compos.* **2009**, *31*, 221–231. [CrossRef]
3. Sasi, E.A.; Peled, A. Three dimensional (3D) fabrics as reinforcements for cement-based composites. *Appl. Sci. Manuf.* **2015**, *74*, 153–165. [CrossRef]
4. Li, V.C. From michromechanics to structural engineering—The design of cementitious composites for civil engineering applications. *Jpn. Soc. Civ. Eng.* **1993**, *10*, 37–48.
5. Li, V.C. On engineered cementitious composites (ECC). *J. Adv. Concr. Technol.* **2003**, *1*, 215–230. [CrossRef]
6. Mechtcherine, V.; Millon, O.; Butler, M.; Thoma, K. Mechanical behaviour of strain hardening cement-based composites under impact loading. *Cem. Concr. Compos.* **2011**, *33*, 1–11. [CrossRef]
7. Hinzen, M.; Brameshuber, W. Improvement of Serviceability and Strength of Textile Reinforced Concrete by Using Short Fibres. In Proceedings of the 4th Colloquium on Textile Reinforced Structures, Dresden, Germany, 3–5 June 2009; pp. 261–272.
8. Butler, M.; Hempel, R.; Schiekel, M. The influence of short glass fibres on the working capacity of textile reinforced concrete. In Proceedings of the 1st International RILEM Symposium Textile Reinforced Concrete, Aachen, Germany, 6–7 September 2006; pp. 45–54.
9. Brameshuber, W.; Hinzen, M. Improving the first crack behaviour of textile reinforced concrete. In Proceedings of the 7th RILEM Workshop on High Performance Fiber Reinforced Cement Composites, Stuttgart, Germany, 1–3 June 2015; pp. 13–20.
10. Gong, T.; Hamza, A.A.; Curosu, I. On the synergetic action between strain-hardening cement-based composites (SHCC) and carbon textile reinforcement. In Proceedings of the 10 th International Conference on Fracture Mechanics of Concrete and Concrete Structures FraMCoS-X, Bayonne, France, 23–26 June 2019. [CrossRef]
11. De Andrade Silva, F.; Butler, M.; Mechtcherine, V.; Zhu, D.; Mobasher, B. Strain rate effect on the tensile behaviour of textile-reinforced concrete under static and dynamic loading. *Mater. Sci. Eng. A* **2011**, *528*, 1727–1734. [CrossRef]
12. Barhum, R.; Mechtcherine, V. Effect of short, dispersed glass and carbon fibres on the behaviour of textile-reinforced concrete under tensile loading. *Eng. Fract. Mech.* **2012**, *92*, 56–71. [CrossRef]
13. Yao, Y.; Silva, F.A.; Butler, M.; Mechtcherine, V.; Mobasher, B. Tension stiffening in textile-reinforced concrete under high speed tensile loads. *Cem. Concr. Compos.* **2015**, *64*, 49–61. [CrossRef]
14. Yao, Y.; Bonakdar, A.; Faber, J.; Gries, T.; Mobasher, B. Distributed cracking mechanisms in textile-reinforced concrete under high speed tensile tests. *Mater. Struct. Constr.* **2016**, *49*, 2781–2798. [CrossRef]
15. Soranakom, C.; Mobasher, B. Modeling of tension stiffening in reinforced cement composites: Part I. Theoretical modeling. *Mater. Struct. Constr.* **2010**, *43*, 1217–1230. [CrossRef]
16. Shi, T.; Leung, C.K.Y. An effective discrete model for strain hardening cementitious composites: Model and concept. *Comput. Struct.* **2017**, *85*, 27–46. [CrossRef]

17. Ranade, R.; Li, V.C.; Heard, W.F. Tensile rate effects in high strength-high ductility concrete. *Cem. Concr. Res.* **2015**, *68*, 94–104. [CrossRef]

18. Shim, V.P.W.; Lim, C.T.; Foo, K.J. Dynamic mechanical properties of fabric armour. *Int. J. Impact Eng.* **2001**, *25*, 1–15. [CrossRef]

19. Zhu, D.; Mobasher, B.; Erni, J.; Bansal, S.; Rajan, S.D. Strain rate and gage length effects on tensile behavior of Kevlar 49 single yarn. *Compos. Part A Appl. Sci. Manuf.* **2012**, *43*, 2021–2029. [CrossRef]

20. Yang, E.-H.; Li, V.C. Tailoring engineered cementitious composites for impact resistance. *Cem. Concr. Res.* **2012**, *42*, 1066–1071. [CrossRef]

21. Curosu, I. Influence of Fiber Type and Matrix Composition on the Tensile Behavior of Strain-Hardening Cement-Based Composites (SHCC) under Impact Loading. Ph.D. Thesis, Technische Universität Dresden, Dresden, Germany, 20 July 2017.

22. Curosu, I.; Mechtcherine, V.; Millon, O. Effect of fiber properties and matrix composition on the tensile behavior of strain-hardening cement-based composites (SHCCs) subject to impact loading. *Cem. Concr. Res.* **2016**, *82*, 23–35. [CrossRef]

23. Curosu, I.; Liebscher, M.; Mechtcherine, V.; Bellmann, C.; Michel, S. Tensile behavior of high-strength strain-hardening cement-based composites (HS-SHCC) made with high-performance polyethylene, aramid and PBO fibers. *Cem. Concr. Res.* **2017**, *98*, 71–81. [CrossRef]

24. Li, V.C.; Leung, C.K.Y. Steady-state and multiple cracking of short random fiber composites. *J. Eng. Mech.* **1992**, *118*, 2246–2264. [CrossRef]

25. Fact Sheet, Ultra High Molecular Weight Polyethylene Fiber Form Dyneema, Eurofibers. Available online: https://issuu.com/eurofibers/docs/name8f0d44,15-11-2010 (accessed on 1 July 2019).

26. Technical Information, PBO Fiber Zylon, Toyobo CO., LTD. Available online: http://www.toyobo-global.com/seihin/kc/pbo/zylon-p/bussei-p/technical.pdf (accessed on 1 July 2019).

27. Allgemeine bauaufsichtliche Zulassung, Verfahren zur Verstärkung von Stahlbeton mit TUDALIT (Textilbewehrter Beton). Available online: http://www.textilbetonzentrum.de/app/download/5806918971/AbZ_Z-31.10-182.pdf (accessed on 30 November 2016).

28. Heravi, A.A.; Mechtcherine, V. Mechanical characterization of strain-hardening cement-based composite (SHCC) under dynamic tensile load. In Proceedings of the 10th International Conference Fracture Mechanics for Concrete and Concrete Structures, Bayonne, France, 24–26 June 2019. [CrossRef]

29. Mobasher, B.; Peled, A.; Pahilajani, J. Distributed cracking and stiffness degradation in fabric-cement composites. *Mater. Struct. Constr.* **2006**, *39*, 317–331. [CrossRef]

30. Peled, A.; Bentur, A. Quantitative description of the pull-out behavior of crimped yarns from cement matrix. *J. Mater. Civ. Eng.* **2003**, *15*, 537–544. [CrossRef]

31. Sueki, S.; Soranakom, C.; Mobasher, B.; Peled, A. Pullout-slip response of fabrics embedded in a cement paste matrix. *J. Mater. Civ. Eng.* **2005**, *19*, 718–727. [CrossRef]

32. Al Ghazali, A.; Schröfl, C.; Mechtcherine, V. Plastic shrinkage of high-performance strain-hardening cement-based composites (HP-SHCC). In Proceedings of the 1st International Conference on Cement & Concrete Technology, Muscat, Oman, 20–22 November 2017; pp. 396–408.

33. Curosu, I.; Mechtcherine, V.; Forni, D.; Cadoni, E. Performance of various strain-hardening cement-based composites (SHCC) subject to uniaxial impact tensile loading. *Cem. Concr. Res.* **2017**, *102*, 16–28. [CrossRef]

34. Jiang, J.; Jiang, C.; Li, B.; Feng, P. Bond behavior of basalt textile meshes in ultra-high ductility cementitious composites. *Compos. Part B Eng.* **2019**, *174*, 107022. [CrossRef]

 *applied sciences*

Article

Long-Term Durability of Carbon-Reinforced Concrete: An Overview and Experimental Investigations

Arne Spelter *, Sarah Bergmann, Jan Bielak and Josef Hegger

Institute of Structural Concrete, RWTH Aachen University, 52074 Aachen, Germany; sbergmann@imb.rwth-aachen.de (S.B.); jbielak@imb.rwth-aachen.de (J.B.); jhegger@imb.rwth-aachen.de (J.H.)
* Correspondence: aspelter@imb.rwth-aachen.de

Received: 20 March 2019; Accepted: 16 April 2019; Published: 21 April 2019

Featured Application: In conventional reinforced concrete structures, corrosion problems often occur due to insufficient concrete cover in exposed structures. Corrosion can be avoided by using carbon textiles as reinforcement. However, the long-term durability behavior of non-metallic reinforcement, e.g., made of carbon filaments and a polymer impregnation, must be considered. This work presents first results and the current state of long-term durability investigations of an epoxy resin impregnated carbon textile.

Abstract: Despite intensive research on material properties of non-metallic technical textiles for internal reinforcement in concrete, the long-term durability is not yet fully understood. In this work, results of preloaded long-term durability tensile tests on carbon-reinforced concrete specimens under environmental factors of stress, temperature, moisture and alkalinity are presented. Based on investigations of non-metallic glass fiber reinforcements with polymer matrices, where strength losses occur over time, it was planned to derive a time to failure curve and to determine a reduction factor for the tensile strength of the carbon textile reinforcement. However, no loss of strength was discovered in residual capacity tests due to the high material resistance and therefore no reduction factor due to the environmental factors could be derived. After more than 5000 h of testing, the residual capacity tests showed an increase in the ultimate failure stress in comparison with the short-term tests. In addition to the long term-durability tests, the influence of the preloading was investigated. The preload was applied to the long-term tests and led to a straighter alignment and loading of the filaments and thus to an increase in the ultimate capacity.

Keywords: alkaline environment; carbon-reinforced concrete; creep; durability; moisture; tensile strength; textile reinforced concrete

---

1. Introduction

Textile reinforced concrete (TRC) is a composite material consisting of non-metallic reinforcement and a concrete matrix adjusted to the requirements of the reinforcement. The grid-like reinforcement consists of impregnated yarns with up to thousands of filaments. (AR-)glass or carbon, for example, are used as filament material, while polymers, such as epoxy resin, styrene-butadiene or vinyl ester, are used for the polymer matrix to improve the utilization of the base material. When using carbon reinforcement, the term carbon-reinforced concrete (CRC) is used.

Due to the non-corrosive carbon filaments, the material is significantly more durable than reinforced concrete and withstands high tensile stresses. AR-glass reinforcement reaches lower ultimate stress levels than carbon reinforcement but is less expensive and still achieves a significantly higher tensile stress than steel reinforcement.

The material behavior of carbon-reinforced concrete has been investigated for almost three decades. During this time various projects with textile reinforcements have been realized [1–4]. Meanwhile, material parameters for design have been investigated [5–8], and standardized testing methods have been derived [9,10].

For the realization of buildings and elements made of CRC, potential users in Germany are dependent on an approval in individual cases or a general building approval. To date, there are no accepted standards, data sheets and guidelines available that make these approvals for dimensioning redundant.

Therefore, the research program C^3—Carbon Concrete Composites—was initiated, which focuses on the economical and marketable application of carbon-reinforced concrete. In individual subprojects of this research program, various aspects for the design and application of carbon-reinforced concrete are investigated [11]. For example, a carbon concrete guideline is drafted within the framework of project C^3-V1.2, while in the project C^3-V2.1, the long-term durability behavior of carbon reinforcement in composite is investigated for both durability and fatigue.

For approvals and eventually design concepts, the question of strength losses due to long-term loading and environmental exposure must be addressed. So far, long-term losses have been considered using conservative reduction and partial safety factors which, however, do not allow for an economical design. For this reason, the Institute of Structural Concrete of RWTH Aachen University is investigating the long-term durability of an epoxy resin impregnated carbon reinforcement combined with a high-strength concrete. The study is part of the collaborative subproject C^3-V2.1. This paper presents the results of current small-scale experiments to determine and evaluate the long-term durability of this material combination and sets the current results in the context of reduction factors from the literature.

To develop a testing concept for investigating the long-term durability of carbon-reinforced concrete, this property must be defined first. According to CSA S807-10 [12] (p. 3), durability is defined as 'the capability of a component, product, or structure to maintain its function for at least a specified period of time without appropriate maintenance'. The properties of fiber-reinforced polymers include alkaline and creep resistance and can both be determined according to CSA S806-12 [13], ACI 440.3R-12 [14] or ISO 10406-1 [15].

The testing methods for alkaline resistance of FRP bars (CSA S806-12 Annex M [13]; ACI 440.3R-12 B.6 [14]; ISO 10406-1 section 11 [15]) are used to investigate the tensile capacity and the weight of an FRP rod before and after immersion in an alkaline solution. These testing methods differentiate between tests with and without loading, as well as immersion in an alkaline solution or exposure in concrete.

The test method for creep rupture of FRP bars (ACI 440.3R-12 [14]) refers to the ASTM D7337/D7337-M [16] 'Standard Test Method for Tensile Creep Rupture of Fiber Reinforced Polymer Matrix Composite Bars', where the creep rupture capacity is defined as 'the force at which failure occurs after a specified period of time from initiation of a sustained force' (p. 1). The testing concepts for the investigation of the creep behavior of the FRP bars are based on the determination of the time to failure due to a constant stress. A semi-logarithmic relationship between time and constant stress is derived [13,15,16].

A testing concept which considers the combined testing of the creep behavior and alkaline resistance was developed by Weber and Baquero [17]. The concept was developed for the approval procedure of the GFRP Schöck ComBar® in Germany. Long-term durability tests at temperature levels of 23, 40 and 60 °C in high-alkaline water-saturated concrete were performed to prove the Arrhenius equation. The equation verifies that the chemical reactions are accelerated by the elevated temperature but are not changed due to a single mechanism that controls the degradation process [18]. Finally, a time-acceleration factor can be determined. In comparison to the previously mentioned testing concepts, the results are plotted in a log-log diagram, where the relationship between the applied stress and the time to failure is linear. Further information on the testing concept can be found in [17,19].

Due to the alkaline environment of the concrete, the presence of moisture and changing temperatures in exterior components as well as loads which lead to stress in the reinforcement, it is necessary to investigate the long-term behavior under combined exposure. In this work, the long-term durability is therefore defined as a constant stress on a textile reinforcement under the influence of the environmental conditions that can be applied during the service life of a building structure without failure of the reinforcement.

2. Long-Term Durability of Non-Metallic Reinforcement

The long-term durability is affected by environmental factors such as stress, temperature, moisture and alkalinity (e.g., [17,20]). These factors are not only used to simulate the environmental conditions, but also for artificial aging, as it is not possible to carry out long-term tests over a service life of up to 100 years [17,21–25]. To determine the residual strength at the end of the life time, an extrapolation of the trend line based on the test data is necessary [16,17,26].

Tensile stress has the greatest influence on the long-term durability of non-metallic reinforcement. Depending on the level of stress, the load leads to a stress fracture of the filaments (creep rupture) after a certain time [27]. Micro cracks in the polymeric matrix appear due to the resulting stress in the yarns, which subsequently enable a chemical attack of the filaments [21]. Furthermore, the matrix can transfer less stress to the neighboring filaments in the area of the micro cracks.

Due to the manufacturing process, the yarns are not exactly aligned, which leads to an uneven load when subjected to stress. However, it is expected that due to the internal composite stresses, the impregnation material creeps. This has a positive effect on the alignment of the filaments. Over time, the micro cracks may grow, which provokes a filament rupture when the ultimate strain or rather the ultimate stress is reached. Consequently, the load has to be transferred to the non-cracked filaments, which can lead to further filament rupture and finally the failure of the textile reinforcement [26].

The ambient temperature reflects the energy level. With increasing temperature, the energy level rises and chemical reactions are accelerated [18]. Litherland et al. [28] carried out fiber strand-in-cement strength tests with Cem-FIL AR glass fibers at temperatures from 20 °C to 80 °C and concluded that a chemical reaction controls the speed over the range of this temperatures.

The long-term durability is also influenced by moisture. Moisture is present in every building component. Outdoor structures such as façade panels and bridges show an increased moisture level due to exposure to rain and melting snow. Many chemical reactions take place in concrete under the influence of moisture [29]. It serves as a transport medium for alkalis and other substances from the concrete or the environment to the reinforcement. According to Orlowsky and Raupach [30], a minimum moisture level must be exceeded before the transportation of moisture and OH-ions on the filament surfaces leads to a loss of strength of unimpregnated AR-glass reinforcement.

For uncracked concrete components, the speed of the damaging process depends on the porosity of the concrete and the diffusion coefficient of the impregnation material of the textile reinforcement. In cracked components, moisture can penetrate faster through the impregnation material to the filaments due to cracks in the concrete matrix and micro cracks in the polymer matrix caused by stress in the area of the cracks [29], as described above.

Besides the environmental factors, the long-term durability also depends on the material parameters of the non-metallic reinforcement. The filament material (e.g., carbon, glass and basalt), the impregnation material (e.g., epoxy resin, styrene-butadiene and vinylester), the cross-sectional area and the shape of the yarns, the transversal yarn spacing as well as the production process affect the long-term durability.

Carbon filaments seem to be resistant under the environmental conditions. Carbon filaments do not undergo corrosion in alkaline environments nor absorb moisture [31]. In addition to that, the influence of the temperature on the durability of the carbon filaments is low. The temperature resistance of carbon even increases with production temperature [32]. However, strength losses may occur due to the impregnation material.

The impregnation material serves to protect the filaments. However, a loss of strength compared to the initial strength due to the impregnation material is possible [33]. Ceroni et al. [34] mentioned the degree of impregnation, the absence of cracks and voids, the resistance to micro-cracking, as well as the degree of curing independent of fiber and impregnation material of the polymeric matrix as key factors for the reduction of durability of FRP materials.

The long-term durability of non-metallic reinforcement is influenced by the environmental factors stress, temperature, moisture and alkalinity. Carbon filaments appear to be resistant to these factors. Therefore, the failure mechanisms that may occur for composites with non-metallic reinforcement must be considered.

3. Failure Mechanisms of Non-Metallic Reinforcement

The environmental factors influence the long-term durability behavior of different types of non-metallic reinforcement to varying degrees and cannot be generalized. For certain materials, these factors inhibit a clear definition of global failure mechanisms. Nevertheless, an attempt will be made to define such mechanisms.

According to Bank et al. [31], the failure occurs in three different phases, or their combination: the fibers, the matrix, and the interphase. A damage of one phase often influences the failure mode of another phase.

Micro cracks in the matrix do not lead to a failure of the composite material (polymeric matrix and fibers) but allow moisture and dissolved chemicals to penetrate the matrix. Under certain circumstances, this leads to an attack on the fibers (e.g., AR-glass) [35]. The failure of the composite material is attributed to the damage of the fiber-matrix interphase [31].

According to Mufti et al. [36], the glass transition temperature is lowered by plastification due to moisture in the matrix of the resin. Due to the moisture absorption and alkalis, the matrix may be damaged because of swelling stresses by cracking, hydrolysis or fiber-matrix debonding. This results in reduction of the stress transfer capability between fiber and matrix [37–39].

The degradation of the fiber-matrix interphase is important for the overall damage of the composite material and, according to Ray [33], the dominating mechanism during environmental aging. Ray [33] examined the effect of temperature on interfaces of fiber-reinforced epoxy composites during humid aging and figured out that the interfacial adhesion is more influenced by hygrothermal aging at higher temperature and longer exposure times. The mechanism of attack depends on the chemistry, structure, morphology and modes of failure at the interface.

Due to environmental factors, failure of the filaments, the matrix or the interphase can lead to failure. A chemical attack of carbon filaments is not expected, but the fiber-matrix interphase or the matrix can be damaged and may lead to long-term failure. To investigate the influences and possible failure mechanisms on an epoxy resin impregnated carbon reinforcement, long-term tests are carried out. The factors stress, temperature and moisture are also used for accelerated aging of the test specimens.

4. Materials and Test Specimens

In this study, an epoxy resin impregnated carbon textile and a high-strength concrete were selected. The chosen biaxial texile solidian GRID Q95/95-CCE-38 from solidian GmbH, whose characteristics where determined in uniaxial single yarn tests [40] (Table 1). A section of the carbon textile reinforcement is displayed in Figure 1.

Table 1. Material characteristics of the carbon textile solidian GRID Q95/95-CCE-38 (properties of one individual yarn [41], test setup according to [42]).

Characteristic	Unit	Warp Direction (0°)	Weft Direction (90°)
Modulus of elasticity	MPa	244,835	243,828
Mean ultimate stress	MPa	3221 (n = 204 tests)	3334 (n = 218 tests)
5% quantile ultimate stress	MPa	2737	2762
95% quantile ultimate stress	MPa	3705	3906
Axial spacing of yarns	mm	38	38
Cross-sectional area per yarn [1]	mm^2	3.62	3.62

[1] Filament area without epoxy impregnation.

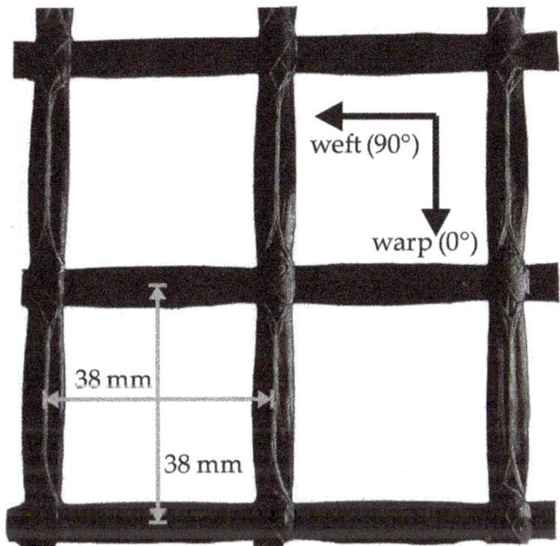

Figure 1. Detail of the tested textile carbon reinforcement.

The requirements on the concrete had changed due to textile reinforcement, so that the concrete composition had to be adapted. The maximum grain size for the ability of penetration through the textile reinforcement [43], the long-term availability of the raw materials, ecological and economic criteria as well as a short-term feasibility of the results in construction practice were taken into account [44] for the development of the concrete, which was part of the basis project B2 of the C^3 program.

The concrete composition used differs slightly from the compositions presented by Schneider et al. [44] because locally available raw materials were used. The maximum grain size is 4 mm, which means that the concrete can be classified as mortar. However, due to the high-strength properties, the term concrete was established. The composition of the high-strength concrete is shown in Table 2.

Table 2. Mix design of cementitious matrix for concrete HF-2-165-4 (mix design adapted from [44]).

Substance	Density [kg/m^3]	Content [kg/m^3]
Binder compound CEM II/C-M Deuna	2962	707
Quartz sand F38 S	2650	294
Quartz sand 0.1–0.5 mm	2630	243.2
Quartz sand 0.5–1.0 mm	2630	201.4
Quartz sand 1.0–2.0 mm	2630	148.9
Quartz sand 2.0–4.0 mm	2630	593.5
Superplasticizer (polycarboxylatether base) MC-VP-16-0205-02	1070	15
Water	1000	165

When concreting, first, the binder and the aggregate are mixed dry for one minute before water and the superplasticizer are added. After a further three minutes of mixing time, the concrete is poured in the formworks. Due to the self-compacting properties of the concrete, no vibration is required. After concreting, the test specimens in the formworks are protected against drying out. After one day, the formworks are removed, and the specimens are stored in water until the 7th day. From the 8th day, the samples are stored at room climate (approximately 20 °C and 65% RH). After reaching a minimum concrete age of 28 days, the test specimens are tested for their short-term, long-term and preloading behavior.

The compressive strength and flexural strength of the concrete were determined on test specimens according to DIN EN 196-1 [45] after 28 to 31 days. As an average of 36 specimens, the compressive strength was 123.7 MPa with a coefficient of variation (COV) of 7.7%. The average flexural strength was 12.2 MPa for 18 samples (COV 15.1%). The average modulus of elasticity was determined to be 43,839 MPa (COV 4.9%) [46] on six cylinders of 30 cm height and 15 cm diameter.

All test specimens are reinforced with a single layer of the carbon textile. The specimens have dimensions of $w/t/l$ = 120/30/1000 mm^3. Each specimen is reinforced with a section of the biaxial fabric with three yarns in test direction (warp direction, Figure 1). The concrete cover is 15 mm.

5. Experimental Investigations

5.1. Introduction

To determine a time to failure curve and a possible reduction factor for a certain service life, e.g., 100 years, for the material combination described in Section 4, long-term tests are carried out. By means of twenty short-term tests with a concrete age of 28 days, the reference load levels for the long-term tests are defined.

The test specimens of the long-term tests were initially preloaded with a final crack pattern for 24 h in tempered water on half of the reference load to ensure a uniform internal load distribution before the constant load is applied. In further tests the influence of the preloading is investigated.

Due to the environmental factors explained in Section 2, the preloading and long-term durability tests are performed under the combined impact of stress, temperature, moisture and alkalinity. The aim of the testing concept developed [19], and originally planned for these tests, was to carry out long-term tests over 200, 1000 and 5000 h. To derive a reduction factor at the end of the service life, a semi-logarithmic relationship between stress and time from variation of the constant stress levels is assumed. To determine a reduction factor, an extrapolation of the time to failure curve is necessary. However, the concept, which is successfully applicable to glass reinforcement, cannot be applied to the long-term tests with the carbon reinforcement presented in this paper, as no failure due to the exposure occurred during the test periods. In residual capacity tests, no reduction could be determined after the end of exposure. Therefore, a different approach will be necessary.

5.2. Experimental Procedure

5.2.1. Short-Term Tests

The tests were carried out in a universal testing machine with a maximum load of 100 kN. Two inductive displacement transducers were pasted on the formwork side and one on the filling side of each specimen to measure the deformations and determine the failure strain of the textile reinforcement.

The load is applied via hydraulic clamping devices through steel plates with clamping length of 200 mm (Figure 2).

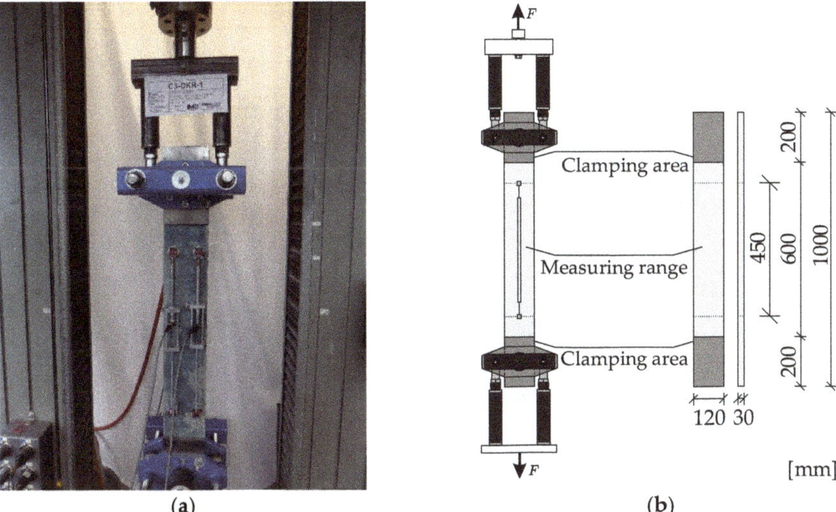

(a)   (b)

Figure 2. (**a**) Specimen in testing machine; (**b**) schematic (according to [10]) of the test setup for short-term tests.

The reference load for the long-term durability tests was determined at room temperature based on two test series, each consisting of ten test specimens. The specimens were tested at an age of 28 days and loaded monotonously with 1 mm/min until failure. The ultimate strain was evaluated at the time of the maximum stress.

5.2.2. Long-Term Tests

The long-term durability tests are carried out in especially developed test rigs (Figure 3a). Those rigs allow a long-term loading of the test specimens as well as a readjustment of the forces by hollow piston cylinders, which are operated hydraulically. In addition to the constant load, the specimens are exposed to tempered water (40 or 60 °C).

The tap water used is not exchanged during the entire test period. The pH-value of the water is 8 and rises rapidly to approximately 11 after dissolution of alkalis from the concrete of the test specimens. This results in an alkaline solution which, besides local attacks by the concrete pore solution on the reinforcement, leads to an attack on the reinforcement material. Accordingly, a simultaneous exposure of stress, moisture, temperature and alkalinity can be examined (Figure 3b).

The load is again applied via clamping devices made of steel plates (clamping length 250 mm). To ensure a permanent clamping of the test specimens, the clamping length increased by 50 mm compared to the short-term tests. The contact pressure is applied by lateral mechanically pretensioned threaded rods. Since the test specimens slipped out during first trial tests, the contact pressure increased to approximately 4.0 MPa (short-term tests 3.0 MPa). The reason for the slipping of the specimens is the thermal expansion of the threaded rods in the tempered water, which leads to a reduced contact pressure compared to the short-term tests.

To measure the creep deformations of the specimens or rather the carbon reinforcement, inductive displacement sensors are placed on the formwork and the filling side of the test specimens. Due to the length of the clamping devices, a test area of 500 mm and a measuring area of 450 mm were chosen.

(a)   (b)

Figure 3. (**a**) Long-term durability test rigs; (**b**) Test specimen under exposure.

While recording the deformations, the test specimens are loaded slowly in the tempered water, up to half of the reference load from the short-term tests. However, due to the high tensile strength of the concrete, no cracks will appear at this load level. Therefore, the specimens are loaded up to a final crack pattern before the load level is reduced back to half of the reference load and kept constant for 24 h.

This constant load serves as a preload for the test specimens, as it was specified in the project C^3-V2.1. The influence of the preloading will be explained in detail in Section 6. After 24 h, the load level is applied and kept constant by manual readjustment over the entire test period until the failure of the specimens. The test time generally depends on the height of the load level.

5.2.3. Preloading Tests

The first part of the preloading tests is performed in the test rigs of the long-term tests (Figure 3a). The clamping devices and the measuring technology are therefore the same.

After 24 h of preloading, the three specimens are removed from the test rigs and dried for 24 h at room temperature. The residual load capacity of the test specimens is determined with the test setup described in Section 5.2.1.

The clamping length is again 250 mm, according to the long-term tests.

The procedure of the preloading is already described in Section 5.2.2. After 24 h of preloading in 60 °C tempered water, the test specimens are removed from the test rigs and dried for 24 h at room temperature. The test specimens are then installed in the testing machine described in Section 5.2.1 and loaded to the ultimate load at 1 mm/min.

6. Results

6.1. Short-Term Tests

The results of the two-test series are presented in Figure 4. The mean failure stress $\sigma_{t,max}$ of the first test series (Figure 4a) was 2969 MPa with a maximum strain ε_t of 10.6‰. The mean failure stress $\sigma_{t,max}$ of the second test series (Figure 4b) was 3130 MPa with a maximum strain ε_t of 11.1‰. This results in a reference failure stress of 3050 MPa with a maximum strain of 10.9‰. The coefficient of variation is 9% for the mean failure stress and 17‰ for the mean strain at failure. The high variance of the strain ε_t can be explained with the indirect measurement of the deformations on the concrete.

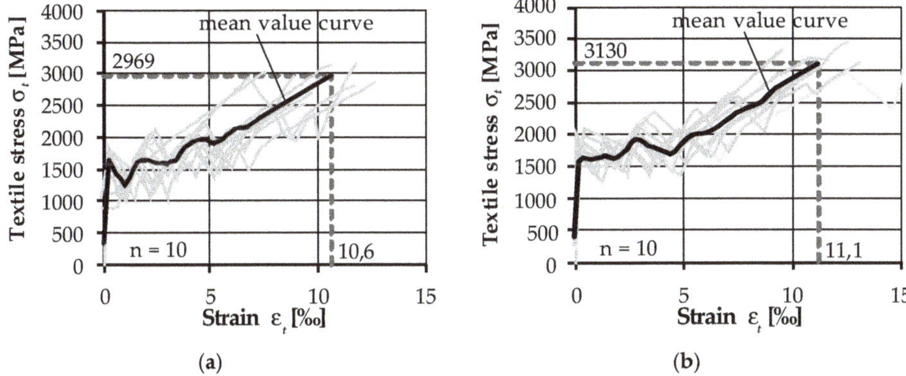

Figure 4. Results of the first (**a**) and second (**b**) short-term test series.

The black line represents the mean value curve from each of the 10 individual curves, which was averaged over strain sections of 0.2‰. The reference load is necessary for the specification of the preloading and long-term durability load levels.

All tests specimens failed due to the rupture of the reinforcement. The minimum failure stress achieved in the first test series was 2532 MPa, while 2692 MPa was achieved in the second test series. The maximum failure stress was 3465 MPa and 3478 MPa, respectively. Two to four cracks occurred in the measuring range during the tests.

6.2. Long-Term Durability Tests

After the reference failure stress (3050 MPa) was determined, constant load levels were selected for the long-term durability tests. For the first load level the aim was to achieve failure within a few hundred hours based on the testing concept [19]. Subsequently, the load levels were to be reduced to reach test times of 1000 and 5000 h.

Table 3 provides an overview of the performed long-term durability tests (CLTT—Carbon Long-Term Tests). Specimens that failed during loading were excluded.

Table 3. Overview of the long-term durability tests.

Tests	Constant Load [MPa]	Test Time [h]	Temperature [°C]	Ratio $\sigma_{t,x}/\sigma_{t,0}$ [-]	Residual Capacity [MPa]
CLTT-1 [1]	2168	11,000	40	0.71	-
CLTT-2	2685	5117	40	0.88	3360
CLTT-3 [1]	2795	7175	40	0.90	-
CLTT-4	2813	5400	40	0.92	3504
CLTT-5	2865	5300	40	0.94	3582 [3]
CLTT-6 [2]	2933	811	60	0.96	-
CLTT-7 [1]	2958	5930	40	0.97	-

[1] Running test; [2] not preloaded, [3] reduced test area.

As displayed in Table 3, the specimens did not fail, with one exception, at loads exceeding 90% of the reference load, so that a few tests were stopped after 5000 to 6000 h to determine the residual capacity of the specimens. The failure of the specimen CLTT-6 is attributed to the non-applied preloading.

No decrease in strength could be observed in the residual load capacity—the ultimate capacity was even higher than the short-term strength. It is assumed that the preloading and the constant loading lead to an uniformization of the individual yarns and of the filaments in each yarn regarding stress and strain levels. This would increase the load capacity of the reinforcement. The effect of a preloading is presented in the next section and discussed in Section 7.

In Figure 5a the results of the long-term durability tests from Table 3 are plotted in a semi-logarithmic diagram. The red arrows in Figure 5a symbolize the ongoing tests. CLTT-1, which was started as a pretest and whose load level is clearly below that of the other tests, is not plotted in the diagram. Figure 5b displays the tests which were stopped after 5000 to 6000 h, as well as their respective residual capacities.

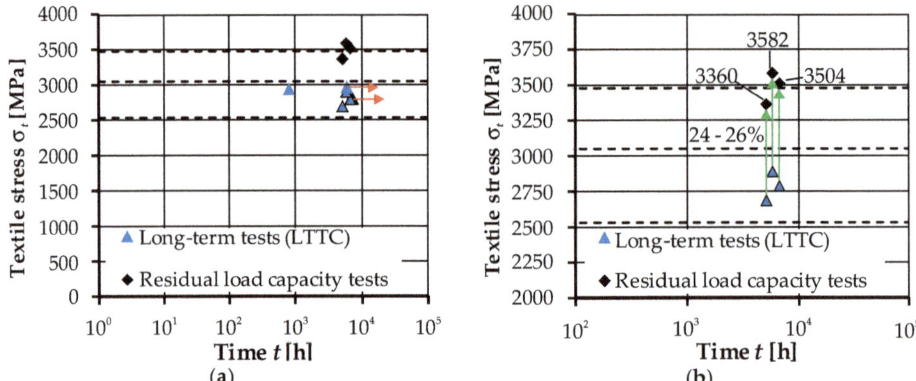

Figure 5. (a) Overview of the long-term durability tests; (b) stopped long-term durability tests and their residual capacity.

The dotted lines represent the maximum textile stress (3476 MPa), the reference stress (3050 MPa) and the lowest textile stress (2532 MPa) achieved in the short-term tests (Section 6.1). The upper and lower lines represent the scatter range of the ultimate stresses of the textile.

As described before, the original aim was the derivation of a time to failure curve for the tested material combination. Based on the test results, however, it is not possible to identify such a curve to determine the strength losses at the end of the service life. For this purpose, a starting point of the time to failure curve must be determined first. Since no failure of the long-term durability tests could be achieved within a few hundred hours due to the constant load levels and environmental factors, preloading tests are carried out.

Figure 5b clearly shows that the stopped tests still had a residual capacity of 24 to 26% above the constant load or 10 to 17% above the reference load, respectively. Therefore, no loss of strength could be detected. However, it is presumed that an alignment of the filaments occurs, which leads to an increase of the load capacity compared to the reference load. The results of the preloading tests are presented in the next section.

6.3. Preloading Tests

The preloading over 24 h in 60 °C tempered water led to an increase of 7% (3270 to 3050 MPa) in the load capacity compared to the reference load. The test specimens had two to three cracks in the measuring area as a result of the preloading. The number of cracks corresponds well with the number of cracks in the short-term (Section 6.1) and long-term durability tests (Section 6.2). When determining the residual capacity, no further cracks occurred so that each test specimen had already achieved a final saturated crack pattern.

A direct comparison of the stress strain curves is not feasible due to the removal of the specimens from the test rigs (Section 5.2.2) and the installation in the testing maschine for the short-term tests (Section 5.2.1).

The higher mean ultimate stress is attributed to the alignment of the filaments (training effect).

7. Discussion

No reduction of the long-term durability for the carbon textile reinforcement could be determined due to the environmental factors of stress, temperature, moisture and alkalinity. The comparison of the residual capacity tests to the short-term tests showed that no loss of strength occurs during the test times. Instead, an increase in the load capacity after more than 5000 to 6000 h was observed. This increase is attributed to a load-oriented alignment of the individual filaments (training effect). The non-exact alignment of the filaments in the production process is reduced because of the pre- and constant long-term load. This leads to an improvement of the ultimate tensile capacity in comparison with the short-term tests, which did not experience any preloading or constant load, respectively. At present, no level of degradation of the textiles exposed to long-term exposure compared to the short-term tests can be determined.

Three preloading tests verified the so-called 'training effect' and will be further investigated in future studies. An increase of 7% in the load capacity compared to the short-term tests was observed. The use of internal fiber sensors [47] could provide information on the exact strain of textile reinforcement in the crack area. This would enable the measurement of the actual strain of the reinforcement in the area of the concrete cracks and hence the development of the strain due to a pre- and constant long-term load. Furthermore, to verify the statement of the 'training effect', the test specimens of the accelerated long-term tests could be compared to test specimens loaded under real conditions.

According to the current state of investigations, it cannot be assumed that the results can directly be transferred to other material combinations of high-strength concrete and carbon textile reinforcement, because many characteristics of the technical textiles influence the long-term durability behavior. Furthermore, it cannot be excluded that the substances of the cementitious matrix will influence the long-term durability of the carbon textiles. At this point, however, it is assumed that investigations with exposure to a high alkaline concrete (pH value 13 to 14) are representative for other cementitious matrices.

Moreover, the investigated load levels will not occur in components. The actual textile stress in a component is significantly lower than the textile stresses at mean value level considered here, as the design of components is based, among other things, on design values. Therefore, the assumed maximum stress is not fully applied to components.

However, the results of the long-term tests of this work correspond well with results from literature. An overview of long-term tests on AFRP, CFRP and GFRP is presented in [27]. For example, Arockiasamy and Amer [48] did not notice any loss of strength at load levels of 65% of the breaking stress during tests on CFRP cables in alkaline solution with a pH value of 13 to 14 over nine months. Micelli and Nanni [21] investigated the durability of three different carbon rods under the influence of alkaline immersion and aging cycles. The test specimens stored for 21 or 42 days in an alkaline solution at 60 °C showed a loss of strength of 1% and 8%, respectively. After alternating combined environmental cycles, strength losses of less than 5% were found.

In accordance with the test results of this work, no or only a slight loss of strength is found in literature for CFRP tests. The durability behavior of CFRP reinforcement and carbon textiles differs due to the geometric properties (cross-sectional area, shape, impregnation material, etc.). However, since no results of long-term durability tests on carbon textiles are available, a reference is made. Due to different production processes of CFRP and carbon textile reinforcement, an increase in strength for the carbon textile reinforcement can be observed.

The test method that was to be used for the experiments could not be applied in this form, so there is a need for further development. Thereby, it must be taken into account that reliable characteristic values of the examination parameter can be derived [49].

8. Conclusions and Outlook

The long-term durability is the combined investigation of creep and alkaline resistance. A test method, which was developed by Weber and Baquero [17] and combines both exposures, was applied to small-scale specimens with a high-strength concrete and a single layer of an epoxy impregnated carbon reinforcement. To determine the long-term durability of the carbon reinforcement, short- and long-term, as well as preloading, tests were performed. Based on the short-term tests, the load levels for the long-term and preload tests were determined.

The results of this work can be summarized as follows:

- The long-term durability behavior of textile reinforcements is generally influenced by environmental factors, such as stress, temperature, moisture and alkalinity (pH-value). For this purpose, test rigs were developed which allow a combined exposure.
- Long-term durability tests with up to 11,000 h test time at temperatures of 40 and 60 °C were carried out at load levels of 70 to 97% of the reference load determined in short-term tests. With one exception, there was no failure due to the applied constant load levels.
- An increase of the ultimate load compared to the short-term tests could be identified in residual capacity tests. A load-oriented alignment of the filaments (training effect) is assumed, which leads to the higher load capacities of 10 to 17% compared to the short-term tests.
- Preloading tests at 60 °C confirmed this observation. After 24 h of preloading at 50% of the ultimate reference load, the residual capacity was 7% higher than the reference load.

Based on the long-term durability tests presented in this paper, it is not yet possible to derive a time to failure curve or a loss of strength for the material combination presented in Section 4. This corresponds to the test results of carbon reinforcement described in literature, where no or only minor strength losses were found. Within the framework of the C^3-V2.1 project, further investigations will be carried out to validate the long-term durability of this and other material combinations with non-metallic carbon reinforcement. The aim is to determine the loss of strength over 100 years for this reinforcement to be able to safely and economically design carbon concrete components. As an outlook, it currently appears possible that no reduction due to external influences is necessary for this type of reinforcement. Moreover, the influence of a preload shall be further investigated.

Author Contributions: Conceptualization, A.S. and S.B.; formal analysis, A.S. and J.B.; experimental investigation, A.S.; data curation, A.S.; writing—original draft preparation, A.S.; writing—review and editing, A.S., S.B. and J.B.; visualization, A.S. and S.B.; supervision, J.H.; project administration, J.H.; funding acquisition, A.S. and J.H.

Funding: This research was funded by BMBF, grant number 03ZZ0321C.

Acknowledgments: The authors thank solidian GmbH for providing the carbon textile reinforcement.

Conflicts of Interest: The authors declare no conflict of interest. The founding sponsors had no role in the design of the study; in the collection, analyses, or interpretation of data; in the writing of the manuscript, and in the decision to publish the results.

References

1. Sharei, E.; Scholzen, A.; Hegger, J.; Chudoba, R. Structural behavior of a lightweight, textile-reinforced concrete barrel vault shell. *Compos. Struct.* **2017**, *171*, 505–514. [CrossRef]
2. Shams, A.; Hegger, J.; Horstmann, M. An analytical model for sandwich panels made of textile-reinforced concrete. *Constr. Build. Mater.* **2014**, *64*, 451–459. [CrossRef]
3. Rempel, S.; Kulas, C.; Will, N.; Bielak, J. Extremely Light and Slender Precast Pedestrian-Bridge Made Out of Textile-Reinforced Concrete (TRC). In *High Tech Concrete: Where Technology and Engineering Meet, Proceedings of the 2017 fib Symposium, Maastricht, The Netherlands, 12–14 June 2017*; Hordijk, D.A., Luković, M., Eds.; Springer International Publishing: Cham, Switzerland, 2017; pp. 2530–2537. ISBN 3319594710.
4. Schumann, A.; Michler, H.; Schladitz, F.; Curbach, M. Parking slabs made of carbon reinforced concrete. *Struct. Concr.* **2018**, *19*, 647–655. [CrossRef]
5. Rempel, S.; Kulas, C.; Hegger, J. Bearing behavior of impregnated textile reinforcement. In *FERRO-11 International Symposium on Ferrocement and 3rd ICTRC International Conference on Textile Reinforced Concrete: Aachen, Germany, 7–10 June 2015*; Brameshuber, W., Ed.; RILEM publications: Paris, France, 2015; pp. 71–78.
6. May, S.; Steinbock, O.; Michler, H.; Curbach, M. Precast Slab Structures Made of Carbon Reinforced Concrete. *Structures* **2018**, *18*, 20–27. [CrossRef]
7. May, S.; Michler, H.; Schladitz, F.; Curbach, M. Lightweight ceiling system made of carbon reinforced concrete. *Struct. Concr.* **2018**, *19*, 1862–1872. [CrossRef]

8. Herbrand, M.; Adam, V.; Classen, M.; Kueres, D.; Hegger, J. Strengthening of Existing Bridge Structures for Shear and Bending with Carbon Textile-Reinforced Mortar. *Materials* **2017**, *10*, 1099. [CrossRef] [PubMed]

9. Brameshuber, W.; Hinzen, M.; Dubey, A.; Peled, A.; Mobasher, B.; Bentur, A.; Aldea, C.; Silva, F.; Hegger, J.; Gries, T.; et al. Recommendation of RILEM TC 232-TDT: Test methods and design of textile reinforced concrete—Uniaxial tensile test: Test method to determine the load bearing behavior of tensile specimens made of textile reinforced concrete. *Mater. Struct.* **2016**, *49*, 4923–4927. [CrossRef]

10. Schütze, E.; Bielak, J.; Scheerer, S.; Hegger, J.; Curbach, M. Einaxialer Zugversuch für Carbonbeton mit textiler Bewehrung. *Beton- Und Stahlbetonbau* **2018**, *113*, 33–47. [CrossRef]

11. Lieboldt, M.; Tietze, M.; Schladitz, F. C^3-Projekt—Erfolgreiche Partnerschaft für Innovation im Bauwesen. *Bauingenieur* **2018**, *93*, 265–273.

12. Canadian Standards Association. *CSA S807-10 (R2015): Specification for Fibre-Reinforced Polymers*; Canadian Standards Association: Mississauga, ON, Canada, 2010.

13. Canadian Standards Association. *S806-12 (R2017): Design and Construction of Building Structures with Fibre-Reinforced Polymers*; Canadian Standards Association: Mississauga, ON, Canada, 2012.

14. American Concrete Institute. *ACI 440.3R-12_ACI_2012_Guide Test Methods for Fiber-Reinforced Polymer (FRP) Composites for Reinforcing or Strengthening Concrete and Masonry Structures*; American Concrete Institute: Farmington Hills, MI, USA, 2012.

15. International Standard. *ISO 10406-1: Fibre-Reinforced Polymer (FRP) Reinforcement of Concrete—Test Methods*; ISO: Geneva, Switzerland, 2015.

16. ASTM International. *D7337/D7337M-12: Test Method for Tensile Creep Rupture of Fiber Reinforced Polymer Matrix Composite Bars*; ASTM International: West Conshohocken, PA, USA, 2012.

17. Weber, A.; Witt Baquero, C. New durability concept for FRP reinforcing bars: Combined durability and creep rupture testing allows for realistic design values. *Concr. Int.* **2010**, *32*, 49–53.

18. Bank, L.C.; Gentry, T.R.; Thompson, B.P.; Russell, J.S. A model specification for FRP composites for civil engineering structures. *Constr. Build. Mater.* **2003**, *17*, 405–437. [CrossRef]

19. Spelter, A.; Rempel, S.; Will, N.; Hegger, J. Testing Concept for the Investigation of the Long-Term Durability of Textile Reinforced Concrete. In *Durability and Sustainability of Concrete Structures (DSCS-2018)*; Falikman, V., Realfonzo, R., Coppola, L., Hàjek, P., Riva, P., Eds.; American Concrete Institute: Farmington Hills, MI, USA, 2018; pp. 55.1–55.9. ISBN 978-1-64195-022-0.

20. Ali, A.H.; Mohamed, H.M.; Benmokrane, B.; ElSafty, A. Effect of applied sustained load and severe environments on durability performance of carbon-fiber composite cables. *J. Compos. Mater.* **2018**, *53*, 677–692. [CrossRef]

21. Micelli, F.; Nanni, A. Durability of FRP rods for concrete structures. *Constr. Build. Mater.* **2004**, *18*, 491–503. [CrossRef]

22. Chen, Y.; Davalos, J.F.; Ray, I. Durability Prediction for GFRP Reinforcing Bars Using Short-Term Data of Accelerated Aging Tests. *J. Compos. Constr.* **2006**, *10*, 279–286. [CrossRef]

23. Sen, R.; Mullins, G.; Salem, T. Durability of E-Glass/Vinylester Reinforcement in Alkaline Solution. *ACI Struct. J.* **2002**, *99*, 369–375.

24. Benmokrane, B.; Wang, P.; Ton-That, T.M.; Rahman, H.; Robert, J.-F. Durability of Glass Fiber-Reinforced Polymer Reinforcing Bars in Concrete Environment. *J. Compos. Constr.* **2002**, *6*, 143–153. [CrossRef]

25. Nkurunziza, G.; Benmokrane, B.; Debaiky, A.S.; Masmoudi, R. Effect of Sustained Load and Environment on Long-Term Tensile Properties of Glass Fiber-Reinforced Polymer Reinforcing Bars. *ACI Struct. J.* **2005**, 615–621.

26. Spelter, A.; Bielak, J.; Will, N.; Hegger, J. Long-term Durability of Textile Reinforced Concrete. In Proceedings of the 2018 fib Congress, Melbourne, Australia, 7–11 October 2018; Foster, S., Gilbert, I.R., Mendis, P., Al-Mahaidi, R., Millar, D., Eds.; Fédération Internationale du Béton fib/International Federation for Structural Concrete: Lausanne, Switzerland, 2018; pp. 1944–1954.

27. International Federation for Structural Concrete. *FRP Reinforcement in RC Structures*; fib: Lausanne, Switzerland, 2007; ISBN 978-2-88394-080-2.

28. Litherland, K.L.; Oakley, D.R.; Proctor, B.A. The use of accelerated ageing procedures to predict the long term strength of GRC composites. *Cem. Concr. Res.* **1981**, *11*, 455–466. [CrossRef]

29. Wang, J.; GangaRao, H.; Liang, R.; Zhou, D.; Liu, W.; Fang, Y. Durability of glass fiber-reinforced polymer composites under the combined effects of moisture and sustained loads. *J. Reinf. Plast. Compos.* **2015**, *34*, 1739–1754. [CrossRef]

30. Orlowsky, J.; Raupach, M. Durability model for AR-glass fibres in textile reinforced concrete. *Mater. Struct.* **2008**, *41*, 1225–1233. [CrossRef]

31. Bank, L.C.; Russel Gentry, T.; Barkatt, A. Accelerated Test Methods to determine the Long-Term Behavior of FRP Composite Structures: Environmental Effects. *J. Reinf. Plast. Compos.* **1995**, *14*, 559–587. [CrossRef]

32. Ehrenstein, G.W. *Faserverbund-Kunststoffe. Werkstoffe-Verarbeitung-Eigenschaften*; 2. Auflage; Hanser Verlag München Wien: München, Germany, 2006; ISBN 3-446-22716-4.

33. Ray, B.C. Temperature effect during humid ageing on interfaces of glass and carbon fibers reinforced epoxy composites. *J. Colloid Interface Sci.* **2006**, *298*, 111–117. [CrossRef] [PubMed]

34. Ceroni, F.; Cosenza, E.; Gaetano, M.; Pecce, M. Durability issues of FRP rebars in reinforced concrete members. *Cem. Concr. Compos.* **2006**, *28*, 857–868. [CrossRef]

35. French, M.A.; Pritchard, G. Environmental stress corrision of hybrid fibre composites. *Compos. Sci. Technol.* **1992**, *45*, 257–263. [CrossRef]

36. Mufti, A.A.; Banthia, N.; Benmokrane, B.; Boulfiza, M.; Newhook, J.P. Durability of GFRP composite rods. *Concr. Int.* **2007**, *29*, 37–42.

37. Drzal, L.T.; Madhukar, M. Fibre-matrix adhesion and its relationship to composite mechanical properties. *J. Mater. Sci.* **1993**, *28*, 569–610. [CrossRef]

38. Wang, Z.; Huang, X.; Xian, G.; Li, H. Effects of surface treatment of carbon fiber: Tensile property, surface characteristics, and bonding to epoxy. *Polym. Compos.* **2016**, *37*, 2921–2932. [CrossRef]

39. Schutte, C.L. Environmental durability of glass-fiber composites. *Mater. Sci. Eng.* **1994**, 255–324. [CrossRef]

40. Rempel, S.; Ricker, M. Ermittlung der Materialkennwerte der Bewehrung für die Bemessung von textilbewehrten Bauteilen. *Bauingenieur* **2017**, *92*, 280–288.

41. Rempel, S. Zur Zuverlässigkeit der Bemessung von biegebeanspruchten Betonbauteilen mit textiler Bewehrung. Ph.D. Dissertation, RWTH Aachen, Aachen, Germany, 2018.

42. Hinzen, M. Prüfmethode zur Ermittlung des Zugtragverhaltens von textiler Bewehrung für Beton. *Bauingenieur* **2017**, *92*, 289–291.

43. Hegger, J.; Voss, S. Investigations on the bearing behaviour and application potential of textile reinforced concrete. *Eng. Struct.* **2008**, *30*, 2050–2056. [CrossRef]

44. Schneider, K.; Butler, M.; Mechtcherine, V. Carbon Concrete Composites C 3—Nachhaltige Bindemittel und Betone für die Zukunft. *Beton- Und Stahlbetonbau* **2017**, *112*, 784–794. [CrossRef]

45. DIN Deutsches Institut für Normung e.V. *Prüfverfahren für Zement—Teil 1: Bestimmung der Festigkeit*; Beuth Verlag GmbH: Berlin, Germany, 2005.

46. DIN Deutsches Institut für Normung e.V. *DIN EN 12390-6:2010-09: Prüfung von Festbeton—Spaltzugfestigkeit von Probekörpern. Deutsche Fassung EN 12390-6:2009*; Beuth Verlag GmbH: Berlin, Germnay, 2010.

47. Schmidt-Thrö, G.; Scheufler, W.; Fischer, O. Kontinuierliche faseroptische Dehnungsmessung im Stahlbetonbau. *Beton- Und Stahlbetonbau* **2016**, *111*, 496–504. [CrossRef]

48. Arockiasamy, M.; Amer, A. *Studies on Carbon FRP (CFRP) Prestressed Concrete Bridge Columns and Piles in Marine Environment*; Florida Department of Transportation: Tallahassee, FL, USA, 1998.

49. Weber, A. Prüfkonzepte für Bewehrungsmaterialien mit zeitabhängigen Widerständen. *Bauingenieur* **2018**, *93*, 323–330.

Article

Fatigue Behaviour of Textile Reinforced Cementitious Composites and Their Application in Sandwich Elements

Matthias De Munck [1],*, Tine Tysmans [1], Jan Wastiels [1], Panagiotis Kapsalis [1], Jolien Vervloet [1], Michael El Kadi [1] and Olivier Remy [2]

[1] Department Mechanics of Materials and Constructions, Vrije Universiteit Brussel (VUB), Pleinlaan 2, 1050 Brussels, Belgium; Tine.Tysmans@vub.be (T.T.); Jan.Wastiels@vub.be (J.W.); Panagiotis.Kapsalis@vub.be (P.K.); Jolien.Vervloet@vub.be (J.V.); Michael.El.Kadi@vub.be (M.E.K.)
[2] CRH Structural Concrete Belgium nv, Marnixdreef 5, 2500 Lier, Belgium; O.Remy@plakagroup.be
* Correspondence: Matthias.De.Munck@vub.be; Tel.: +32-(0)2-629-2927

Received: 19 February 2019; Accepted: 26 March 2019; Published: 28 March 2019

Abstract: Using large lightweight insulating sandwich panels with cement composite faces offers great possibilities for the renovation of existing dwellings. During their lifetime, these panels are subjected to wind loading, which is equivalent to a repeated loading. To guarantee the structural performance of these panels during their entire lifetime, it is necessary to quantify the impact of these loading conditions on the long term. The fatigue behaviour was, therefore, examined in this paper both at the material level of the faces and at the element level as well. plain textile reinforced cementitious composite (TRC) specimens were subjected to 100,000 loading cycles by means of a uniaxial tensile test, while sandwich beams were loaded 100.000 times with a four-point bending test. Results show that the residual behaviour is strongly dependent on the occurrence of cracks. The formation of cracks leads to a reduction of the initial stiffness. The ultimate strength is only affected in a minor way by the preloading history.

Keywords: textile reinforced cementitious composites (TRC), sandwich elements; fatigue; uniaxial tensile tests; four-point bending tests; digital image correlation (DIC)

---

1. Introduction

The energy and thermal insulation regulations for both new buildings and renovations become stricter year after year, leading directly to a growing demand for low-energy insulating building solutions. Particularly for renovation, the installation time on site needs to be reduced to the minimum to limit the inconvenience for the current residents. In this context, large lightweight prefabricated sandwich panels offer great possibilities. Renovating and insulating existing dwellings by placing panels with the dimensions of one story facilitates the installation process and reduces the total renovation time to a couple of days.

Nowadays sandwich panels are already widely spread in construction. They are characterized by a large stiffness to weight ratio thanks to the composite action between the two stiff faces and the insulating core. Different materials can be used, both for the core as for the skins [1]. Steel, wood, concrete, etc., have been used as facing material. Precast concrete sandwich panels are commonly used for walls of (industrial) buildings. Typically, these panels consist of steel reinforced concrete faces with a thickness of 60 mm or more, which ensure the load-bearing capacity. The total weight of such panels can be substantially reduced by omitting the non-structural concrete, i.e., the concrete cover necessary to protect the steel rebars against corrosion. This can be achieved by using technical textiles as an alternative reinforcement to steel. Since these textiles are not sensitive to corrosion, no concrete cover is needed, and lightweight sandwich panels with skins of only a few millimeter's thickness can be achieved. The combination of a cementitious matrix with a textile reinforcement has been widely investigated for different kinds of applications: reinforcing and/or strengthening of concrete and masonry structures [2–5], construction of pedestrian bridges [6], and as skins for the fabrication of sandwich panels. In addition to an experimental characterization of the bending behaviour of TRC sandwich panels [7–10], various work has been done on the analytical [11,12] and numerical [13–15] modelling of this behavior.

Textile reinforced cementitious composite (TRC) is characterized by a linear behavior in compression and a non-linear behavior in tension. This non-linear tensile behavior was observed in various experiments [16–20]. Parallel to this experimental campaign, different models were elaborated to predict the structural behavior of TRC [21–24]. Generally, the tensile behavior of the composite can be divided into three different stages (Figure 1). During the first linear stage, fibers and matrix are working together, and the modulus is given by the law of mixtures. For composites with a low fiber volume fraction, the modulus of the matrix is determining for the resulting initial modulus E1. Cracks start forming in the matrix when the ultimate tensile stress of the matrix is exceeded, resulting in a reduction of the modulus. Once a crack is formed, the load is redistributed. This process of the formation of cracks and subsequently redistribution of the load is called the multiple cracking stage and repeats itself until the matrix is fully saturated with cracks. The third and last stage is called the post-cracking stage. In this linear stage, the additional load is only carried by the fibers. The tangential modulus of the composite E3 is determined by the modulus of the fibers and the fiber volume fraction. Finally, failure of the composite material is induced by tensile rupture of the textile at a strain largely exceeding the tensile failure strain of the matrix.

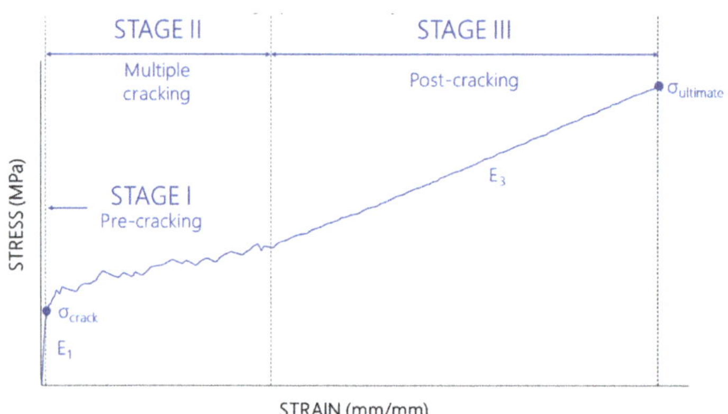

Figure 1. Characteristic tensile behavior of textile reinforced cementitious composite (TRC).

Applied in sandwich panels, the TRC faces are subjected to different loading conditions. One of the determining loading conditions is wind loading, comparable to a repeated loading. The influence of TRC sandwich panels subjected to a repeated loading has been studied little before. Cuypers et al. [25] cyclically loaded sandwich panels with E-glass fiber reinforced cementitious faces up to 2/3rd of their ultimate load. The panels were subjected to ten loading cycles and subsequently loaded up to failure. During the repeated loading an accumulation of the residual deformation was observed. Literature data on the fatigue behavior of TRC itself are limited. Hegger [6] and Mesticou [26] performed cyclic loading tests on TRC coupons, but the number of loading cycles was limited. Remy [27] and Cuypers [28] studied the behavior of TRC specimens, combining an inorganic phosphate cement with E-glass fibers. The specimens were subjected up to 10^7 cycles for different maximum cyclic loads. By assessing the evolving modulus, it was concluded that the accumulation of damage was not stabilized after 10^7 cycles. In all the above-mentioned research, samples were loaded and unloaded using a uniaxial tensile test up to a load at which the multiple cracking fully took place. An evaluation of the fatigue behavior of TRC at low load levels, i.e., below the matrix cracking stress is lacking in the literature.

Applied as thin faces in a sandwich panel loaded in bending, TRC will subjected to (nearly) uniform tension or compression and to loading cycles at relatively low stress levels originating from the characteristic wind loadings. The investigation of repeated loading conditions at the lower stress range is crucial and differs from the work done in literature. The formation of cracks at lower stress levels needs to be evaluated. This occurrence of cracks is directly linked to the ingress of aggressive substances and, thus, to durability measures of the façade panels. In addition, the modulus of TRC in the cracked state is significantly reduced. This lowered modulus has a direct impact on the displacements of the panels. A proper comprehension of the tensile fatigue behavior of TRC is, thus, indispensable to evaluate the long-term behavior of the resulting sandwich panels, and this already at low stress levels to account for the serviceability limit state.

This paper describes an extensive experimental study on 27 TRC coupons and 13 sandwich beams with TRC faces. Nine coupons were tested statically to identify the reference tensile behavior. The other 18 samples were divided into three different series. Each series was subjected to 100,000 tensile loading–unloading cycles up to a different predefined stress level, based on the expected loading conditions in serviceability limit state: 0.5 MPa, 1.0 Mpa, and 2.0 MPa. Afterwards, a static tensile test was performed to quantify the residual behavior. From the 13 sandwich beams, five sandwich beams were used as reference beams and loaded up to failure. The eight other beams, divided into two series, were subjected to 100,000 loading–unloading cycles and subsequently loaded up to failure. The maximum cycle load of the first series was equivalent to an elastic tensile stress of 1.0 MPa in the TRC skin, for the second series this was equivalent to an elastic tensile stress of 2.0 MPa. For both the coupons and the sandwich beams, an extensive analysis was performed on the hysteresis curves of the repeated loading tests and on the residual static behavior. In addition to a comparison of the structural behavior, the cracking behavior was also investigated in detail. Conclusions were drawn on the evolution of the parameters of the hysteresis curves and on the degradation of the static behavior.

2. Materials and Methods

2.1. Material Characteristics

To allow a clear comparison between the investigations on the component and the element level, the same materials were used for the coupons and the sandwich faces. For the TRC coupons, a premix mortar was reinforced with multiple layers of alkali-resistant (AR) glass fiber textiles. For the sandwich beams, an expanded polystyrene (EPS) core was covered with the same textile, embedded in the premix mortar. The material properties of both the EPS and the TRC constituents (mortar and textile) are described below.

2.1.1. Mortar

A commercially available Portland cement-based shrinkage-compensated mortar was chosen as the matrix of the TRC. Its maximum grain size was 0.5 mm, and a water to binder mass ratio of 0.15 was considered. Six flexural and twelve compression tests were performed according to the European Standards [29], to characterize the flexural strength $f_{ct,f}$ and compressive strength f_{cc} (Table 1).

Table 1. Mechanical properties of the used mortar.

	$f_{ct,f}$ MPa	f_{cc} MPa
Average	6.35	23.17
Standard Deviation	0.45	1.59

2.1.2. Textile Reinforcement

A technical textile made of AR glass rovings is embedded in the mortar. The textile is polymer coated and woven into an orthogonal mesh. The textile has a nominal tensile strength of 2500 N per 50 mm, a total surface weight of 200 g/m² (165 g/m² glass fibers), and a mesh opening of 5 mm in both directions [30] (Figure 2).

Figure 2. An alkali-resistant (AR) glass textile with a mesh size of 6 mm used.

2.1.3. Expanded Polystyrene

For the fabrication of the sandwich beams, expanded polystyrene (EPS) was chosen as a rigid insulating core. EPS is not the most performant thermal insulating material, but this is largely compensated by its low cost and low density (15–20 kg/m³). Both cost and density are key parameters for the considered application. The properties of the used EPS 200 are listed in Table 2.

Table 2. Properties expanded polystyrene (EPS) 200 [31].

Density kg/m³	E-Modulus MPa	Bending Strength kPa
20	10	250

2.2. Specimen Preparation

The experiments on the material level were carried out on prismatic TRC coupons, with a nominal length of 500 mm, a nominal width of 75 mm, and a nominal thickness of 10 mm. All specimens had an identical build-up, reinforced with two layers of textile equally distributed over the height (Figure 3). To do so, they were made separately using a hand lay-up technique; the mortar was cast three times. After spreading out the first layer of mortar, a reinforcement fiber net was placed and impregnated in the mortar (Figure 4). Subsequently, a second layer of mortar and second reinforcement grid were placed. Once the third layer of mortar was cast, a plastic sheet was used to seal the mold and to prevent premature evaporation of the water. All the coupons were demoulded after 24 h and had the same curing process for 28 days; stored at ambient temperature (approximately 20 °C) and a relative humidity of between 45% and 60% for at least 28 days. The resulting fiber volume fraction of the samples was equal to 1.29% (0.65% in the loading direction).

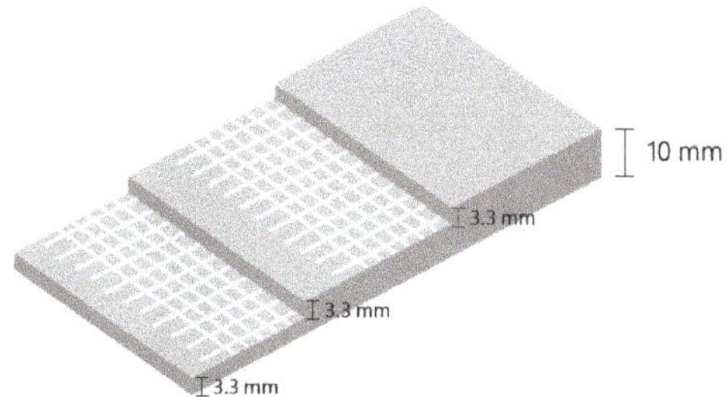

Figure 3. Stacking sequence TRC coupons.

(a)   (b)

Figure 4. A hand lay-up technique was used for the preparation of the specimens: (**a**) TRC coupons and (**b**) sandwich beams with TRC faces.

For the element level, sandwich beams with a total length of 2500 mm and a width of 200 mm were fabricated and tested. An EPS core with a thickness of 200 mm was covered on both sides with a 5 mm thick TRC layer using a hand lay-up technique. After spreading out a layer of mortar, the textile grid was embedded in the mortar. The glass fiber volume fraction was equal to that of the coupons, 0.65% in the loading direction. The faces were sealed with a plastic cover to prevent premature evaporation. All beams were stored for at least 28 days at ambient temperature (approximately 20 °C) and relative humidity of between 45% and 60%.

2.3. Test Set-Up

2.3.1. Uniaxial Tensile Tests

For the material level, uniaxial tensile tests on TRC coupons were preferred over bending tests, since they are more representative for the considered application: the thin faces of a sandwich panel subjected to bending are loaded under (nearly) uniform tension or compression. In the uniaxial test set-up, the load was introduced via bolt through aluminum end-plates, which were glued to the TRC coupons with a two-component glue. Stress concentrations were avoided by tapering the end-plates (Figure 5). The specimens were loaded by a servo-hydraulic actuator with a capacity of 25 kN, using a 10 kN load cell. The static tests were displacement-controlled with a rate of 1 mm/min. The cyclic loading was load controlled at a frequency of 10 Hz. Displacements were measured with a dynamic extensometer at one side and digital image correlation (DIC) at the other side.

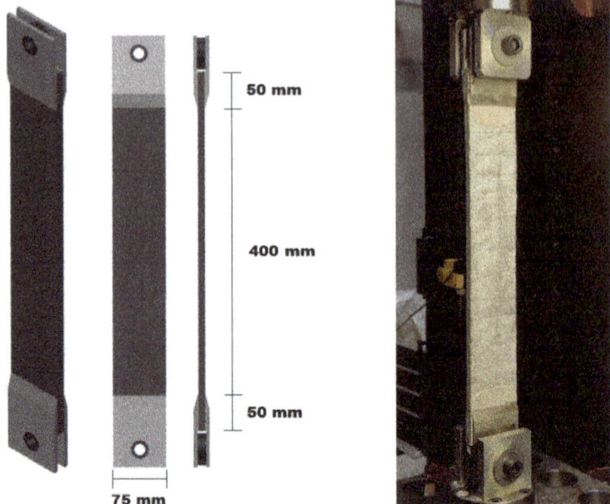

Figure 5. Uniaxial test set-up and dimensions.

2.3.2. Four-Point Bending Tests

To assess the fatigue behavior on the element level, sandwich beams were both statically and cyclically loaded using a four-point bending test set-up. The distance between the roller supports was 2200 mm. The load was induced by means of a 500 mm span dividing beam, connected through a 10 kN load cell to a servo-hydraulic actuator with a capacity of 25 kN (Figure 6). The static tests were displacement-controlled with a rate of 1 mm/min. The cyclic loading was load-controlled at a frequency of 2 Hz. To avoid local stress concentrations, aluminum distribution plates were placed at the supports and at the loading areas. Mid-span displacements were monitored with an LVDT (Linear Variable Differential Transducer). DIC was used to monitor the cracking behavior of the tensile face partly in the zone of the constant moment.

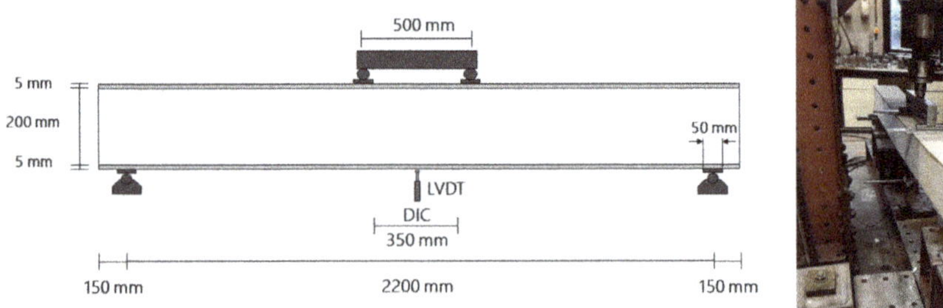

Figure 6. Four-point bending test set-up.

2.3.3. Digital Image Correlation

To measure strains and displacements and to visualize cracks, digital image correlation (DIC) was used. It is an optical, non-contacting method to measure displacement- and strain-fields of a specimen. The measurement is based on the comparison of a reference image (generally unloaded condition) with images taken at different load steps. The settings of the DIC systems are specified in Table 3. The features of the used cameras are listed in Table 4. For the static tests, images were acquired at a frequency of 0.3 1/s. During the cyclic tests, one image every 10 cycles was taken during the first 100 cycles. Further, one image was acquired every 100 cycles up to cycle 1000 and every 1000 cycles up to cycle 100,000. A more extensive description of the working principle of DIC can be found in the literature [32].

Table 3. Settings digital image correlation (DIC) analysis.

	TRC Coupons	Sandwich Beams
subset	21 pxs	21 pxs
step	7 pxs	7 pxs
filter size	11	11
area of interest	250×45 mm	250×190 mm

Table 4. Camera features.

Type of Camera	CCD
lenses size	8 mm
resolution	2546×2048 pxs

3. Results and Discussion

3.1. Investigations on TRC Coupons

In total 27 specimens were tested. First, the reference behavior (TRC REF) was characterized by testing nine specimens in a quasi-static way. The stresses were calculated using the nominal dimensions, strains were measured using DIC over a length of 320 mm. The average stress–strain curve was determined from the different curves; the standard deviation is presented as the shaded area (Figure 6). In the first linear stage, the modulus E1 was 11.38 GPa. A modulus E3 of 411.34 MPa was determined in the third stage. On average, the first crack appeared at a stress equal to 1.65 MPa.

As mentioned before, building elements, such as façade panels, will be prone to repeated loading conditions originating from the wind. Under characteristic wind loading, the stress level in the faces will not exceed 2 MPa (Figure 7). To have a clear insight on the influence of such loading conditions, three series of six specimens were cyclically loaded up to three different stress levels; 0.5 MPa, 1.0 Mpa, and 2.0 MPa. These series were nominated as: TRC CYCL 0.5 MPa, TRC CYCL 1.0 Mpa, and TRC CYCL 2.0 MPa.

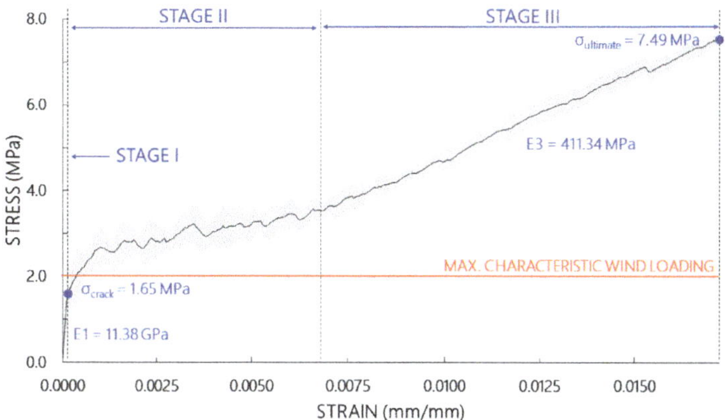

Figure 7. Reference behavior of TRC coupons.

All specimens were loaded 100,000 cycles up to the specific stress level and unloaded to a stress level above 0 MPa to avoid loading the specimens in compression. To compare the different series, focus was put on the evolution of some distinctive fatigue parameters during the cyclic loading: the cycle modulus and the dissipative energy. The modulus per cycle was determined from the hysteresis curve by linear regression and expressed relatively compared to the modulus of the first cycle. The dissipative energy was calculated as the area enclosed by the loading–unloading curve of each cycle. As for the modulus, this was expressed relatively compared to the energy dissipation of the first cycle. The evolution of these different fatigue parameters was observed to be strongly related to the formation of cracks. Without the occurrence of cracks, both the modulus and the dissipative energy remained more or less constant (Figure 8a). Regardless of the applied maximum cycle stress (1.0 MPa or 2.0 MPa), each formation of a crack led to a decrease of the modulus, together with an increase in the dissipative energy (Figure 8b).

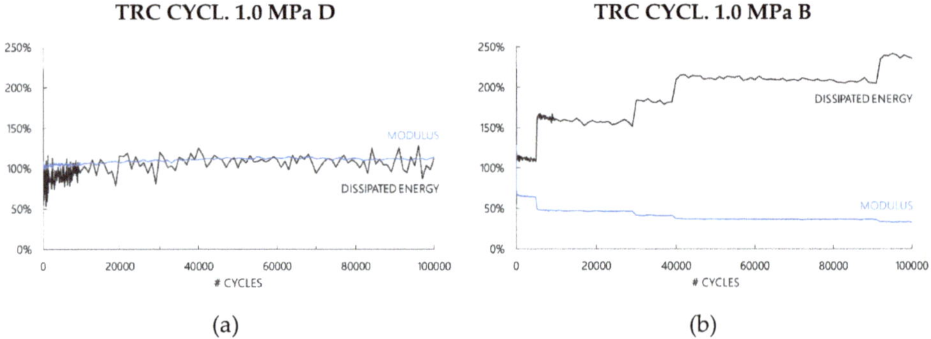

Figure 8. The evolution of the modulus and the dissipated energy of representative TRC coupons during the loading–unloading cycles: (**a**) an uncracked sample and (**b**) a cracked sample.

Using DIC enabled tracking and visualizing the crack patterns during the repeated loading tests. The crack widths or crack openings were calculated using the displacement fields measured by the DIC. The formation of cracks was related to the maximum cycle stress. The higher the maximum cycle stress the higher the probability of forming cracks. TRC CYCL 0.5 MPa remained uncracked for all six tested coupons. This stress level did not come near the cracking stress of the TRC equal to the average reference cracking stress of 1.65 MPa, which was observed during the static tests. The cracking phenomena of TRC have a very stochastic nature, resulting in a large variation on the first cracking strength. A standard deviation of 0.54 MPa was determined, leading to a smeared range of 1.11 MPa and 2.19 MPa. TRC CYCL 1.0 MPa and TRC CYCL 2.0 MPa were loaded closer to this range which explains the larger probability of the occurrence of cracks. Two specimens of TRC CYCL 1.0 MPa and one specimen of TRC CYCL 2.0 MPa remained uncracked (Table 5). Overall, more cracks were observed for TRC CYCL 2.0 MPa. The outlier is specimen A of TRC CYCL 1.0 MPa, in which 22 cracks were observed after the cyclic preloading. No clear explanation could be found for this; damage caused during manufacturing is the most probable cause for this distorted behavior.

Table 5. Number of cracks formed during cyclic preloading of the TRC coupons.

Specimen	TRC CYCL 0.5 MPa	TRC CYCL 1.0 MPa	TRC CYCL 2.0 MPa
A	0	22	7
B	0	8	12
C	0	0	10
D	0	0	0
E	0	5	16
F	0	1	1

Once a crack was formed, its width tended to stabilize. Both for TRC CYCL. 1.0 MPa as for TRC CYCL. 2.0 MPa, the crack width increased during the first hundred cycles but stabilized afterwards. As an example, the evolution of the crack width was shown for a sample subjected to a maximum cycle stress of 1.0 MPa (Figure 9).

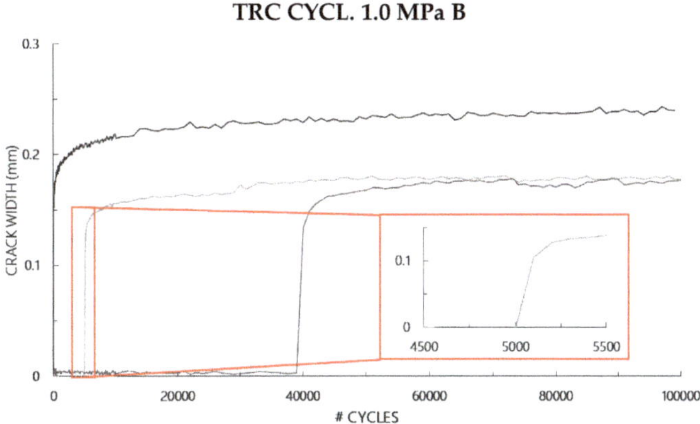

Figure 9. Evolution of the crack width of some representative cracks formed during cyclic loading of TRC coupons.

After passing all loading cycles, a static test as described in Section 2.3.1 was performed to determine the residual capacity of the TRC coupons. The stresses were calculated using the nominal dimensions of the coupons, the strains were measured using DIC over a length of 320 mm. The obtained stress–strain curves were grouped per series. In Figure 10 one can see that the residual behavior was only slightly affected for series TRC CYCL 0.5 MPa. As shown in Table 5 no cracks were induced during the cyclic loading. The formation of cracks was found as the only degradation fatigue mechanism, leading directly to the explanation of the observed residual behavior: both the modulus, as the strength were comparable to the reference behavior (Table 6). The only difference between the residual curves and the reference one was found in the multiple cracking stage. As can be seen in Table 6, more cracks were formed for TRC CYCL 0.5 MPa samples compared to REF, which could explain the limited stress increase during the multiple cracking stage, and, thus, the shifted stress–strain curves.

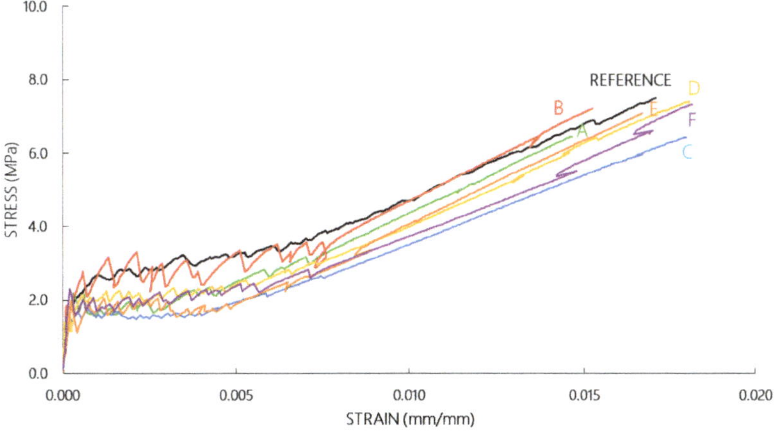

Figure 10. Residual behavior of TRC CYCL. 0.5 MPa.

Table 6. Quantitative comparison of some distinctive parameters of the static tensile stress–strain curve: the average reference versus the specimens cyclically preloaded to 0.5 MPa.

Specimen		E_1 GPa	E_3 GPa	σ_{crack} MPa	$\sigma_{ultimate}$ MPa	# Cracks After Cyclic Preloading	At Failure
REF	avg	11.38	0.41	1.65	7.49	-	13
	st dev	1.94	0.018	0.54	0.52	-	2
CYCL. 0.5 MPa	A	8.99	0.45	2.19	6.44	0	14
	B	15.32	0.49	1.85	7.22	0	11
	C	11.74	0.37	1.49	6.43	0	19
	D	14.60	0.43	1.38	7.40	0	19
	E	9.22	0.48	1.96	7.07	0	14
	F	12.64	0.43	2.30	7.32	0	17
	avg	12.08	0.44	1.86	6.98	-	16
	st dev	2.42	0.038	0.34	0.40	-	3

As the maximum stress to which the TRC coupons were cyclically loaded was increased for series TRC CYCL 1.0 MPa and TRC CYCL 2.0 MPa, so did the probability to the formation of cracks. Once a crack was formed, this directly resulted in a decrease of E_1 measured in the residual stress–strain behavior. The more cracks are formed, the lower E_1. Since this initial modulus was very dependent on the occurrence of cracks, a large scatter exists on the results and the average value of E_1 was not representative and, therefore, not displayed in Tables 7 and 8. A specimen fully saturated with cracks showed a linear behavior in the static tests, for example, specimen A in Figure 11 and specimen E in Figure 12. For these specimens, few extra cracks were formed during the static tests and the residual behavior was characterized with a modulus equal to E_3 (Tables 7 and 8). The cyclic preloading did not affect the modulus in the last branch, E_3 (Tables 7 and 8). Overall, the ultimate strength was lowered after subjection to loading cycles up to 1.0 MPa (Table 7) and 2.0 MPa (Table 8). However, no clear link was found between the number of cracks and the ultimate strength.

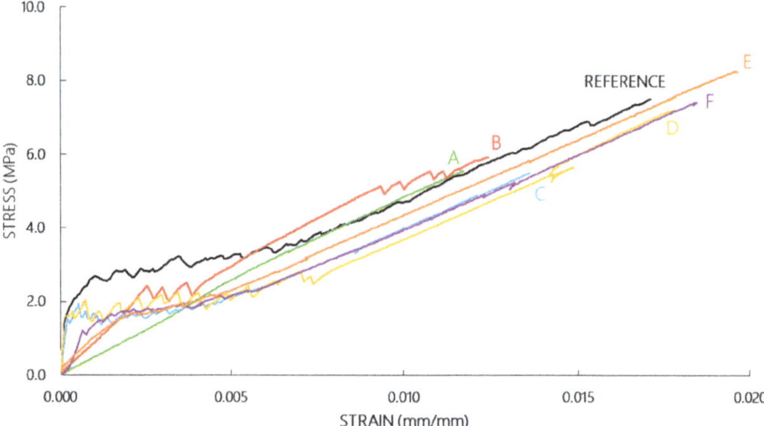

Figure 11. Residual behavior of TRC CYCL. 1.0 MPa.

Table 7. Quantitative comparison of some distinctive parameters of the static tensile stress–strain curve: the average reference versus the specimens cyclically preloaded to 1.0 MPa.

Specimen		E_1 GPa	E_3 GPa	σ_{crack} MPa	$\sigma_{ultimate}$ MPa	# Cracks After Cyclic Preloading	# Cracks At Failure
REF	avg	11.38	0.41	1.65	7.49	-	13
	st dev	1.94	0.018	0.54	0.52	-	2
	A	0.48	0.48	-	5.55	22	24
	B	0.92	0.49	2.41	5.92	8	14
	C	8.37	0.39	1.53	5.49	0	22
CYCL.	D	12.27	0.45	1.63	7.20	0	14
1.0 MPa	E	0.91	0.41	1.70	8.25	5	29
	F	2.59	0.41	1.20	7.41	1	28
	avg	-	0.44	1.69	6.64	-	22
	st dev	-	0.037	0.40	1.04	-	6

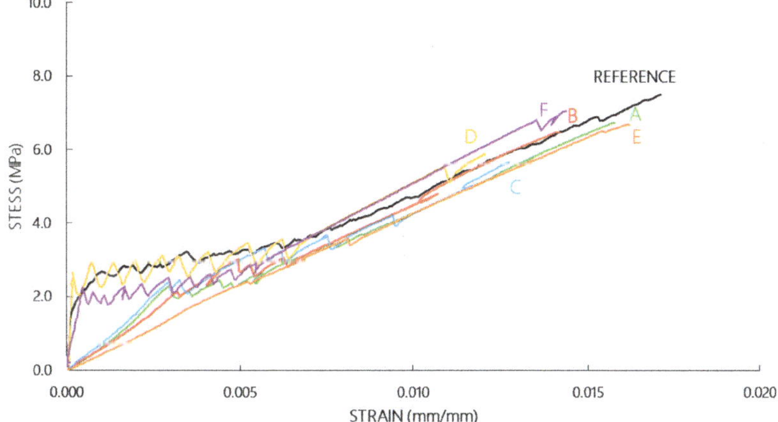

Figure 12. Residual behavior of TRC CYCL. 2.0 MPa.

Table 8. Quantitative comparison of some distinctive parameters of the static tensile stress–strain curve: the average reference versus the specimens cyclically preloaded to 2.0 MPa.

Specimen		E_1 GPa	E_3 GPa	σ_{crack} MPa	$\sigma_{ultimate}$ MPa	# Cracks After Cyclic Preloading	# Cracks At Failure
REF	avg	11.38	0.41	1.65	7.49	-	13
	st dev	1.94	0.018	0.54	0.52	-	2
	A	0.84	0.42	2.30	6.74	7	22
	B	0.65	0.48	2.13	6.48	12	19
	C	0.81	0.43	2.41	5.65	10	20
CYCL.	D	13.06	0.47	2.66	5.88	0	10
2.0 MPa	E	0.48	0.42	2.42	6.68	16	24
	F	4.76	0.48	2.24	7.13	1	11
	avg	-	0.45	2.36	6.43	-	18
	st dev	-	0.029	0.17	0.51	-	5

Monitoring the static tests up to failure with DIC measurements enabled to map cracking patterns and measure crack widths and spacings. The average, maximum and total crack widths were calculated from the measured displacement fields for each specimen separately at different stress levels. In addition to the crack spacing itself, the cumulative frequency was also determined to quantify the degree of saturation, i.e., the ratio of actually formed cracks to the total number of cracks at failure. To compare the different series to the reference behavior, the average per series was calculated out of the average and total crack width, the crack spacing and the cumulative frequency. For the maximum crack width, the absolute maximum crack width observed over all specimens within one series was determined. The maximum crack width is an important parameter for designing concrete building elements regarding durability measures, decisive for the ingress of aggressive materials.

Looking at the cumulative frequency one could observe that more cracks were formed at lower stress levels for TRC CYCL 0.5 MPa, TRC CYCL 1.0 Mpa, and TRC CYCL 2.0 MPa., while for TRC REF 20% of the total amount of cracks were still formed at stress levels above 4.0 MPa (Figure 13a). The crack spacing showed a discrepancy between TRC REF and TRC CYCL 0.5 MPa versus TRC CYCL 1.0 MPa and TRC CYCL 2.0 MPa (Figure 13b) again. For TRC CYCL 1.0 MPa and TRC CYCL 2.0 MPa cracks originated at lower stress levels leading to a diminished crack spacing. As observed for the cumulative frequency, the crack patterns of the cyclically preloaded series were nearly complete at the stress level of 2.5 MPa. At all stress levels, the crack spacing of these series was lower compared to TRC REF, leading to the conclusion that more cracks were formed for TRC CYCL 0.5 MPa, TRC CYCL 1.0 Mpa, and TRC CYCL 2.0 MPa.

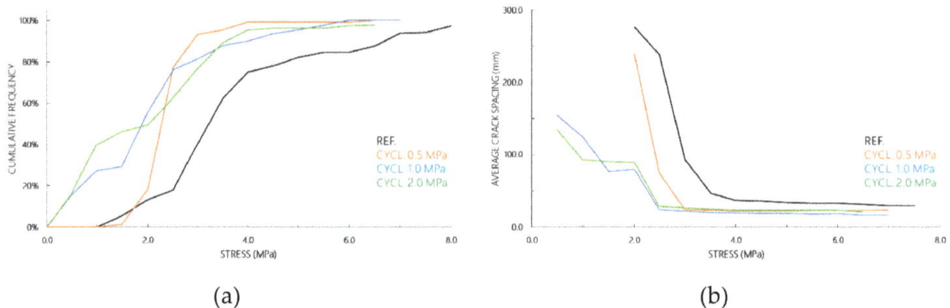

(a)   (b)

Figure 13. The number of cracks present in the different samples was used to compare: (**a**) the average cumulative frequency and (**b**) the average crack spacing.

Other than for TRC REF and TRC CYCL 0.5 MPa, TRC CYCL 1.0 Mpa, and TRC CYCL 2.0 MPa showed cracks already at stress levels lower than 1.0 MPa (Figure 14), as a consequence of the fact that cracks were already formed during the cyclic preloading. This is displayed for all studied parameters. Looking at the average crack width, a discrepancy was observed for stress levels below 2.0 MPa and above 2.0 MPa: at lower stress levels, TRC CYCL 1.0 MPa and TRC CYCL 2.0 MPa had a higher average crack width, while for higher stress levels, wider cracks were measured for TRC REF and TRC CYCL. 0.5 MPa (Figure 14a). The same tendency was seen for the maximum crack width but less pronounced (Figure 14b). Since this maximum crack width was strongly related to the durability requirements, one could state that the impact of repeated loading on durability measurements is little. The lowest total crack width was observed for TRC REF and this at all stress levels. (Figure 14c).

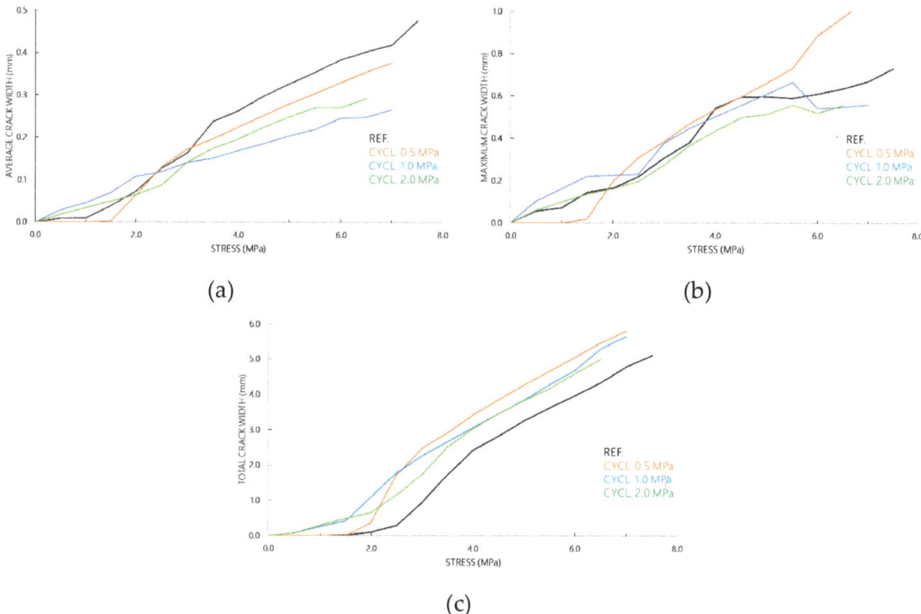

$$(a) \qquad (b)$$

$$(c)$$

Figure 14. The cracking patterns of the different series were compared using the: (**a**) the average crack width, (**b**) the maximum crack width and (**c**) the total crack width.

3.2. Investigations on Sandwich Beams

In addition to the fatigue behavior of TRC coupons under tensile loading, the fatigue behavior in bending of sandwich (SW) beams with TRC faces was investigated. In total 13 beams were tested. To determine the virgin, reference behavior of the panels (SW REF), five samples were tested with a quasi-static four-point bending test, as described in Section 2.3.2. The other beams were subjected to a cyclic preloading, divided into two different series analogous to the experiments on the TRC coupons. Since no influence was observed for a cyclic preloading of the coupons up to 0.5 MPa this was excluded from further experiments. The stress levels of 1.0 MPa and 2.0 Mpa, respectively, in the tensile face were converted to equivalent loads for the sandwich beams using a validated numerical model [33], respectively, 0.5 kN and 1.0 kN. These beams will be referred to as SW CYCL 1.0 MPa and SW CYCL 2.0 MPa. Similar to the coupons, the typical hysteresis curve was analyzed by comparing some fatigue parameters. The cycle stiffness and the dissipated energy were calculated similarly as explained in Section 3.1.

The evolution of the stiffness and the dissipated energy show the same trends as for the coupons. The stiffness and the dissipated energy evolved in the opposite direction and directly linked to the formation of cracks. For uncracked specimens, they remained constant (Figure 15a). The occurrence of a crack was accompanied by a drop in stiffness and a jump in dissipated energy, as was seen for sample SW CYCL. 2.0 MPa A at cycle 93.000 in Figure 15b.

Overall, fewer cracks were observed for the sandwich bending tests (Table 9) compared to the coupon uniaxial tensile tests. A possible reason for this was the different anchorage length in the different configurations. When applied in the sandwich beam, the textile was embedded over a longer length in the matrix, leading to a better anchorage. In addition, the presence of the insulating core could have had a beneficial effect, the bond between the EPS-core and the TRC faces restricted the crack widths.

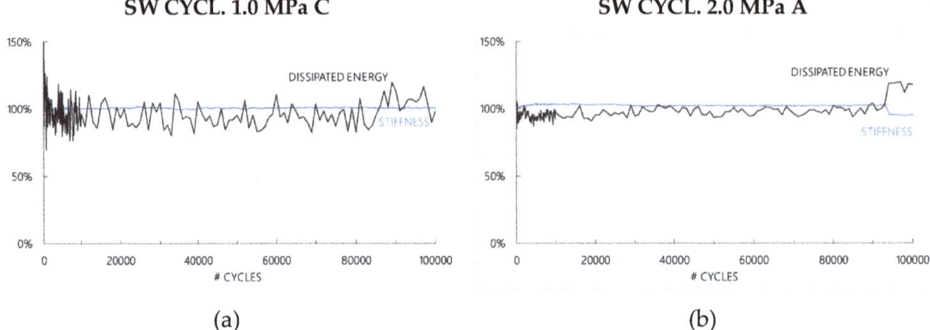

Figure 15. The evolution of the stiffness and the dissipated energy of sandwich beams during the loading–unloading cycles: (**a**) an uncracked sample and (**b**) a cracked sample.

Table 9. Number of cracks formed during cyclic preloading of the sandwich beams.

Specimen	SW CYCL 1.0 MPa	SW CYCL 2.0 MPa
A	0	2
B	0	1
C	0	2
D	0	6

Similar to the TRC coupons, once a crack was formed in the tensile face of the sandwich beam, its width increased during the first subsequent thousand cycles but stabilized gradually afterwards (Figure 16). The occurrence of cracks was the only observed damage mechanism, the core and the face-core interface were not affected. The fatigue behavior of the sandwich beams was dependent on the behavior of the TRC faces.

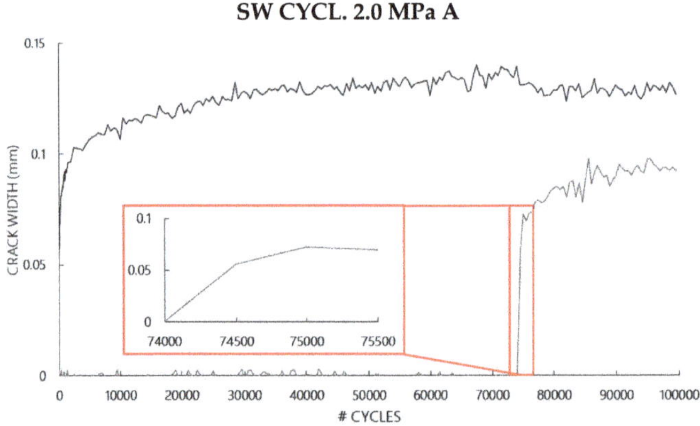

Figure 16. Evolution of the crack width of some representative cracks formed during cyclic loading of sandwich beams.

After the loading–unloading cycles, the residual static behavior was quantified by means of a quasi-static four-point bending test. As for the TRC coupons, the only observed degradation phenomenon was the formation of cracks. For SW CYCL 1.0 MPa no cracks were observed during the cyclic loading and, thus, the residual behavior was very similar to the behavior of SW REF (Figure 17). As can be seen in Table 10 both the stiffness and strength of SW REF and SW CYCL. 1.0 MPa were situated in the same range.

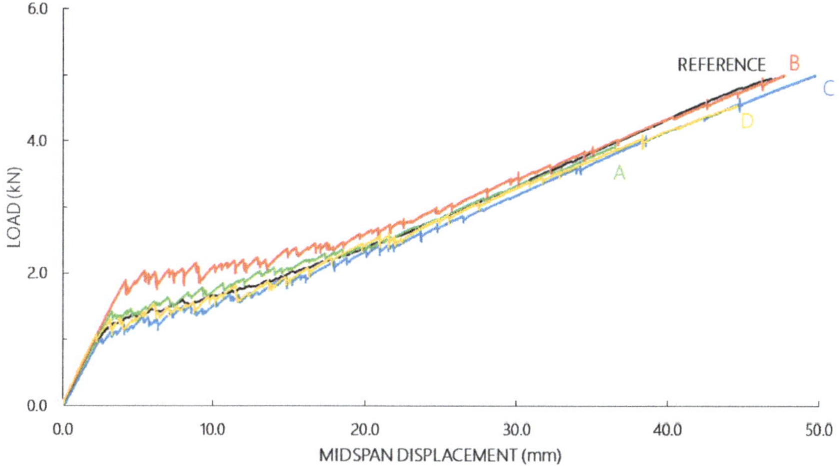

Figure 17. Residual behavior of SW CYCL. 1.0 MPa.

Table 10. Quantitive comparison of some distinctive parameters of the static load-displacement curve: the average reference versus the beams cyclically preloaded to 1.0 MPa.

Specimen		S_1 kN/mm	S_3 kN/mm	P_{crack} kN	$P_{ultimate}$ kN	# Cracks	
						After CyclicPreloading	At Failure
REF	avg	0.451	0.098	1.12	4.99	-	17
	st dev	0.009	0.002	0.27	0.45	-	4
SW CYCL. 1.0 MPa	A	0.437	0.091	1.42	3.95	0	12
	B	0.474	0.090	1.89	5.14	0	13
	C	0.418	0.091	0.98	5.35	0	13
	D	0.488	0.090	1.04	4.62	0	13
	avg	0.454	0.090	1.34	4.76	-	13
	st dev	0.028	0.000	0.36	0.54	-	0

All specimens of SW CYCL 2.0 MPa showed cracks after subjection to the loading cycles (Table 11). However, the occurrence of cracks had a smaller impact on the initial stiffness S_1 compared to cracked TRC coupons. The stiffness in the last branch S_3 was not affected by the cyclic preloading. The average ultimate capacity was 23% lower for the SW CYCL sandwich beams compared to the SW REF beams (Figure 18).

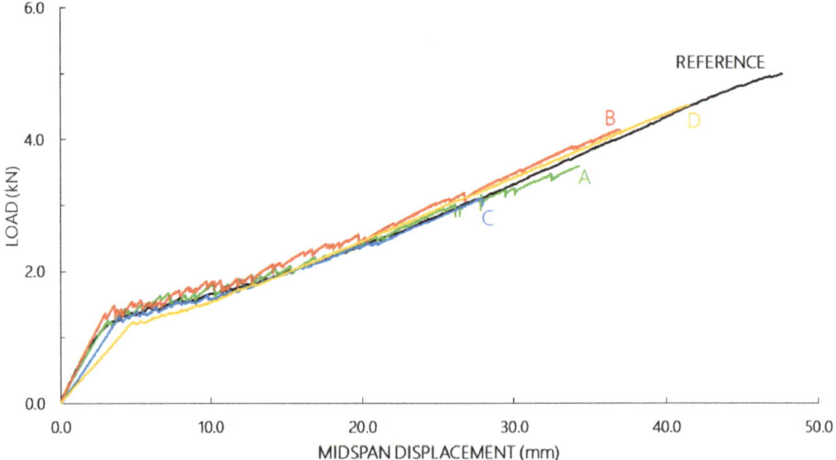

Figure 18. Residual behavior of SW CYCL. 2.0 MPa.

Table 11. Quantitive comparison of some distinctive parameters of the static load-displacement curve: the average reference versus the beams cyclically preloaded to 2.0 MPa.

Specimen		S_1 kN/mm	S_3 kN/mm	P_{crack} kN	$P_{ultimate}$ kN	# Cracks	
						After CyclicPreloading	At Failure
REF	avg	0.451	0.098	1.12	4.99	-	17
	st dev	0.009	0.002	0.27	0.45	-	4
SW CYCL. 2.0 MPa	A	0.412	0.082	1.25	3.60	2	9
	B	0.455	0.097	1.36	4.15	1	10
	C	0.336	0.086	1.32	3.11	2	15
	D	0.254	0.096	1.22	4.52	6	24
	avg	-	0.090	1.29	3.84	-	15
	st dev	-	0.006	0.055	0.54	-	6

As shown in Figure 5, a part of the tensile face was monitored using DIC. An analysis of the cracks was performed as for the TRC coupons. However, due to the narrow area of interest of the DIC a limited amount of the cracks was captured, and, thus, only the maximum crack width was looked at. The evolution of the maximum crack width of SW REF and SW CYCL 1.0 MPa was very similar (Figure 19). SW CYCL 2.0 MPa showed larger crack widths for all load levels, which could be attributed to the occurrence of cracks during the cyclic preloading and the inherent debonding between fibers and matrix. This observation did not match the results of the TRC coupons, no increase of the maximum crack width was seen for TRC CYCL. 2.0 MPa. For the moment no clear reasoning could be made regarding this particular discrepancy between TRC coupons and SW beams.

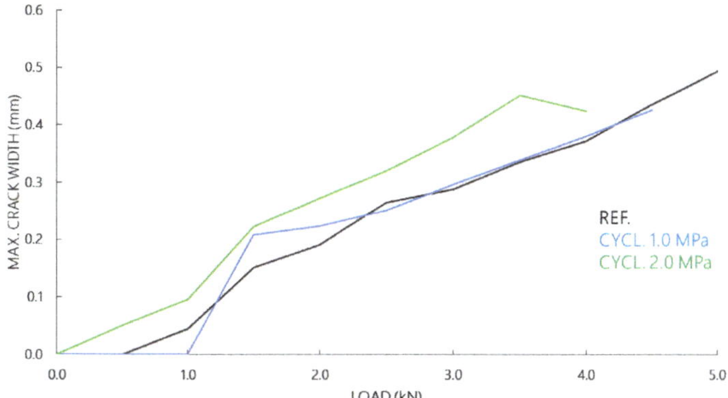

Figure 19. The maximum crack width observed in the sandwich beams per series.

4. Conclusions

This paper investigated the fatigue behavior of TRC and of sandwich panels with faces made of TRC. Uniaxial tests were performed on rectangular TRC coupons, in total 27 specimens were fabricated and tested, divided into four series: one reference series and three series which were cyclically preloaded up to different stress levels (0.5 MPa, 1.0 Mpa, and 2.0 MPa). In addition, 13 sandwich beams were tested by means of a four-point bending test. Three different series were considered: one reference series and two series cyclically preloaded up to the corresponding stress level (1.0 MPa and 2.0 MPa) in the tensile face of the sandwich beam. Afterwards, all specimens (incl. the reference specimens) were loaded up to failure. All tests, both cyclic and static, were monitored with DIC to map the cracking patterns and to measure the actual crack widths.

A large similarity was observed between the experiments on the material level and the experiments on the element level. One could conclude that the fatigue behavior of the sandwich panels was strongly dependent on the fatigue behavior of the TRC faces. No degradation was observed in the core, nor in the interface between core and faces. The sandwich beams were less sensitive to the formation of cracks and degradation of the mechanical behavior compared to the TRC coupons. Possible reasons could be found in the different anchorage length of the textiles and in the presence of EPS-core; the bond between EPS and TRC restricted the crack widths.

Fatigue parameters as cycle modulus and dissipative energy were investigated. All of them were related to the occurrence of cracks, which was the only observed damage mechanism: if no cracks were formed during the loading–unloading cycles, the parameters remained constant. In the presence of cracks, a decrease of the modulus and an increase of the dissipative energy and residual accumulative strains was observed. The modulus/stiffness and dissipative energy evolved in the opposite direction. Once a crack originated during the cyclic loading, its width grew during the first subsequent cycles but evolved asymptotically. The formation of cracks was governed by the maximum cycle stress. The higher the latter, the larger the probability of the occurrence of cracks.

The residual capacity of the cyclically preloaded specimens was compared to a virgin reference behavior, obtained with identical experimental set-ups. Samples which remained uncracked after being subjected to the repeated loading conditions had a residual behavior equal to the reference behavior. The presence of cracks was reflected in the residual behavior by a lower initial modulus/stiffness. After being fully saturated with cracks, the remaining modulus is only dependent on the modulus of the fibers. This modulus remained unaffected leading to the conclusion that the fibers were not degraded by the cyclic preloading. The ultimate capacity was degraded little after subjection to repeated loading conditions. No clear relation was observed between the occurrence and number of cracks and the loss of ultimate capacity.

Author Contributions: Conceptualization, M.D.M, T.T. and O.R.; methodology, M.D.M., J.W. and T.T.; formal analysis, M.D.M.; investigation, M.D.M.; writing—original draft preparation, M.D.M.; writing—review and editing, T.T., J.W., J.V.; M.E.K.; P.K.; visualization, J.V.; supervision, T.T., J.W. and O.R.; project administration, M.D.M. and T.T.; funding acquisition, M.D.M, T.T. and O.R.

Funding: This research was funded by Agentschap voor Innovatie en Ondernemen (VLAIO) and CRH Structural Cngrant number 150251.

Acknowledgments: The authors gratefully acknowledge Agentschap voor Innovatie en Ondernemen (VLAIO) and CRH Structural Concrete Belgium nv for funding the research of the first author through a Baekeland mandate.

Conflicts of Interest: The authors declare no conflict of interest.

References

1. Davies, J.M. *Lighweight Sandwich Construction;* John Wiley and Sons Ltd.: Hoboken, NJ, USA, 2001.
2. Larbi, A.S.; Contamine, R.; Ferrier, E.; Hamelin, P. Shear strengthening of RC beams with textile reinforced concrete (TRC) plate. *Constr. Build. Mater.* **2010**, *24*, 1928–1936. [CrossRef]
3. Verbruggen, S.; Tysmans, T.; Wastiels, J. TRC or CFRP strengthening for reinforced concrete beams: An experimental study of the cracking behaviour. *Eng. Struct.* **2014**, *77*, 49–56. [CrossRef]
4. Papanicolaou, C.G.; Triantafillou, T.C.; Papathanasiou, M.; Karlos, K. Textile reinforced mortar (TRM) versus FRP as strengthening material of URM walls: Out-of-plane cyclic loading. *Mater. Struct. Constr.* **2008**, *41*, 143–157. [CrossRef]
5. Parisi, F.; Lignola, G.P.; Augenti, N.; Prota, A.; Manfredi, G. Rocking response assessment of in-plane laterally-loaded masonry walls with openings. *Eng. Struct.* **2013**, *56*, 1234–1248. [CrossRef]
6. Hegger, J.; Voss, S. Investigations on the bearing behaviour and application potential of textile reinforced concrete. *Eng. Struct.* **2008**, *30*, 2050–2056. [CrossRef]
7. Nguyen, V.A.; Jesse, F.; Curbach, M. Experiments about load bearing behaviour of lightweight sandwich beams using textile reinforced and expanded polystyrene concrete. *Struct. Concr.* **2015**, *17*, 760–767. [CrossRef]
8. Colombo, I.G.; Colombo, M.; Prisco, M. Bending behaviour of Textile Reinforced Concrete sandwich beams. *Constr. Build. Mater.* **2015**, *95*, 675–685. [CrossRef]
9. Portal, N.W.; Flansbjer, M.; Zandi, K.; Wlasak, L.; Malaga, K. Bending behaviour of novel Textile Reinforced Concrete-foamed concrete (TRC-FC) sandwich elements. *Compos. Struct.* **2017**, *177*, 104–118. [CrossRef]
10. Cuypers, H.; Wastiels, J. Analysis and verification of the performance of sandwich panels with textile reinforced concrete faces. *J. Sandw. Struct. Mater.* **2011**, *3*, 589–603. [CrossRef]
11. Junes, A.; Larbi, A.S. An indirect non-linear approach for the analysis of sandwich panels with TRC facings. *Constr. Build. Mater.* **2016**, *112*, 406–415. [CrossRef]
12. Shams, A.; Hegger, J.; Horstmann, M. An analytical model for sandwich panels made of textile-reinforced concrete. *Constr. Build. Mater.* **2014**, *64*, 451–459. [CrossRef]
13. Finzel, J.; Häussler-Combe, U. Textile reinforced concrete sandwich panels: Bending tests and numerical analyses. In Proceedings of the Euro-C 2010, Schladming, Austria, 15–18 March 2010; pp. 789–795.
14. Miccoli, L.; Fontana, P. Numerical modelling of UHPC and TRC sandwich elements for building envelopes Façade element components. *Int. Assoc. Bridge Struct. Eng.* **2015**, *105*, 1–9.
15. Djamai, Z.I.; Bahrar, M.; Salvatore, F.; Larbi, A.S.; El Mankibi, M. Textile reinforced concrete multiscale mechanical modelling: Application to TRC sandwich panels. *Finite Elem. Anal. Des.* **2017**, *135*, 22–35. [CrossRef]
16. Contamine, R.; Larbi, A.S.; Hamelin, P. Contribution to direct tensile testing of textile reinforced concrete (TRC) composites. *Mater. Sci. Eng. A* **2011**, *528*, 8589–8598. [CrossRef]
17. De Andrade Silva, F.; Butler, M.; Mechtcherine, V.; Zhu, D.; Mobasher, B. Strain rate effect on the tensile behaviour of textile-reinforced concrete under static and dynamic loading. *Mater. Sci. Eng. A* **2011**, *528*, 1727–1734. [CrossRef]
18. Barhum, R.; Mechtcherine, V. Effect of short, dispersed glass and carbon fibres on the behaviour of textile-reinforced concrete under tensile loading. *Eng. Fract. Mech.* **2012**, *92*, 56–71. [CrossRef]

19. Hartig, J.; Jesse, F.; Schicktanz, K.; Häußler-Combe, U. Influence of experimental setups on the apparent uniaxial tensile load-bearing capacity of Textile Reinforced Concrete specimens. *Mater. Struct. Constr.* **2012**, *45*, 433–446. [CrossRef]
20. Hegger, J.; Will, N.; Curbach, M.; Jesse, F. Tragverhalten von textilbewehrtem Beton: Verbund, Ribbildung und Tragverhalten. *Beton-Und Stahlbetonbau* **2004**, *99*, 452–455. [CrossRef]
21. El Kadi, M.; Tysmans, T.; Verbruggen, S.; Vervloet, J.; de Munck, M.; Wastiels, J.; van Hemelrijck, D. A layered-wise, composite modelling approach for fibre textile reinforced cementitious composites. *Cem. Concr. Compos.* **2018**, *94*, 107–115. [CrossRef]
22. Bertolesi, E.; Carozzi, F.G.; Milani, G.; Poggi, C. Numerical modeling of Fabric Reinforce Cementitious Matrix composites (FRCM) in tension. *Constr. Build. Mater.* **2014**, *70*, 531–548. [CrossRef]
23. Promis, G.; Gabor, A.; Hamelin, P. Analytical modeling of the bending behavior of textile reinforced mineral matrix composite beams. *Compos. Struct.* **2011**, *93*, 792–801. [CrossRef]
24. Tysmans, T.; Wozniak, M.; Remy, O.; Vantomme, J. Finite element modelling of the biaxial behaviour of high-performance fi bre-reinforced cement composites (HPFRCC) using Concrete Damaged Plasticity. *Finite Elem. Anal. Des.* **2015**, *100*, 47–53. [CrossRef]
25. Cuypers, H. Analysis and Design of Sandwich Panels with Brittle Matrix Composite Faces for Building Applications. Ph.D. Thesis, Vrije Universiteit Brussel, Brussels, Belgium, 2002.
26. Mesticou, Z.; Bui, L.; Junes, A.; Larbi, A.S. Experimental investigation of tensile fatigue behaviour of Textile-Reinforced Concrete (TRC): Effect of fatigue load and strain rate. *Compos. Struct.* **2017**, *160*, 1136–1146. [CrossRef]
27. Remy, O. Lightweight Stay Formwork: A Concept for Future Building Applications. Ph.D. Thesis, Vrije Universiteit Brussel, Brussels, Belgium, 2012.
28. Cuypers, H.; Gu, J.; Croes, K.; Dumortier, S.; Wastiels, J. Evaluation of fatigue and durability properties of E-glass fibre reinforced phosphate cementitious composite. *Int. Symp. Brittle Matrix Compos.* **2000**, *6*, 127–136.
29. Belgian Bureau for Standardisation (NBN). *NBN EN 196-1:2016 Methods of Testing Cement-Part 1: Determination of Strength*; Belgian Bureau for Standardisation (NBN): Brussels, Belgium, 2016.
30. Knauf. Gitex, Glasvezelwapening Technische Fiche. 2018. Available online: http://www.knauf.be/nl/product/gitex-glasvezelwapening (accessed on 18 January 2018).
31. Kemisol. Technische Documentatie EPS. 2004. Available online: http://www.kemisol.be/ (accessed on 15 March 2018).
32. Sutton, M.A.; Orteu, J.J.; Schreier, H. *Image Correlation for Shape, Motion and Deformation Measurements*; Springer: New York, NY, USA, 2009.
33. De Munck, M.; Vervloet, J.; El Kadi, M.; Verbruggen, S.; Wastiels, J.; Remy, O.; Tysmans, T. Modelling and experimental verification of flexural behaviour of textile reinforced cementitious composite sandwich renovation panels. In Proceedings of the 12th fib International PhD Symposium in Civil Engineering, Prague, Czech Republic, 29–31 August 2018; pp. 179–186.

 *applied sciences*

Article

Bond Fatigue of TRC with Epoxy Impregnated Carbon Textiles

Juliane Wagner * and Manfred Curbach

Institute of Concrete Structures, 01062 TU Dresden, Germany; manfred.curbach@tu-dresden.de
* Correspondence: Juliane.Wagner1@tu-dresden.de; Tel.: +49-351-463-39419

Received: 29 March 2019; Accepted: 8 May 2019; Published: 15 May 2019

Abstract: For the economical construction of fatigue loaded structures with textile reinforced concrete (TRC), it is necessary to investigate the fatigue behavior of the materials. Since next to the tensile load-bearing behavior, the bond behavior of a material is crucial as well, the present paper deals with the bond fatigue of TRC with epoxy-impregnated carbon textiles. First, static tests are carried out to determine the sufficient anchorage length of the investigated material combination. Afterwards, the influence of cyclic loading on the necessary anchorage length, deformation, stiffness, and residual strength is investigated. The results of the cyclic tests are summarized in stress-number of cycles to failure (S-N) diagrams. In the end, it can be said that the cyclic loading has no negative impact on the necessary anchorage length. If specimens withstand the cyclic loading, there is no difference between their residual strength and the reference strength. The failure of specimens occurs only at high load levels, provided that the anchorage length is sufficient.

Keywords: textile reinforced concrete; carbon reinforced concrete; TRC; CRC; bond; fatigue; carbon textile; epoxy impregnation; test setup

1. Introduction

Textile reinforced concrete (TRC) has been under investigation for about two decades now. During this time, it was used—among other applications—for the construction of several bridges, e.g., [1–9]. Whilst the pedestrian bridges were built without prior separate fatigue investigations, the fatigue resistance of the road bridges was tested in the laboratory on true scaled structures. However, it would be uneconomical to perform a fatigue test on an entire structure every time. Therefore, it is necessary to have a closer look at the fatigue behavior of textile reinforced concrete. This is among other things the objective of the research projects C^3-V1.2 and C^3-V2.1, belonging to the research program C^3–Carbon Concrete Composite [10].

Currently, there are already some investigations on the tensile fatigue behavior of TRC. In [11–14], tensile fatigue tests with different carbon textiles were carried out, and in [15], the tensile fatigue behavior of an alkali-resistant glass textile was investigated. Since the high tensile strength of technical textiles is only advantageous when the occurring forces can be transmitted from the concrete to the textile, the bond behavior of TRC should not be ignored. As the authors don't know any research concerning the topic of bond fatigue of TRC, our own investigations were carried out and are presented in this paper.

For the investigation of bond fatigue, a suitable test setup was developed [16]. Whilst here flexible carbon textiles with a styrene-butadiene impregnation were tested, the applicability of the test setup also for stiff carbon textiles with an epoxy impregnation was proved in [17]. On the basis of this research, more precise investigations on the influence of load level and anchorage length on deformation, stiffness, and number of cycles to failure were done. The results are presented in the following sections.

2. Materials and Methods

2.1. Materials

The investigations were made with a material combination that is usually used for newly built structures out of TRC. As reinforcement, a stiff carbon textile impregnated with epoxy resin was used and embedded in a high-strength concrete. Figure 1 shows the biaxial textile (solidian GRID Q95/95-CCE-38) from solidian GmbH with a 38-mm axial fibre strand distance in both directions [18]. More information about the textile is shown in Table A1 in Appendix A.

Figure 1. Carbon textile impregnated with epoxy resin.

The related high-strength concrete (HF-2-145-5) was specially developed for use in TRC [19]. The maximum grain size was 5 mm, and was realized by grit instead of gravel. Further information on the composition can be found in Table A2. The average compressive strength and flexural strength of tested specimens at the age of 28 days was 127 N/mm^2 and 12 N/mm^2, respectively. The values were determined on three prisms (40 × 40 × 160 mm) according to [20].

2.2. Samples

As no standard for TRC exists yet, the geometry of the specimens was chosen according to testing recommendations for TRC [21], which were created in a previous research project of the research programme C^3. The specimens were 110 mm wide and 30 mm thick. The anchorage length was varied, so that the specimens had lengths between 76–304 mm. The samples were centrally reinforced with one textile layer with three fiber strands in the longitudinal direction.

They were manufactured in coated timber formworks (Figure 2a). First, huge concrete panels with a size of 990 mm × 380 mm × 30 mm were casted. During the casting process, the textile was clamped at two sides to fix it in its position. After casting, the panels were left in the formwork for one day covered with foil. Afterwards, they were stored in water until the seventh day after casting. From here until testing, they were stored in a climate chamber at a temperature of 20 °C and 65% relative humidity. Before the testing, at a minimum age of 28 days, the specimens were sawed out of the panels (Figure 2b).

2.3. Test Setup

The test setup for the investigation of the bond bearing behavior of the chosen material combination is known as the 'double-sided textile pull-out (DPO) test' (Figure 3a). Initially, the test was developed at RWTH Aachen University, e.g., [22], for the determination of the needed anchorage length. Stuck steel plates define the investigated anchorage length on both ends of the specimen. A small gap between the steel plates works as a predetermined breaking point in the middle of the specimen. By pulling the steel plates apart, tension is initiated in the specimen and hence in the textile as well. If the textile ruptures, the chosen anchorage length was long enough to transfer the complete tensile load. If bond failure occurs and the textile, e.g., pulls out, the anchorage length was too short. The applicability of the test setup for fatigue tests was investigated e.g., in [16].

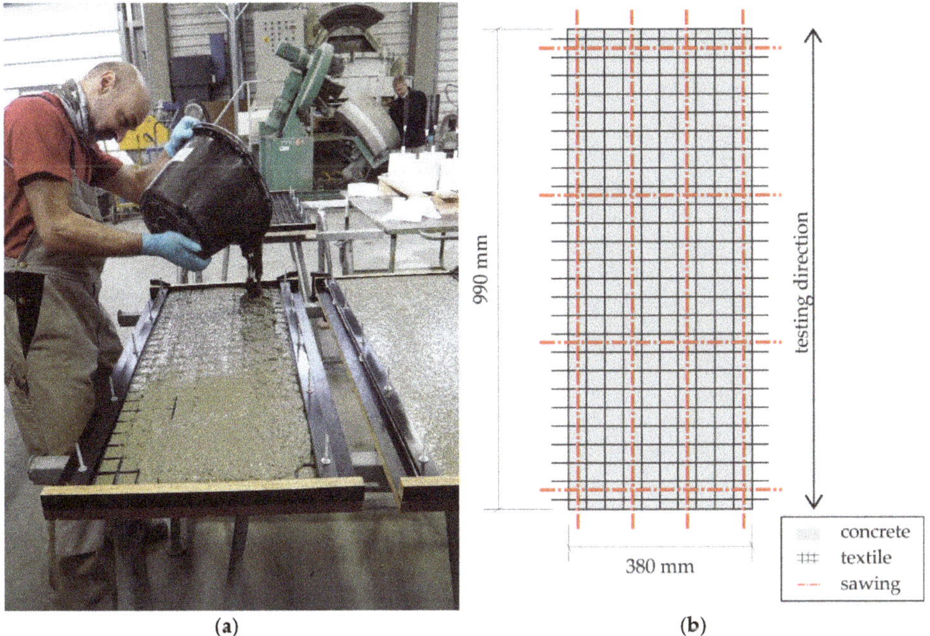

Figure 2. Manufacturing of the specimens: (**a**) Casting of huge panels; (**b**) Sawing out of specimens (exemplary cut).

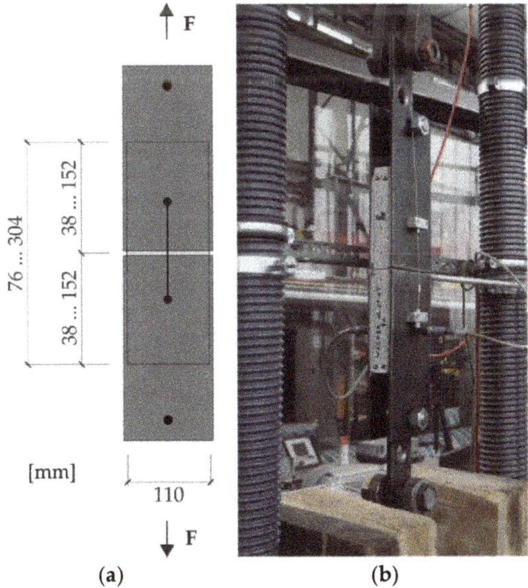

Figure 3. Test setup for static and cyclic double-sided textile pull-out (DPO) tests: (**a**) Specimen geometry; (**b**) Testing machine with specimen.

The tests were carried out in a servo-hydraulic tension testing machine with accuracy class 1 and a load capacity of 100 kN for cyclic tests. During the tests, the machine force was recorded by a load cell. The deformation of the specimen at the predetermined breaking point was measured by two extensometers, which were fixed on the steel plates on both sides of the specimens. In the cyclic tests, the number of load cycles was recorded as well.

2.4. Load Regime and Experimental Program

In order to define the load levels of the cyclic tests, first the reference strength had to be determined. As in the cyclic tests, the influence of cyclic loading on the necessary anchorage length ought to be investigated. The static reference tests were done with different anchorage lengths as well, which were defined as a multiple of the fiber strand distance, *a*. There were four different anchorage lengths varying between 1*a*–4*a*. The minimum number of specimens per length was 10. The average value per length was set as the reference strength for the cyclic tests.

The load regime for the cyclic tests was the following. First, the specimen was loaded path-controlled up to a defined mean stress. Thereby, it was important that the crack in the middle of the specimen had completely developed. Afterwards, the cyclic loading was started, force-controlled with a sinusoidal oscillation and constant amplitude. The load frequency was 12 Hz. The maximum number of load cycles was set to 2×10^6 to limit the duration of the test, and because this is a usual choice for cyclic tests with concrete or steel. Afterwards, runouts that withstood the cyclic loading were tested path-controlled until failure to determine their residual strength.

To reduce the amount of data, measurements, e.g., the force and deformation during the cyclic loading, were done in intervals. After a defined period of time, 800 measuring values were recorded with a measuring rate of 400 Hz. The duration of this period increased over the course of the cyclic testing: during the first load cycles, every measuring point was recorded, until approximately the 5000th load cycle, the period was 10 seconds; until 30,000 load cycles, it was three minutes, and afterwards, it was 15 min.

The experimental program is shown in Table 1. Per anchorage length, two different minimum stresses σ_{min} were investigated with different related maximum stresses σ_{max}. The load levels are specified as a percentage of the reference strength. As one of the goals of the research project C^2-V2.1 is to create stress-number of cycles to failure (S-N) diagrams for TRC, load levels had to be chosen that lead to the failure of the specimens. Since preliminary tests showed that only high amplitudes cause failure, a maximum σ_{min} of 50% was set. The second σ_{min} of 30% was chosen as a compromise between the distance of the two investigated minimum stresses and the limit of the testing machine (regarding the combination of amplitude and frequency). The related maximum stress was increased after each test series, with the aim to force the failure of the specimens. Each test series contained four specimens. However, especially in test series with short anchorage lengths, some specimens failed already during the static loading up to the mean stress. So, these specimens could not be taken into account in the evaluation in the following section.

Table 1. Experimental program.

Anchorage Length	Number of Tested (Evaluated *) Specimens at Load Level $\sigma_{min}/\sigma_{max}$ [%]				
	30/90	30/95	50/70	50/85	50/90
1*a*	4 (3)	4 (2)	4 (2)	4 (2)	4 (3)
2*a*	4	4 (3)	4	4	4 (3)
3*a*	4	4	4	4	4
4*a*	4	4	4	4	4

* If different to number of tested specimens.

3. Results

3.1. Reference Tests

The results of the reference tests with different anchorage lengths are displayed in Figure 4. Here, one can see the reached maximum stress and the related measured anchorage length per specimen.

Additionally, the average tensile strength (grey line) and the range of variation of this material combination (grey area) can be seen. These values were determined in tensile tests according to [23].

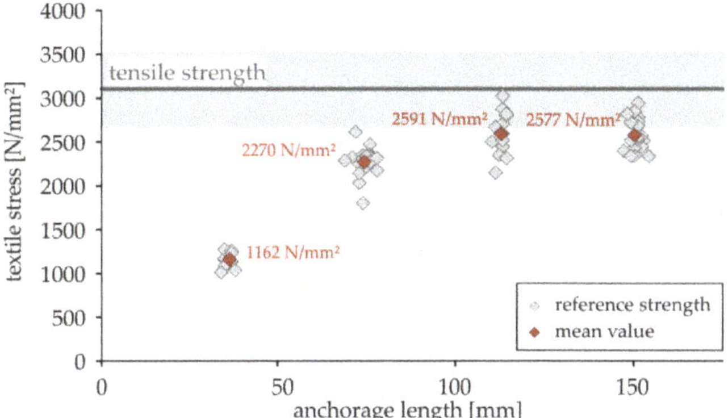

Figure 4. Maximum reached stresses in static double-sided textile pull-out (DPO) tests with different anchorage lengths.

It is clearly visible that the maximum textile stress increases with increasing anchorage length. However, the increase stops at the anchorage length of 3*a*, and the specimens with 4*a* anchorage length reach nearly the same values as the ones with 3*a*. One may assume that an anchorage length of 3*a* is sufficient to transfer the maximally possible bond force of this material combination and that longer anchorage lengths do not lead to higher loads. None of the specimens failed due to textile rupture, but rather by spalling in the reinforcement layer. Therefore, the maximally reached values are lower than the tensile strength. However, the values are quite close to the range of variation of the tensile strength, and therefore, the results are considered acceptable for the moment.

3.2. Cyclic Stress–Strain Behavior

In Figure 5, exemplary stress–deformation curves of a non-failed and a failed specimen are shown. Furthermore, the curves of the reference tests are displayed in the background. Both specimens were tested at the same load level, which was located in the transition area to the fatigue strength. This area is usually characterized by the simultaneous appearance of failed and non-failed specimens, as well as a large scattering in the number of cycles to failure. In the present diagrams, specimens with an anchorage length of 3*a* were chosen. Examples for cyclic stress–deformation curves of specimens with shorter or longer anchorage lengths are shown in Appendix B, Figures A1–A3.

Generally, the cyclic stress–deformation curves consist of three different sections. The first section includes the static loading up to the required mean stress and the formation of the crack in the predetermined breaking point. Therefore, the shape of this section of the curve follows that of the reference tests. The second part of the curve starts with the beginning of the cyclic loading. The increase and decrease of the stress during the cyclic loading are clearly visible. At the same time, the deformation increases with an increasing number of load cycles. In the case that the specimen fails due to the cyclic loading (in the present investigations, failure occurred in the form of splitting in the reinforcement layer), the cyclic stress–deformation curve ends up at this point (Figure 5b). If the specimen survived (Figure 5a), there is a third section of the stress–deformation curve. Due to the stopping of the testing machine after reaching the required number of load cycles, the third section of the curve begins with a decrease of stress and deformation. Afterwards, the residual strength of the specimen is determined statically. Therefore, the stress–deformation curve increases steeply until the failure of the specimen. Hereby, this part of the curve approximates the curves of the reference tests (grey in Figure 5), and its slope—and thus the stiffness of the specimens—was nearly the same or even steeper (higher stiffness), compared to the reference tests.

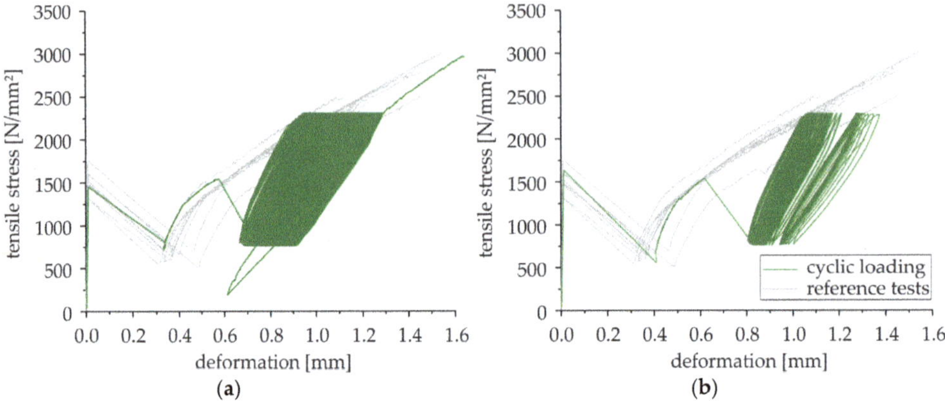

Figure 5. Cyclic stress–deformation curves (here: anchorage length = $3a$; $\sigma_{min}/\sigma_{max}$ = 30%/90%): (**a**) non-failed specimen; (**b**) failed specimen (load cycles: 57733).

3.3. Development of Deformation

3.3.1. General Remarks

The development of deformation during the cyclic loading can be shown in cyclic creep curves. According to e.g., [24,25], these curves are divided into three sections. In the first section, the deformation increases rapidly and non-linearly. Afterwards, in section two, the curve increases linearly and less steeply. Section three is also called the beginning of fatigue failure; it is characterized by a non-linear and rapid increase again. The end of section three and thereby also the cyclic creep curve is marked by the failure of the specimen.

To investigate the deformation behavior of the tested specimens, their cyclic creep curves were compared. Therefore, the average deformation of the two extensometers at mean stress is shown as a function of normalized load cycles, whereby "0" marks the beginning of the cyclic loading. Due to the measurement at intervals, curves of specimens with a very small number of cycles to failure consist of just a few measuring points, and in most cases, their failure was not recorded. For this reason, these curves are shown just until the last measuring point during the cyclic loading.

Exemplary, Figure 6 displays the cyclic creep curves of a failed and a non-failed specimen with the same anchorage length and load level. As one can see, the first two sections of the cyclic creep curve are clearly visible. Remarking on section three is not that simple with the present investigations, because the failure of the specimens mostly was not recorded. Assessing specimens that failed at a quite high number of load cycles, nearly none, or if any, just a low non-linear increase in deformation can be noticed before failure, which means that failure occurs quite abruptly.

As one can see also in Figure 6, the absolute value of deformation is no indicator of impending failure, because the non-failed specimen reached higher deformations than the failed one with the same anchorage length and load level.

3.3.2. Dependence on Load Level

To investigate the influence of the load level on the deformation of the specimens, the cyclic creep curves of specimens with the same anchorage lengths are compared in Figure 7. Here, specimens with an anchorage length of $3a$ are shown as examples. Figure 7a shows the creep curves of specimens with a minimum stress of 30% and maximum stresses of 90% and 95%, respectively. Figure 7b shows curves of specimens with 50% minimum stress and different maximum stresses between 70–90%. Failed specimens are illustrated by dashed lines. Curves of specimens with other anchorage lengths are shown in Figure A4 in Appendix B.

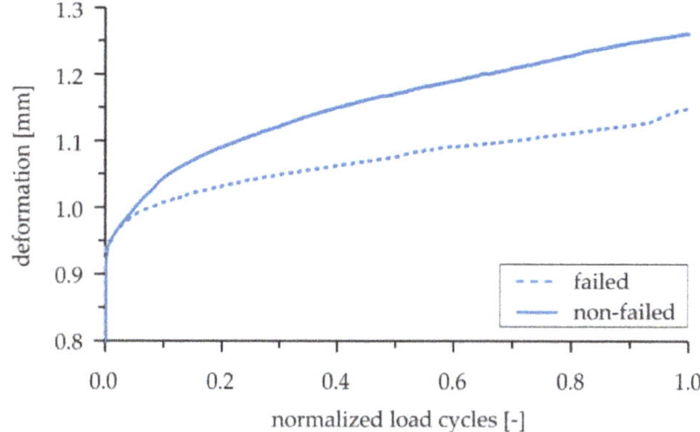

Figure 6. Comparison of cyclic creep curves of a failed and a non-failed specimen (here: anchorage length = $4a$; $\sigma_{min}/\sigma_{max}$ = 30%/95%).

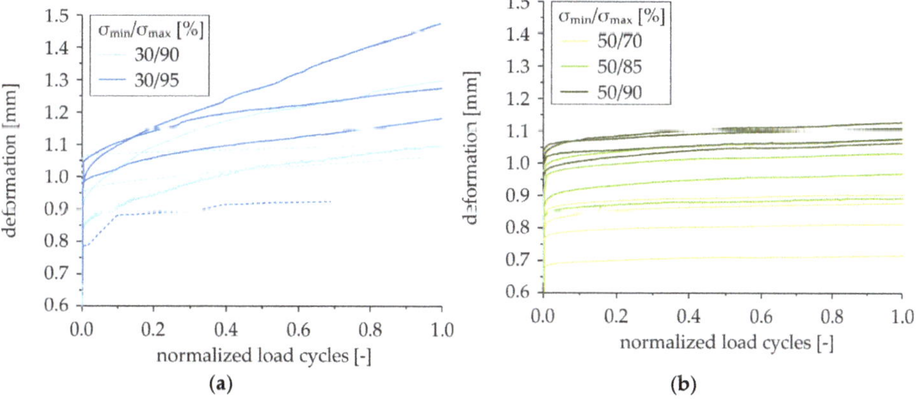

Figure 7. Comparison of normalized cyclic creep curves depending on the load level (here: anchorage length = $3a$): (**a**) 30% minimum stress; (**b**) 50% minimum stress.

One can see that deformations become larger with increasing maximum stress. However, not only the maximum stress itself, but also the related amplitude σ_a affects the amount of deformation. That means, regarding e.g., load levels 30/90 and 50/90 in Figure 8 (showing the same curves as Figure 7 but depending on the amplitude), specimens with the same maximum stress but a higher amplitude (viz. a lower minimum stress) show higher deformations than the ones with a lower amplitude. In addition, it can be seen that not only do deformations become larger with higher amplitudes, but the slope of section two of the cyclic creep curve also becomes steeper.

3.3.3. Dependence on Anchorage Length

Now, the influence of the anchorage length on the deformation is regarded. Therefore, the four different investigated anchorage lengths were compared at several load levels. Figure 9 shows an example of the cyclic creep curves at a load level of 50/85. The comparison of the creep curves at other load levels is shown in Figure A5. Again, failed specimens are marked by dashed lines.

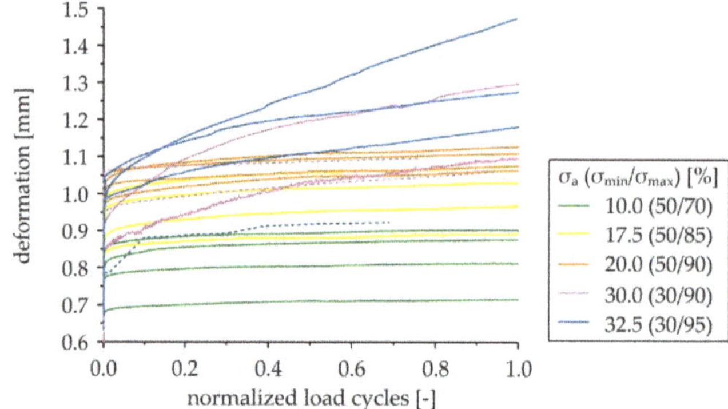

Figure 8. Comparison of normalized cyclic creep curves depending on the amplitude (here: anchorage length = 3*a*).

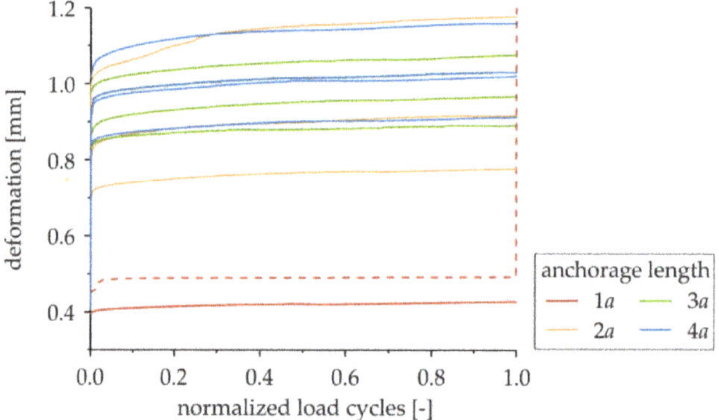

Figure 9. Comparison of normalized cyclic creep curves of different anchorage lengths at similar load level (here: $\sigma_{min}/\sigma_{max}$ = 50%/85%).

In Figure 9, it can clearly be seen that the deformation increases with increasing anchorage length. However, the difference between the deformation of specimens with 1*a* and 2*a* as anchorage lengths is larger than the difference between specimens with 2*a* and 3*a*. Finally, the deformations of specimens with anchorage lengths of 3*a* and 4*a* show no differentiation anymore. The reason for this is to be found in the absolute reference strengths, increasing non-linearly with increasing anchorage length (see Section 3.1).

3.4. Development of Stiffness

3.4.1. General Remarks

The development of the stiffness of a specimen can be described by regarding the development of the secant modulus. According to [26], the secant modulus describes the slope of the secant between the maximum and minimum point of a hysteresis loop, and can be determined separately for every single load cycle (Figure 10) [27]. In contrast to [26], in the present investigations, deformations instead of elongations are indicated. Therefore, the physical unit of the secant modulus is $\frac{\frac{N}{mm^2}}{mm}$ instead of $\frac{N}{mm^2}$ (see also [16]).

To compare the absolute stiffnesses of several specimens, the development of their secant modulus is shown as a function of normalized load cycles. Whereby, similar to the deformations, "0" marks the beginning of the cyclic loading. As already explained in Section 3.3.1, due to the measurement in intervals, curves of failed specimens are only drawn until the last measuring point before failure.

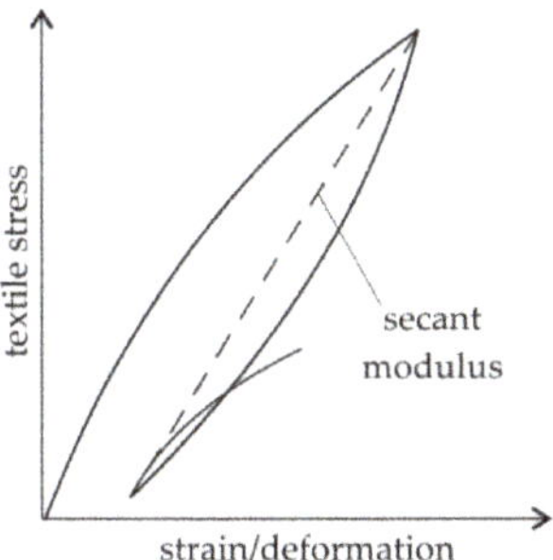

Figure 10. Determination of the secant modulus.

3.4.2. Dependence on Load Level

The dependence of the stiffness on the load level is shown in Figure 11. As an example, here, the curves of specimens with $3a$ as an anchorage length are shown. The evaluations for the other anchorage lengths can be seen in Figure A6. Similar to the deformations, there is a clear relation between the amplitude and the secant modulus, which becomes lower with increasing amplitude. When calculating the secant modulus, the deformation is the denominator; hence, it is mathematically justified that the secant modulus has to become lower with increasing amplitude (see also [16]), because the deformation increases with increasing amplitude (see Section 3.3.2). Comparing the curves with high amplitudes (blue and purple) with the ones with lower amplitudes (orange, yellow, and green), one can see that—similar to the deformations—there is a difference in their decrease, which is lower in the curves with lower amplitudes.

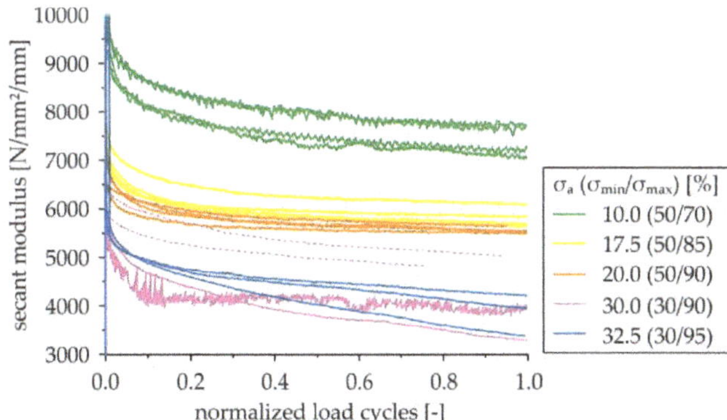

Figure 11. Comparison of the normalized development of the secant modulus depending on the amplitude (here: anchorage length = $3a$).

3.4.3. Dependence on Anchorage Length

Similar to the investigation of deformations, now the dependence of the absolute stiffness on the anchorage length is regarded. Figure 12 shows the development of the stiffness of specimens with the four investigated anchorage lengths at a load level of 50/85. The curves at the other load levels can be found in Figure A7. As one can see, the absolute stiffness decreases with increasing anchorage length. However, there is just a small increase between the specimens with $2a$ and $3a$, and no increase between the specimens with an anchorage length of $3a$ and $4a$. The reason is also a mathematical one, remembering that the stiffness depends on the deformation, and that there was also nearly no increase of deformation between specimens with an anchorage length of 2*a* and longer ones (see Section 3.3.3).

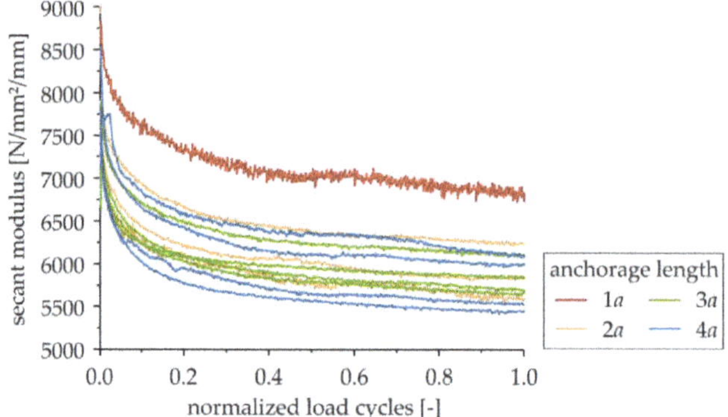

Figure 12. Comparison of the normalized development of the secant modulus of different anchorage lengths at similar load level (here: $\sigma_{min}/\sigma_{max}$ = 50%/85%).

3.5. S-N Diagram

The most common way of evaluating fatigue tests is to create S-N diagrams. Here, the number of cycles to failure *N* at a defined load level *S* can be read off. Figure 13 shows the S-N diagram for the present investigations. The two different minimum stresses are marked by two different colors. The related maximum stresses can be seen at the y-axis. The different anchorage lengths are defined by different symbols. Runouts are edged by a black line, and marked by an arrow with the number of non-failed specimens. For a better understanding, the S-N diagram is broken down by the anchorage lengths in Figure A8, and the results are listed in Table A3.

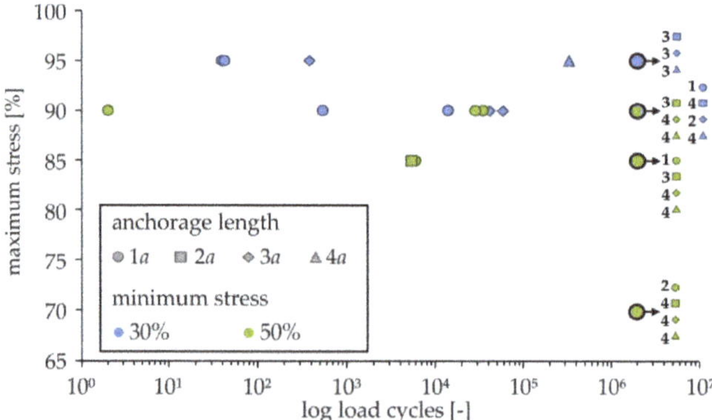

Figure 13. S-N diagram with relative maximum stresses for DPO tests with different anchorage lengths and two different minimum stresses.

As one can see in Figure 13, there is only a very small number of failed specimens. Only at high loads ($\sigma_{max} \geq$ 85%) and with 1*a* as the anchorage length did failure occur rapidly and assuredly. Specimens with anchorage lengths of 2*a* or longer failed seldom, and even at high loads (e.g., σ_{max} = 95%), runouts occurred.

In the S-N diagram in Figure 14, the maximum stresses are shown as absolute stresses, ignoring the different minimum stresses. Furthermore, the mean tensile strength and its range of variation and the reference stresses of the different anchorage lengths are shown. It can clearly be seen that with shorter anchorage lengths (1*a* and 2*a*), the transmittable load is quite low. Only with anchorage lengths of 3*a* or 4*a* can loads close to the tensile strength be reached. However, from 3*a* to 4*a*, no significant increase in transmittable load can be noticed.

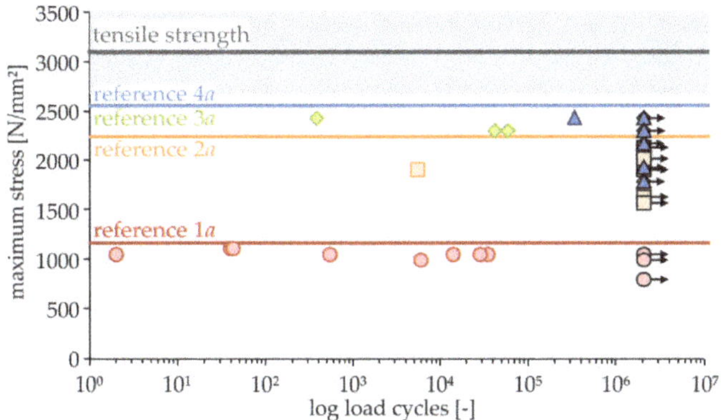

Figure 14. S-N diagram with absolute maximum stresses for DPO tests with different anchorage lengths.

3.6. Residual Strength

If specimens withstood the cyclic loading, their residual strength was tested subsequently. In Figure 15, the residual strengths are compared to the reference strengths from Section 3.1. The results above the grey line mean a higher residual strength compared to the reference strength, and the results below the line mean a lower residual strength. Figure 15a displays the results of the specimens with a minimum stress of 30%, and Figure 14b displays the ones with 50% minimum stress. Different symbols stand for the different anchorage lengths, and the darker the color of the symbols, the higher the related maximum stress. The single values of the results can also be found in Table A3.

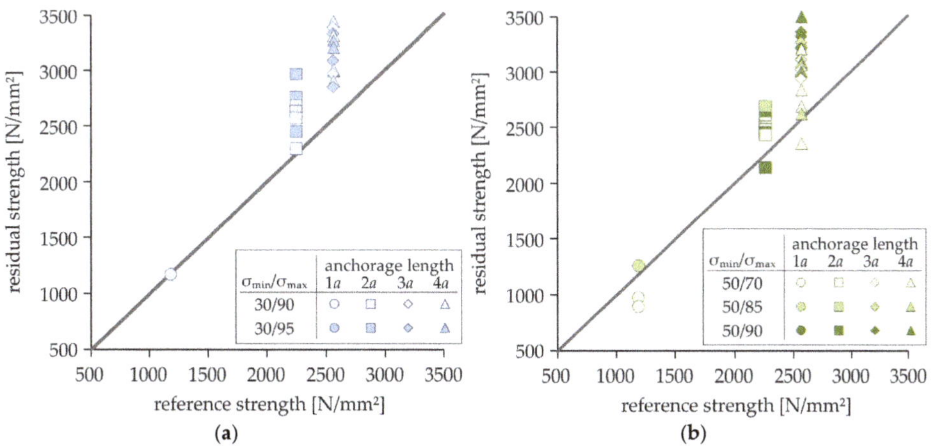

Figure 15. Comparison of residual and reference stresses: (**a**) 30% minimum stress; (**b**) 50% minimum stress.

As one can see in Figure 15, the residual strengths are higher or at least at the same level than the reference strengths, and even high maximum stresses seem not to cause damage, leading to lower residual strengths. That means that there is no negative impact on the load-bearing capacity of the material by fatigue loading. A load increase after cyclic loading is not unusual, and is often noticed in fatigue experiments with several materials. A final explanation for this phenomenon has not been found yet. In the present investigation, the reason for the load increase could be the activation of more filaments in a fiber strand due to the cyclic loading. However, this theory still has to be proven.

In Figure 16, the dependence of the textile stress on the anchorage length, which is known from Section 3.1, is supplemented by the determined residual stresses. The test results of the reference stresses, all of them failed by splitting, are marked by grey rhombs, whilst orange rhombs mark the splitted specimens in residual strength tests. If specimens in residual strength testing failed by textile rupture, the results are displayed by red crosses.

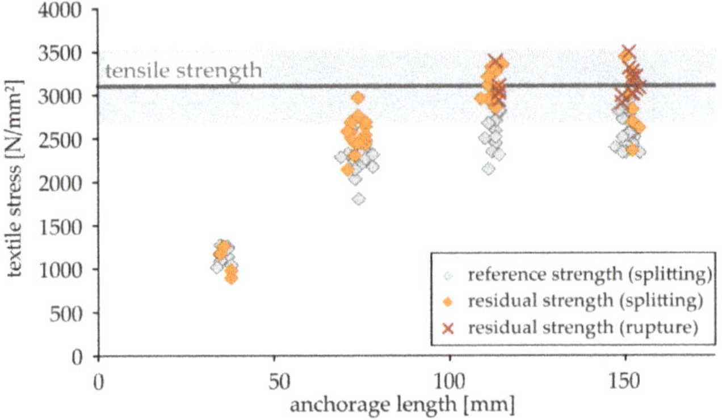

Figure 16. Comparison of textile stresses and failure mechanisms in reference and residual strength tests with different anchorage lengths.

As one can see, textile rupture only occurred at the long anchorage lengths of 3*a* or higher, where the residual strengths reach the value of the mean textile stress. However, there was no dependence of the failure mode on the applied load level. Furthermore, as splitting and rupture occur at the same stresses, the failure mode is also not dependent on the value of the residual strength.

4. Summary, Conclusions, and Outlook

Regarding the results from Section 3, it can be stated that the cyclic loading has no negative impact on the required anchorage length for the investigated material combination. This assertion is justified by the following. First, there was no further increase in deformation when the anchorage length was increased from 3*a* to 4*a* (Section 3.3.3). Additionally, there was no decrease in absolute stiffness regarding these specimens (Section 3.4.3). Finally, in testing the residual strength, there was no load increase from an anchorage length of 3a, and some of these specimens failed by textile rupture at the level of the tensile strength, which is an indicator that the tested anchorage length is sufficient to transfer the complete tensile load, even after cyclic loading.

In conclusion, it can be said that the bond fatigue behavior of the investigated material combination is quite good, provided that the anchorage length is sufficient. The development of the deformation and stiffness of a specimen during cyclic loading depends on the applied maximum load as well as on the related minimum load (Sections 3.3.2 and 3.4.2). However, even at high maximum loads (e.g., σ_{max} = 95%), runouts occurred (Section 3.5). The tested residual strengths of runouts were generally higher or at least at the same level than the reference strengths (Section 3.6). Regarding the stress–deformation curves, the curve of the residual strength meets the curves of the reference tests again (Section 3.2); thus—regarding runouts—no negative impact of the cyclic loading can be detected. However, as there was nearly no third section of the observed cyclic creep curves of the failed specimen (Section 3.3.1), one may assume that cyclic failure occurs quite abruptly without advance notice. For this reason, it is strongly recommended to not extrapolate the S-N curves over the experimentally proven load levels.

Whilst doing the above described investigations, the following concerning the use of DPO tests was found by SCHÜTZE AND CURBACH in [28]: a material combination (carbon textile with polyacrylate-based impregnation and fine grain concrete) usually used for the strengthening of existing reinforced concrete structures was tested in static DPO tests. Independently of the anchorage length, all the specimens failed due to splitting at quite low loads, and no increase of load with increasing anchorage length could be detected. Firstly, it is assumed that due to eccentricity between the stuck steel plates and the textile layer, there is a moment leading to splitting, and secondly, that the stuck steel plates block the formation of cracks, except for the one in the middle of the specimen. At this point, the whole deformation is concentrated, and the critical deformation, leading to splitting, is reached quite early. The investigations with this material combination have shown that the DPO test only represents the bond behavior of some specific situations, and cannot be used to determine the necessary anchorage length for thin layers of this material combination. For this material combination, much more realistic results were reached with an overlap test (e.g., according to [29]). Even though the findings in [28] were made with a different material combination than in the research presented in Section 3, comparative studies using overlap tests with the material combination described in Section 2 of the present paper should be done to prove the obtained results.

Author Contributions: Conceptualization, M.C.; Formal analysis, J.W.; Funding acquisition, M.C.; Investigation, J.W.; Project administration, M.C.; Supervision, M.C.; Validation, J.W. and M.C.; Visualization, J.W.; Writing—original draft, J.W.; Writing—review & editing, M.C.

Funding: This research was funded by the German Federal Ministry of Education and Research. The experiments were carried out in the projects C3-V1.2 'Verification and testing concepts for standards and approvals' (funding period: 01.2016–04.2018, grant number: 03ZZ0312A) and C3-V2.1 'Long-term behaviour of carbon reinforced concrete' (funding period: 09.2017–02.2020, grant number: 03ZZ0321A) as part of the project consortium 'C3—Carbon Concrete Composites' with circa 170 partners all over Germany.

Acknowledgments: First of all, we would like to thank the colleagues in the Otto Mohr Laboratory, where the specimens were produced and the experiments took place. Further thanks for the good cooperation go to the colleagues and partners in the research projects. For the provision of the carbon textile free of charge, we would also like to thank solidian GmbH!

Conflicts of Interest: The authors declare no conflict of interest.

Appendix A. Materials

Textile "solidian GRID Q95/95-CCE-38"

Table A1. Characteristics of the textile (values according to [18]).

		Longitudinal	Transversal
Fiber strand distance	[mm]	38	38
Cross-section of the strand	[mm^2]	3.62	3.62
Average tensile strength	[N/mm^2]	3200	3300
Modulus of elasticity	[N/mm^2]	>220000	>205000

Concrete "HF-2-145-5"

Table A2. Composition of the concrete (values according to [19]).

Ingredients	Quantity [kg/m^3]
Binder	621
Quartz fine sand	250
Sand 0/2	530
Granite grit 2/5	837
Super-plasticizer	16
Water	145

Appendix B. Results

Cyclic stress–deformation curves

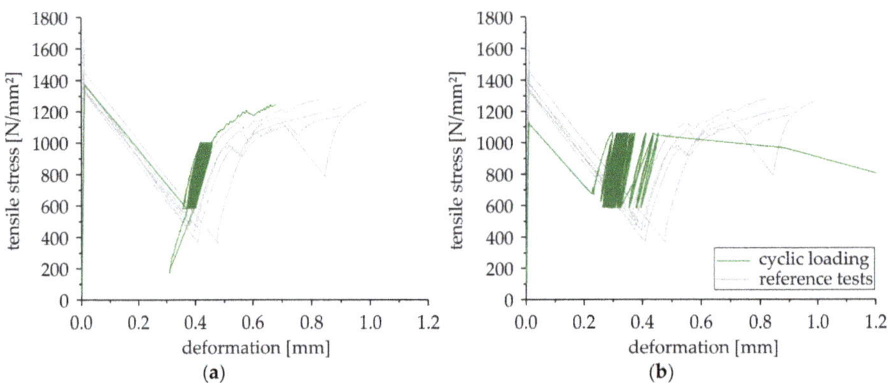

Figure A1. Cyclic stress–deformation curves (here: anchorage length = 1*a*): (**a**) non-failed specimen; (**b**) failed specimen (load cycles: 34323).

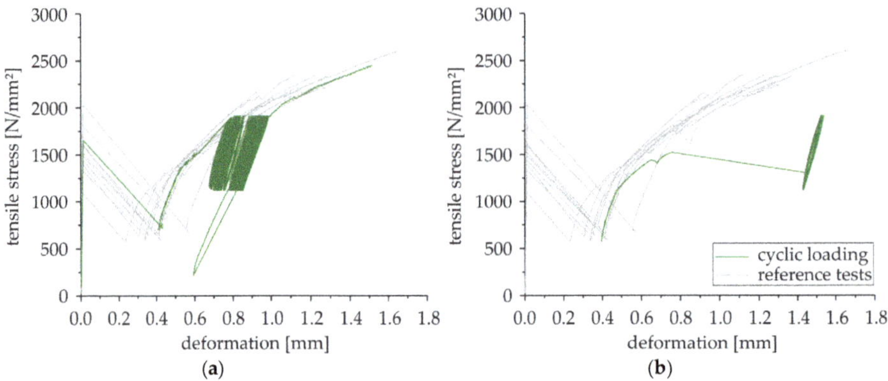

Figure A2. Cyclic stress–deformation curves (here: anchorage length = 2*a*): (**a**) non-failed specimen; (**b**) failed specimen (load cycles: 5308).

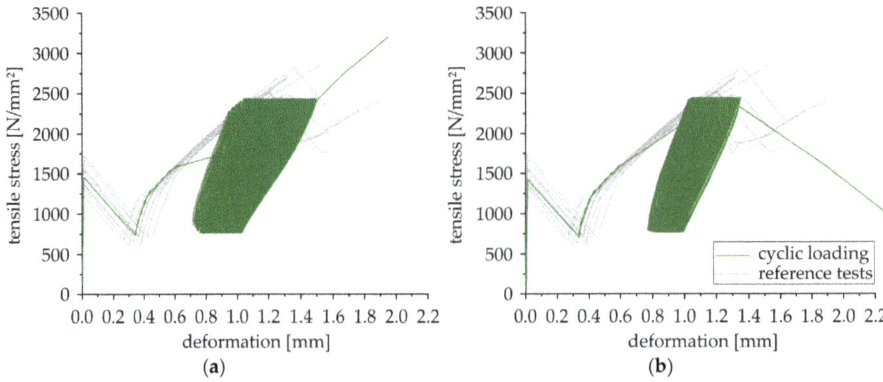

Figure A3. Cyclic stress–deformation curves (here: anchorage length = 4*a*): (**a**) non-failed specimen; (**b**) failed specimen (load cycles: 328198).

Cyclic creep curves depending on the load level

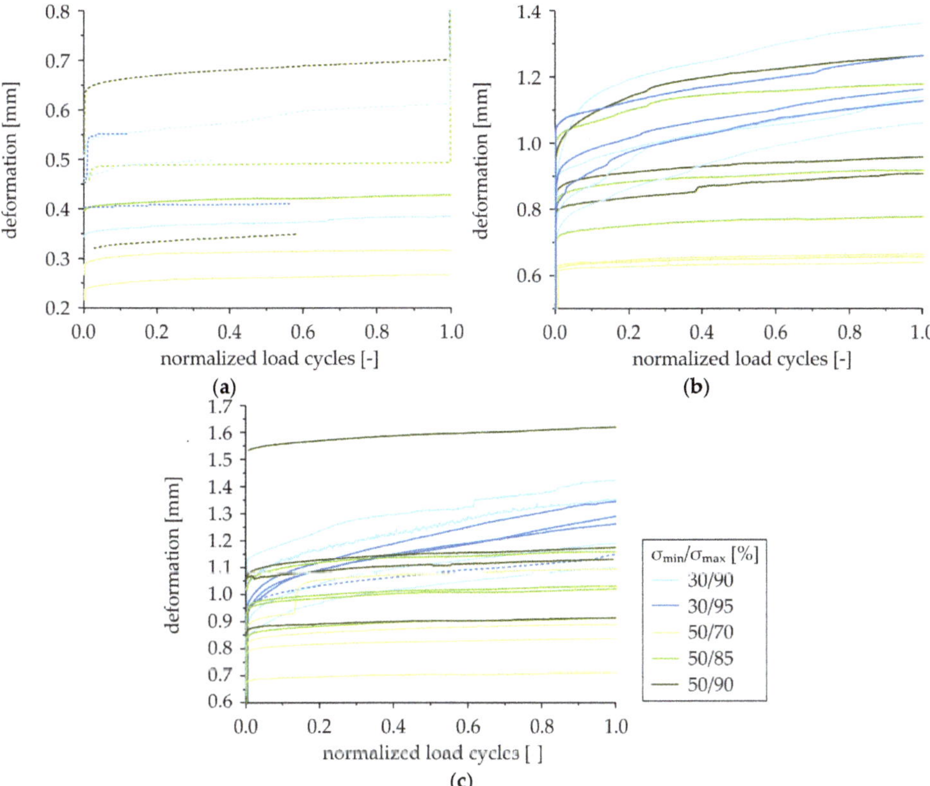

Figure A4. Comparison of normalized cyclic creep curves depending on the load level: (**a**) anchorage length = 1*a*; (**b**) anchorage length = 2*a*; (**c**) anchorage length = 4*a*.

Cyclic creep curves depending on the anchorage length

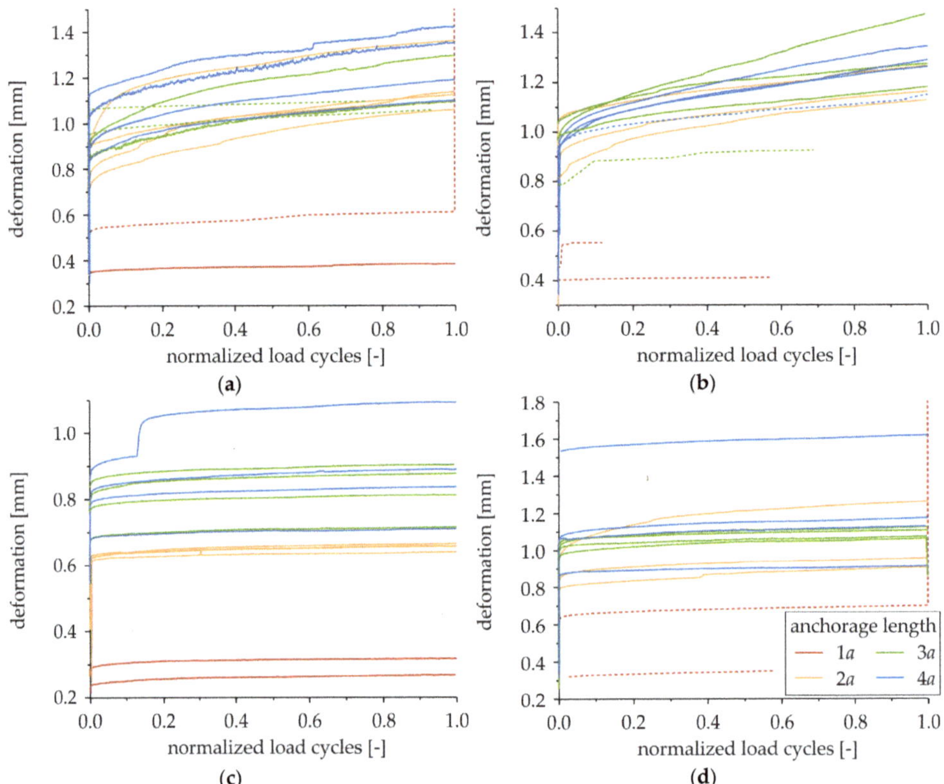

Figure A5. Comparison of normalized cyclic creep curves depending on the anchorage length: (a) $\sigma_{min}/\sigma_{max}$ = 30%/90%; (b) $\sigma_{min}/\sigma_{max}$ = 30%/95%; (c) $\sigma_{min}/\sigma_{max}$ = 50%/70%; and (d) $\sigma_{min}/\sigma_{max}$ = 50%/90%.

Development of secant modulus depending on the load level

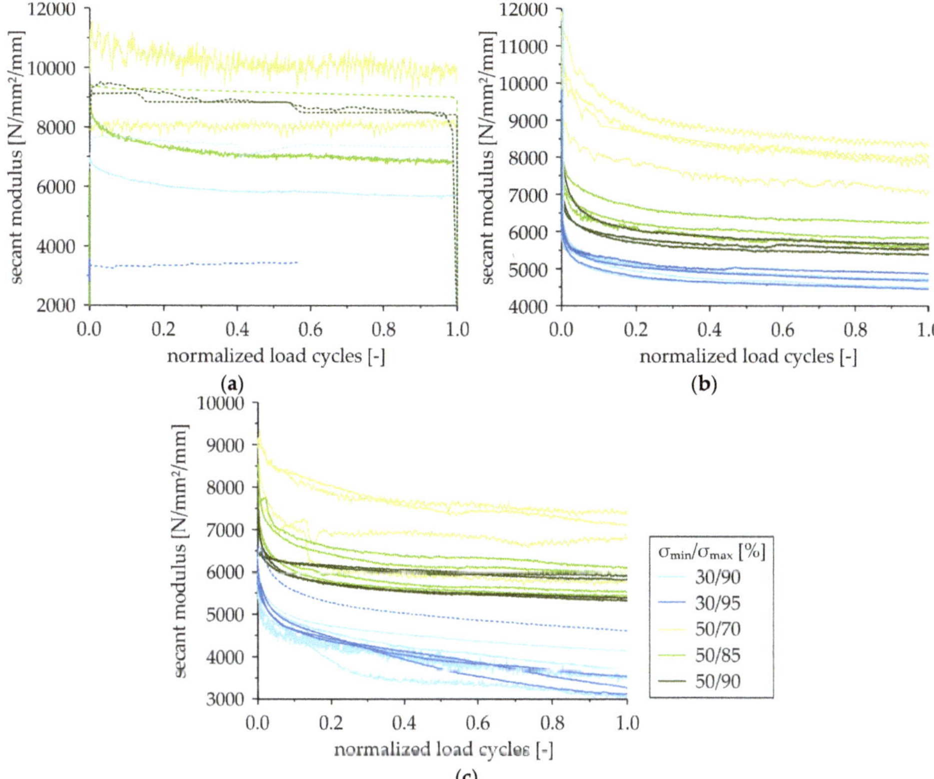

Figure A6. Comparison of the normalized development of the secant modulus depending on the load level: (**a**) anchorage length = 1*a*; (**b**) anchorage length = 2*a*; and (**c**) anchorage length = 4*a*.

Development of secant modulus depending on the anchorage length

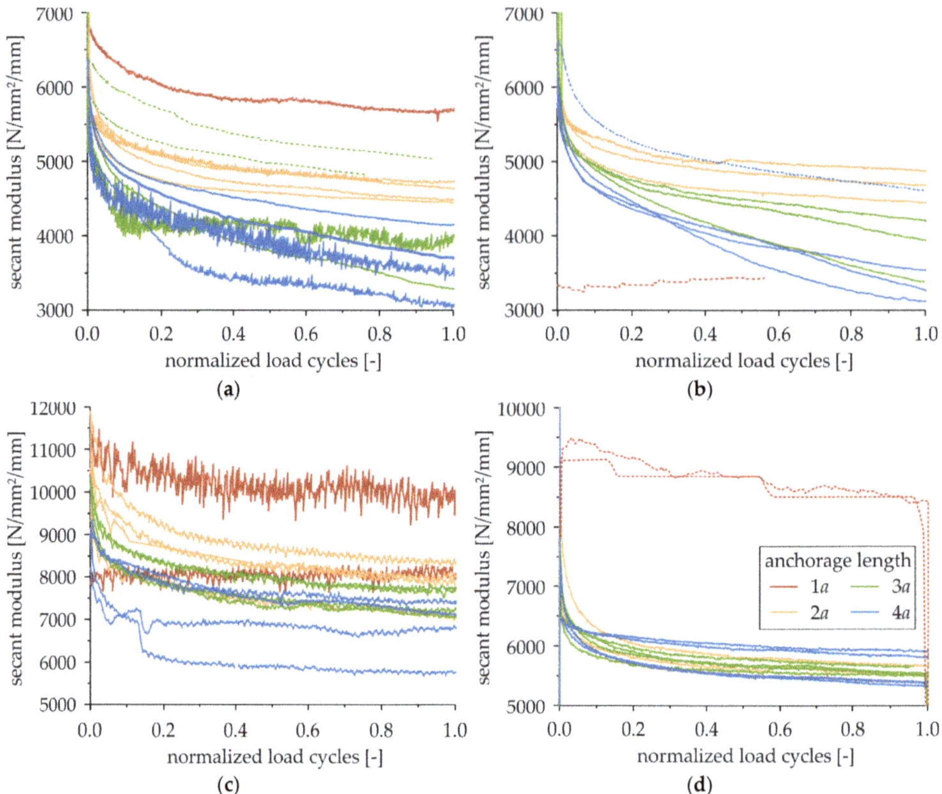

Figure A7. Comparison of the normalized development of the secant modulus depending on the anchorage length: (**a**) $\sigma_{min}/\sigma_{max}$ = 30%/90%; (**b**) $\sigma_{min}/\sigma_{max}$ = 30%/95%; (**c**) $\sigma_{min}/\sigma_{max}$ = 50%/70%; and (**d**) $\sigma_{min}/\sigma_{max}$ = 50%/90%.

S-N diagram broken down by the anchorage lengths

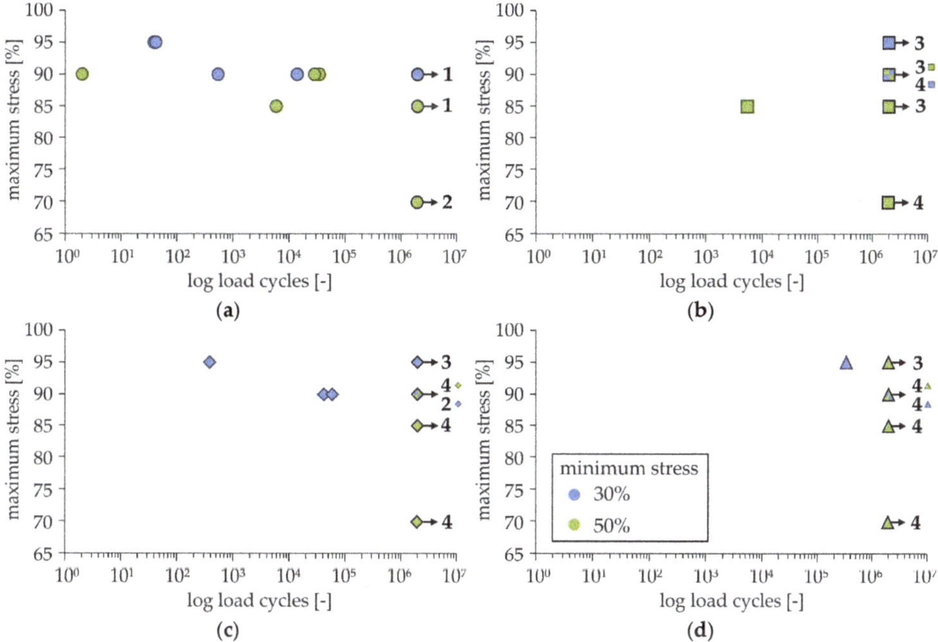

Figure A8. S-N diagrams with relative maximum stresses and two different minimum stresses: (**a**) anchorage length = 1*a*; (**b**) anchorage length = 2*a*; (**c**) anchorage length = 3*a*; and (**d**) anchorage length = 4*a*.

Test results number of cycles to failure and residual strength

Table A3. Test results.

Anchorage Length	Load Level $\sigma_{min}/\sigma_{max}$ [%]	Number of Cycles to Failure	Residual Strength [N/mm^2]	Anchorage Length	Load Level $\sigma_{min}/\sigma_{max}$ [%]	Number of Cycles to Failure	Residual Strength [N/mm^2]
1*a*	30/90	532	-	2*a*	30/90	2009000 *	2296
		13801	-			2009000 *	2567
		2009000 *	1169			2009000 *	2677
	30/95	39	-			2009000 *	2629
		42	-		30/95	2009000 *	2447
	50/70	2009000 *	976			2009000 *	2753
		2009000	895			2009000 *	2961
	50/85	5865	-		50/70	2009000 *	2501
		2009000 *	1259			2009000 *	2536
	50/90	2	-			2009000 *	2478
		28112	-			2009000 *	2429
		34323	-		50/85	5308	-
3*a*	30/90	41078	-			2009000 *	2683
		57733	-			2009000 *	2449
		2009000 *	2965			2009000 *	2447
		2009000 *	3379		50/90	2009000 *	2137
	30/95	378				2009000 *	2443
		2009000 *	3083			2009000 *	2576
		2009000 *	3326	4*a*	30/90	2009000 *	2897

Table A3. *Cont.*

Anchorage Length	Load Level $\sigma_{min}/\sigma_{max}$ [%]	Number of Cycles to Failure	Residual Strength [N/mm^2]	Anchorage Length	Load Level $\sigma_{min}/\sigma_{max}$ [%]	Number of Cycles to Failure	Residual Strength [N/mm^2]
		2009000 *	2844			2009000 *	3000
	50/70	2009000 *	3015			2009000 *	3433
		2009000 *	2930			2009000 *	3306
		2009000 *	3101		30/95	328198	-
		2009000 *	2944			2009000 *	3262
	50/85	2009000 *	3107			2009000 *	3201
		2009000 *	3039			2009000 *	2990
		2009000 *	2964		50/70	2009000 *	2353
		2009000 *	3257			2009000 *	2828
	50/90	2009000 *	3311			2009000 *	2683
		2009000 *	2949			2009000 *	3197
		2009000 *	3206		50/85	2009000 *	3057
		2009000 *	3348			2009000 *	3007
						2009000 *	2615
						2009000 *	3261
					50/90	2009000 *	2994
						2009000 *	3204
						2009000 *	3077
						2009000 *	3485

* Runout.

References

1. Scheerer, S.; Chudoba, R.; Garibaldi, M.P.; Curbach, M. Shells made of Textile Reinforced Concrete—Applications in Germany. *J. Int. Assoc. Shell Spat. Struct. JIASS* **2017**, *58*, 79–93. [CrossRef]
2. Curbach, M.; Graf, W.; Jesse, D.; Sickert, J.-U.; Weiland, S. Segmentbrücke aus textilbewehrtem Beton—Konstruktion, Fertigung, numerische Berechnung. *Beton-und Stahlbetonbau* **2007**, *102*, 342–352. [CrossRef]
3. Michler, H. Segmentbrücke aus textilbewehrtem Beton—Rottachsteg Kempten im Allgäu. *Beton-und Stahlbetonbau* **2013**, *108*, 325–334. [CrossRef]
4. Kulas, C.; Goralski, K. Die weltweit längste Textilbetonbrücke—Technische Details und Praxiserfahrungen. *Beton-und Stahlbetonbau* **2014**, *109*, 812–817. [CrossRef]
5. Rempel, S.; Hegger, J.; Kulas, C. A pedestrian bridge made of textile reinforced concrete. In Proceedings of the 16th European Bridge Conference, Endinburgh, UK, 23–25 June 2015.
6. Helbig, T.; Rempel, S.; Unterer, K.; Kulas, C.; Hegger, J. Fuß- und Radwegbrücke aus Carbonbeton in Albstadt-Ebingen—Die weltweit erste ausschließlich carbonfaserbewehrte Betonbrücke. *Beton-und Stahlbetonbau* **2016**, *111*, 676–685. [CrossRef]
7. Rempel, S.; Kulas, C.; Will, N.; Bielak, J. Extremely Light and Slender Precast Pedestrian-Bridge Made Out of Textile-Reinforced Concrete (TRC). In Proceedings of the fib Symposium, Maastricht, The Netherlands, 12–14 June 2017. [CrossRef]
8. Rempel, S. Erste Straßenbrücke aus Carbonbeton. In Proceedings of the 9th Carbon- und Textilbetontage, Dresden, Germany, 26–27 September 2017.
9. Bielak, J.; Bergmann, S.; Hegger, J. Querkrafttragfähigkeit von Carbonbeton-Plattenbrücken mit C-förmiger Querkraftbewehrung—Theoretische und experimentelle Untersuchungen für zwei Straßenbrücken in Gaggenau. *Beton-und Stahlbetonbau* **2019**, *114*. [CrossRef]
10. Homepage of the Project Carbon Concrete Composite. Available online: www.bauen-neu-denken.de (accessed on 21 March 2019).
11. Jesse, F. Ermüdet Textilbeton? Verhalten unter nicht-ruhender Beanspruchung. In Proceedings of the 4th Anwendertagung Textilbeton, Dresden, Germany, 27–28 September 2012.

12. Feix, J.; Hansl, M. Pilotanwendungen von Textilbeton für Verstärkungen im Brückenbau. In Proceedings of the 25th Dresdner Brückenbausymposium, Dresden, Germany, 9–10 March 2015; Curbach, M., Ed.; pp. 99–110.

13. Schütze, E.; Lorenz, E.; Curbach, M. Static and Dynamic Fatigue Strength of Textile Reinforced Concrete. In Proceedings of the IABSE Conference, Nara, Japan, 13–15 May 2015; IABSE, Ed.; pp. 332–333.

14. Holz, K.; Schütze, E.; Garibaldi, P.; Curbach, M. Determination of Material Properties of TRC under Cyclic Loads. *ACI Spec. Publ. SP-ACI 549-01* **2018**, *324*, 1–16.

15. De Munck, M.; Tysmans, T.; Wastiels, J.; Kapsalis, P.; Vervloet, J.; El Kadi, M.; Remy, O. Fatigue Behaviour of Textile Reinforced Cementitious Composites and Their Application in Sandwich Elements. *Appl. Sci.* **2019**, *9*, 1293. [CrossRef]

16. Wagner, J.; Holz, K.; Curbach, M. Zyklische Verbundversuche mit Carbonbeton. *Beton-und Stahlbetonbau* **2018**, *113*, 525–534. [CrossRef]

17. Wagner, J.; Curbach, M. Tensile load bearing and Bond Behaviour of Carbon Reinforced Concrete under cyclic Loading. In Proceedings of the fib Congress, Melbourne, Australia, 7–11 October 2018; Foster, S., Gilbert, R., Mendis, P., Al-Mahaidi, R., Millar, D., Eds.;

18. Solidian GmbH. Technical Data Sheet Solidian GRID Q95/95-CCE-38. 2017. Available online: https://www.solidian.com/fileadmin/user_upload/pdf/TDS/solidian_GRID_Q95.95-CCE-38.pdf (accessed on 14 May 2019).

19. Schneider, K.; Butler, M.; Mechtcherine, V. Carbon Concrete Composites C^3—Nachhaltige Bindemittel und Betone für die Zukunft. *Beton-und Stahlbetonbau* **2017**, *112*, 784–794. [CrossRef]

20. DIN EN 196-1. *Prüfverfahren für Zement–Teil 1: Bestimmung der Festigkeit*; Deutsche Fassung EN 196-1:2016; Beuth: Berlin, Germany, November 2016.

21. Bielak, J.; Scholzen, A.; Chudoba, R.; Schütze, E.; Schmidt, J.; Reichel, S. Prüfempfehlung zur Verwendung in C^3—Beidseitiger Textilauszugversuche/Double Sided Textile Pull-Out (DPO). In *Ergebnisbericht Vorhaben B3—Konstruktionsgrundsätze, Sicherheits-und Bemessungskonzepte sowie standardisierte Prüfmethoden für Carbonbeton*; Research Report; TU Dresden: Dresden, Germany, 2016.

22. Bielak, J.; Li, Y.; Hegger, J.; Chudoba, R. Numerical and Experimental Characterization of Anchorage Length for Textile Reinforced Concrete. In *RILEM Bookseries 15, Proceedings of Strain-Hardening Cement-Based Composites (SHCC4), Dresden, Germany, 18–20 September 2017*; Mechtcherine, V., Slowik, V., Kabele, P., Eds.; Springer: Berlin/Heidelberg, Germany, 2018; pp. 409–417.

23. Schütze, E.; Bielak, J.; Scheerer, S.; Hegger, J.; Curbach, M. Einaxialer Zugversuch für Carbonbeton mit textiler Bewehrung/Uniaxial tensile test for carbon reinforced concrete with textile reinforcement. *Beton-und Stahlbetonbau* **2018**, *113*, 33–47. [CrossRef]

24. Klausen, D. Festigkeit und Schädigung von Beton bei häufig wiederholter Beanspruchung. Ph.D. Dissertation, TU Darmstadt, Darmstadt, Germany, 1978.

25. Balázs, G.L. *Deformation based fatigue failure criterion. In, Localized Damage III—Computer-Aided Assessment and Control*; Aliabadi, M.H., Carbinteri, A., Kaliszky, S., Cart-Wright, D.J., Eds.; Computational Mechanics Publications: Southampton, UK, 1994; pp. 631–638.

26. Holmen, J.O. Fatigue of Concrete by Constant and Variable Amplitude Loading. Ph.D. Dissertation, University of Trondheim, Norwegian Institute of Technology, Division of Concrete Structures, Trondheim, Norway, 1979.

27. Oneschkow, N. *Analyse des Ermüdungsverhaltens von Beton anhand der Dehnungsentwicklung*, 2nd ed.; Berichte aus dem Institut für Baustoffe, Heft 13; Institut für Baustoffe, Leibniz Universität Hannover: Hannover, Germany, 2016.

28. Schütze, E.; Curbach, M. Zur experimentellen Charakterisierung des Verbundverhaltens von Carbonbeton mit Spalten als maßgeblichem Versagensmechanismus. *Bauingenieur* **2019**, *94*, 133–141.

29. Lorenz, E. Endverankerung und Übergreifung textiler Bewehrungen in Betonmatrices. Ph.D. Dissertation, TU Dresden, Dresden, Germany, 2014.

 *applied sciences*

Article

The Effect of Elevated Temperatures on the TRM-to-Masonry Bond: Comparison of Normal Weight and Lightweight Matrices

Paraskevi D. Askouni *, Catherine (Corina) G. Papanicolaou and Michael I. Kaffetzakis

Structural Materials Laboratory, Department of Civil Engineering, University of Patras, 26504 Rio, Greece; kpapanic@upatras.gr (C.G.P.); mkaffetzakis@upatras.gr (M.I.K.)
* Correspondence: askounip@upatras.gr; Tel.: +30-2610-996-587

Received: 3 April 2019; Accepted: 21 May 2019; Published: 27 May 2019

Abstract: Textile Reinforced Mortar (TRM) is a composite material that has already been successfully used as an externally bonded strengthening means of existing structures. The bond of TRM with various substrates is of crucial importance for determining the degree of exploitation of the textile. However, little is known on the effect of elevated/high temperatures on the TRM-to-substrate bond characteristics while relevant testing protocols are also lacking. This study focuses on the experimental assessment of the TRM-to-masonry bond after exposure of masonry wallettes unilaterally furnished with TRM strips at 120 °C and 200 °C for 1 h. The shear bond tests on cooled-down specimens were carried out using the single-lap/single-prism set-up. Two TRM systems were investigated sharing the same type of textile, which is a dry AR glass fiber one (either in a single-layer or in a double-layer configuration) and different matrices: one normal weight (TRNM) and another lightweight (TRLM) of equal compressive strengths. At control conditions (non-heated specimens) and after exposure at a nominal air temperature of 120 °C, both single layer TRM systems exhibited similar bond capacities. After exposure at a nominal air temperature of 200 °C single-layer and double-layer TRNM overlays outperformed their TRLM counterparts. A critical discussion is based on phenomenological evidence and measured response values.

Keywords: textile reinforced mortar; bond; masonry; normal weight/lightweight aggregates; elevated temperatures

---

1. Introduction

TRM is an innovative composite material suitable for use as externally bonded strengthening means for strengthening or rehabilitation of existing concrete [1] and masonry [2] structures. In the case of masonry structures, and especially those of a monumental character, TRMs comprise an appealing choice since they: (i) allow for minimal geometry and self-weight change of the strengthened structural members, (ii) are chemically, physically, and mechanically compatible with various substrates, (iii) can be reversible (to a certain extent), (iv) can be applied under low temperatures and/or high humidity conditions, and (v) allow for vapor permeability of the substrate.

The quality of bond at both the TRM-to-substrate interface and the textile-to-matrix one is a crucial parameter for the efficient use of this composite material and has received appreciable attention from academia through a number of experimental studies that investigate the effect of individual parameters on bond behavior [3–6]. Nevertheless, there is only a limited number of studies addressing the effect of elevated or high temperatures (ranges being rather arbitrarily defined in each study) on the bond characteristics of TRM systems. To the authors' knowledge, only two studies exist that focus on the experimental investigation of the residual bond capacity of TRM-to-masonry interfaces after exposure to elevated/high temperatures. In the study of Maroudas and Papanicolaou [7], a TRM system consisting of a dry AR glass fiber textile embedded in a cementitious matrix was applied on masonry wallettes made of solid clay fair-faced bricks. Single-lap/single-prism tests were conducted after specimens' exposure at 100 °C, 200 °C, and 300 °C for 1 h. It was concluded that the residual shear load that can be undertaken by this type of TRM/masonry interfaces after exposure at 100 °C, 200 °C, and 300 °C amounts, respectively, to 65%, 60%, and 50% of the respective load reached ambient conditions. In the study of Ombres et al. [8], various TRM systems were applied on masonry wallettes: one made of PBO fiber textile and a cement-based matrix while two other were made of dry basalt fiber textiles (of different aerial weight and grid spacing) and a lime-based mortar. Single-lap/single-prism tests were conducted after specimens' exposure at 100 °C, 150 °C, and 200 °C for 3 h. According to the results, the bond response after specimens' heating was affected by the fibers' type and textiles' density. The residual shear bond load decreased by up to 50% and 90% in the case of PBO-TRM and 'heavy' basalt-TRM, respectively, while it remained almost unaffected in the case of the 'light' basalt-TRM ('light' having half the aerial weight of 'heavy'). Furthermore, a numerical model for the simulation of TRMs' bond behavior under different temperatures was proposed by Donnini et al. [9]. For the calibration of the model, double-lap/single-prism shear bond tests were conducted (prisms, in this case, standing for single bricks). The TRM systems comprised textiles with either uncoated (dry) or epoxy-impregnated carbon fiber yarns embedded in a cement-based mortar. Specimens were tested under two different regimes: (i) while, at 120 °C, being conditioned to the same temperature for a duration of 100 min, and (ii) after exposure at 120 °C for 60 min. It was shown that—when tested under the former heating regime—TRMs with impregnated yarns lost more than half of their bond capacity while the TRMs with dry yarns remained almost unaffected. However, TRMs with impregnated yarns retained almost all of their initial bond capacity when tested under the second heating regime. Lastly, there exist two more studies presenting the results of shear bond tests carried out on concrete specimens furnished with TRM strips both under and after their exposure to elevated/high temperatures. In the study of Ombres [10], single-layered or double-layered TRM overlays consisting of PBO textile and a cement-based matrix were applied on concrete prisms. Single-lap/single-prism shear bond tests were conducted after specimens' exposure at 50 °C and 100 °C for 8 h. The bond capacity of the double-layered TRM strips was reduced for both exposure temperatures while the capacity of the single-layered strip was negatively affected only for the higher exposure temperature. In the study of Raoof and Bournas [11], TRM strips with three or four dry carbon textile layers were applied on concrete prisms through a polymer-modified cementitious mortar for the realization of single-lap/double-prism tests. Two testing procedures were followed. According to the first one, specimens were heated at a target temperature (50 °C, 75 °C, 100 °C, 150°C, 200 °C, 300 °C, 400 °C, and 500 °C) for 60 min and they were then tested under the same temperature. These specimens were not seriously affected in terms of TRM bond capacity up to a temperature of 400 °C while their bond load was less than or equal to half of the untreated specimens' load for a temperature of 500 °C. According to the second procedure, specimens were loaded up to a certain percentage of their bond strength (25%, 50%, and 75%) at ambient conditions and they were then heated up to failure (temperature at failure: approximately 300 °C when stressed to 25% and 50% of their bond strength and approximately 75 °C when stressed to 75% of their bond strength).

Review of the existing literature on the topic reveals the lack of commonly accepted testing protocols for the investigation of the heating effect on the TRMs' shear bond performance (either during or post heating) when bonded on various substrates. Temperatures examined are subjectively characterized as 'elevated' or 'high' and heating regimes are arbitrarily selected, which renders comparisons between different experimental campaigns difficult or even invalid. Specimens' treatment prior to testing with a special focus on the moisture content of the TRM overlay (but also of the substrate) is yet another parameter that deserves special attention should rational testing procedures be drafted. Once there, the role of bond critical parameters like the number of textile layers, the textile geometry, the type of the matrix, and the type of substrate on heat-struck TRM/substrate joints remains to be investigated.

Although there is no clear definition of what is considered to be the range of "elevated temperatures" as opposed to that of "high temperatures", one would identify (from the available literature) 300 °C as the boundary temperature between them. Realistic exposure conditions to higher (or much higher) temperatures should be simulated with fire-resembling time-temperature histories (since these temperatures are frequently the result of fire events). The "elevated temperatures" range is of relevance when the aim is to assess the properties of concrete-like materials (TRM, in this case) with surface temperatures close to the ones reached in this paper. Practical examples of situations resulting into such surface concrete (TRM) temperatures are: (i) hot weather exposure (where concrete surface temperatures close to 50 °C or higher can be recorded, [12]) and (ii) exposure to climatic conditions inside nuclear reactor containments (where the maximum allowed—accident-inflicted or short-term—surface temperatures on concrete elements is set to approximately 175 °C, [13]).

This paper presents the experimental assessment of the residual bond capacity of two externally bonded TRM systems on masonry substrates post their exposure to elevated temperatures. To this purpose, the single-lap/single-prism shear bond test set-up was employed due to its simplicity in terms of specimens' preparation and specimens' handling during the heating treatment. Two TRM systems were investigated to share the same type of textile, namely a dry fiber AR glass textile (either in a single-layer or in a double-layer configuration) and different matrices: one normal weight and another lightweight of equal compressive strengths. It is highlighted that this is one of the first publications involving lightweight matrices for the design of TRM strengthening systems.

2. Materials, Specimens, and Experimental Program

2.1. Materials

2.1.1. Masonry

The masonry substrate was simulated by a wallette made of stack-bonded ridge-faced perforated fired clay bricks and a cement/lime-based mortar. Bricks had a compressive strength of 5 MPa and nominal dimensions 190 mm × 83 mm × 58 mm (as in length × width × height). Masonry joints were approximately 10 mm thick and were made of an M10 general purpose masonry mortar, according to the EC 6 classification [14], containing cement (CEM II 32.5N), lime, and sand in proportions 1:0.5:5, by volume. The compressive strength and the elastic modulus of the wallettes normal to the mortar joints were equal to 5.8 MPa (CoV 10%) and 3.2 GPa (CoV 19%), respectively, as determined by testing and complying with the recommendation LUMB1 of RILEM TC 76 Reference [15].

2.1.2. TRM

The reinforcement was a woven textile consisting of dry AR glass fiber yarns equally arranged in two orthogonal directions with a 17 mm mid-yarn spacing (yarns' text being equal to 2450 g/km in both directions) and an aerial weight of 300 g/m^2. The textile's tensile strength (f_{tex}) and elastic modulus (E_{tex}) were determined through lab testing (partly as per standard EN ISO 13934-1 [16] provisions) using seven yarns' strips and were found equal to 505 MPa (CoV 11%) and 83 GPa (CoV 17%), respectively. Two cement-based mortars were used as matrices. The normal weight mortar (N) contained Portland cement (CEM II 42.5N), fine limestone sand (d_{max} = 2 mm), silica fume (d_{max} = 1 μm), limestone filler (d_{max} = 120 μm), and effective water in proportions 1:2.07:0.11:0.28:1.82, by volume. The lightweight mortar (L) contained the same materials' exception being the normalweight sand, which was replaced with pumice sand (cement, pumice sand, silica fume, limestone filler, and effective water in proportions 1:2.07:0.11:0.28:1.61, by volume). The density of the normal weight and lightweight mortar was equal to 2113 kg/m^3 and 1760 kg/m^3, respectively. The compressive and flexural strengths of the mortars were determined at 28 days, according to standard EN 1015-11 [17], and were, respectively, found to be equal to 55 MPa (CoV 11%) and 5 MPa (CoV 18%) for the normal weight mortar and 55 MPa (CoV 9%) and 3 MPa (CoV 14%) for the lightweight one. It is noted that the prisms for the mortars' mechanical characterization were cast using a part of the total mortar batches mixed for the cast of the TRM strips and they were cured next to the reinforced masonry prisms (see Section 2.2). The tensile response of the Textile Reinforced Normal Weight Mortar (TRNM) and the Textile Reinforced Lightweight Mortar (TRLM) was determined for both single-layer and double-layer coupons, according to the procedure described in AC434 ICC-ES [18] (three identical specimens per coupons' configuration) and is depicted in Figure 1 while their tensile strength (f_{TRM}) and the corresponding axial strain (ε_{TRM}) is presented in Table 1. From the experimental axial tensile stress versus axial tensile strain curves shown in Figure 1, it is deducted that the response of the single-layer specimens does not qualify as that of a strain-hardening material. This is attributed to the low (longitudinal) fibers' volume fraction. Lightly reinforced cementitious matrices (like the single-layer coupons used herein) exhibit: (i) substantial drops of the load corresponding to the formation of the first (and of any other) crack with this loss of the load carrying capacity being irrecoverable after the formation of each crack, (ii) limited crack formation capacity, and (iii) zero or slight strain hardening (as the result of the fibers' inadequacy to effectively bridge the cracks). On the other hand, double-layer specimens present a strain-hardening response, which could be idealized as trilinear up to failure consisting of an initial uncracked stage followed by a crack development one and a final post-cracked stage. In both cases of single-layered and double-layered coupons, the failure mode was due to load-aligned fibers' slippage from within the mortar and simultaneous fibers' fracture during the enlargement of a previously created mortar crack.

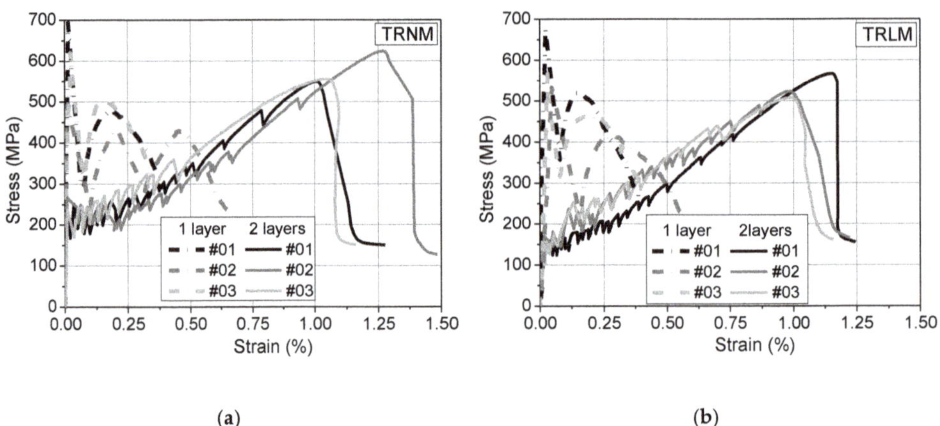

(a)   (b)

Figure 1. Axial tensile stress versus axial tensile strain curves of: (**a**) TRNM (Textile Reinforced Normal Weight Mortar) and (**b**) TRLM (Textile Reinforced Lightweight Mortar) coupons (stress is calculated by dividing with the load-aligned fibers' cross section).

Table 1. Mechanical characteristics of TRM (Textile Reinforced Mortar) coupons.

Coupon's Configuration	First Crack Stress f_{FCR} (MPa) {CoV}	Tensile Strength f_{TRM} (MPa) {CoV}	Axial Strain Corresponding to f_{TRM} ε_{TRM} (%) {CoV}
TRNM 1 layer	604 {17%}	604 {17%}	0.02 {19%}
TRNM 2 layers	241 {12%}	576 {7%}	1.11 {13%}
TRLM 1 layer	594 {13%}	594 {13%}	0.03 {20%}
TRLM 2 layers	151 {17%}	532 {6%}	1.04 {9%}

2.2. Specimens

Shear bond test specimens were designed and constructed (for the most part), according to the recommendation of RILEM TC 250-CSM [19]. Each specimen comprised one wall prism (wallette) reinforced with a single-layer or double-layer TRM overlay on one side, which was cast according to the wet lay-up process (Figure 2). Each mortar layer (2 and 3 for single-layer and double-layer TRMs, respectively) was approximately 4 mm-thick and was toweled flush to the prescribed thickness by means of a 4 mm-thick rubber-cork frame. The TRM overlay (and, hence, the textile strip) was centrally bonded on the wallettes' width-wise. The length and width of each TRM overlay were equal to 250 mm and 120 mm (7 yarns), respectively, while its distance from the wall's top edge was equal to 40 mm so that stress concentration phenomena would be avoided. The textile(s) projected from both sides of the TRM overlay, that is from the top and bottom part of the overlay designated herein as "loaded end" and "free end", respectively (Figure 2). Specimens were moist-cured for 7 days (covered with wet burlaps) and then stored in lab conditions (20 °C ± 2 °C, 65% RH) for 21 more days until pre-heating treatment, heating, and final testing. Prior to testing, the extremity/extremit(ies) of the projecting textile(s) from the loaded end of the TRM overlays were sandwiched between either two or three glued-on fiber-reinforced polymer (FRP) tabs depending on the number of textile layers (i.e., one or two, respectively).

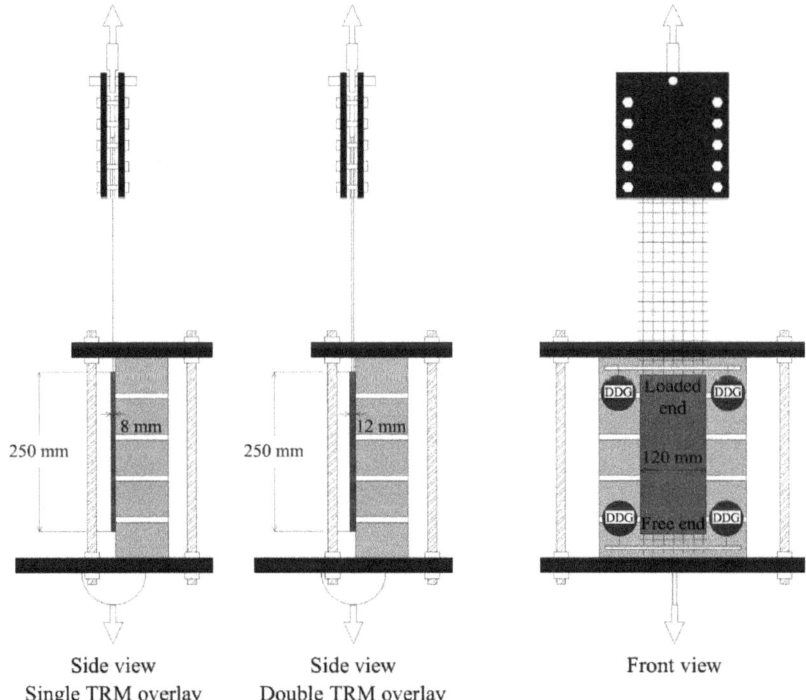

Figure 2. Single-lap/single-prism shear bond test set-up.

2.3. Experimental Program

2.3.1. Test Plan

In total, 30 specimens were constructed: half of them (15) were furnished with Textile Reinforced Normal Weight Mortar overlays and the rest (15) with Textile Reinforced Lightweight Mortar ones. Specimens were tested at ambient temperature (20 °C—control specimens) and, after exposure, at 120 °C and 200 °C. For both minimum and maximum temperatures (20 °C and 200 °C), prisms furnished with both single-layer and double-layer TRM strips were tested. For 120 °C, prisms tested bore only single-layer TRM strips. Three identical specimens were tested per case. The specimens' notation has the form of Tx_Sy_z_n (see also Table 2), where x is the nominal exposure temperature (T) in °C, y is the number of textile strips (S), z is the type of mortar used as a matrix (N or L), and n is the specimen number in a group of identical specimens. The bond length of the TRM strip (250 mm) was selected so as to ensure its adequate anchoring on the substrate.

Table 2. Experimental results.

Specimen	Temperature (°C)	Tex Strip #	TKLE (°C) {CoV}	TKFE (°C) {CoV}	TKM (°C) {CoV}	σ_{max} (MPa)	$\sigma_{max,average}$ (MPa) {CoV}	$d_{r,max}$ (mm)	$d_{r,max,average}$ (mm) {CoV}	s_{max} (mm)
T20S1N01	20	1	-	-	-	313.28	344.36 {11%}	0.26	0.28 {29%}	0.005
T20S1N02	20	1				335.91		0.22		
T20S1N03	20	1				383.88		0.38		
T20S2N01	20	2	-	-	-	340.72	330.53 {7%}	0.48	0.66 {32%}	0.07
T20S2N02	20	2				305.58		0.90		
T20S2N03	20	2				345.30		0.61		
T20S1L01	20	1	-	-	-	375.32	377.57 {6%}	0.13	0.11 {17%}	0.005
T20S1L02	20	1				355.49		0.09		
T20S1L03	20	1				401.90		0.11		
T20S2L01	20	2	-	-	-	341.48	369.49 {7%}	0.66	0.69 {18%}	0.04
T20S2L02	20	2				391.14		0.83		
T20S2L03	20	2				375.86		0.59		
T120S1N01	120	1	51 {8%}	45 {9%}	63 {20%}	337.66	341.74 {2%}	0.28	0.38 {27%}	0.07
T120S1N02	120	1				348.36		0.36		
T120S1N03	120	1				339.19		0.49		
T120S1L01	120	1	49 {1%}	49 {17%}	53 {11%}	316.27	331.55 {6%}	0.39	0.46 {22%}	0.03
T120S1L02 *	120	1				-		-		
T120S1L03	120	1				346.83		0.54		
T200S1N01	200	1	110 {14%}	99 {13%}	109 {2%}	345.30	317.29 {9%}	0.49	0.48 {2%}	0.10
T200S1N02	200	1				319.33		0.47		
T200S1N03	200	1				287.24		0.48		
T200S2N01	200	2	121 {7%}	102 {9%}	118 {4%}	300.23	322.13 {10%}	0.74	0.56 {32%}	0.12
T200S2N02	200	2				306.34		0.57		
T200S2N03	200	2				359.82		0.38		
T200S1L01	200	1	117 {4%}	105 {7%}	118 {3%}	238.35	251.59 {11%}	0.49	0.40 {21%}	0.06
T200S1L02	200	1				282.66		0.37		
T200S1L03	200	1				233.77		0.33		
T200S2L01	200	2	125 {12%}	111 {3%}	133 {4%}	233.00	271.63 {13%}	0.79	0.59 {32%}	0.08
T200S2L02	200	2				276.30		0.55		
T200S2L03	200	2				305.58		0.42		

* Machine stopped working unexpectedly during testing.

2.3.2. Pre-Heating Treatment and Heating Regime

Prior to subjecting the specimens to the prescribed heating regime, all of them were inserted in an electrical oven at a constant temperature of 40 °C for 24 h with the aim of bringing them to a state of approximately equal moisture content. Upon removal from the oven and until their exposure to elevated temperatures, specimens were wrapped in a PVC membrane in order to prevent moisture exchange with the atmosphere. For heating to 120 °C and 200 °C, the electrical oven (the same used for the drying sequence) was preheated prior to specimens' placement at a temperature higher than the target one in order to counterbalance the heat loss due to door opening. Each specimen rested in the oven for 1 h under the target temperature and remained undisturbed after turning the oven off until its temperature reached the ambient one. All heated specimens were tested upon completion of the heating-cooling regime. Figure 3 presents the time history of the specimens' preparation and treatment.

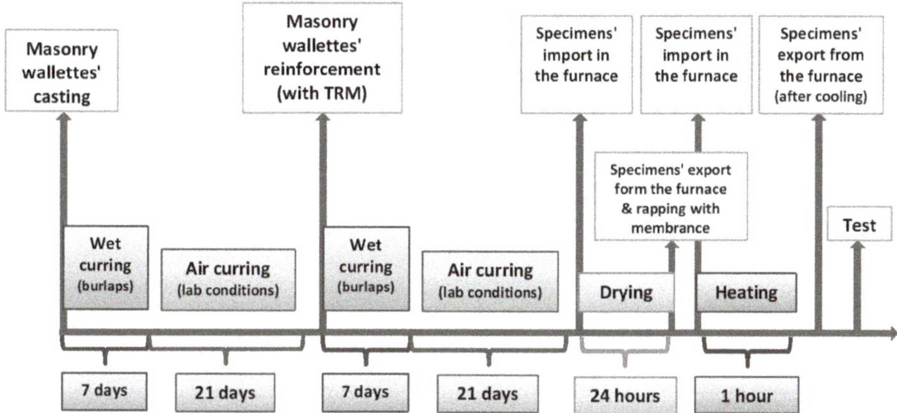

Figure 3. Time history of specimens' preparation and treatment.

For specimens subjected to the previously described heating-cooling regime, the free part of the wallette surface furnished with the composite overlay had been covered with the same mortar used for the TRM so that real-life conditions were simulated during heating. For these specimens, the protection of the projecting bare textile was achieved using a combination of ceramic wool insulation and rock mineral wool insulation with aluminum coating, as depicted in Figure 4a. It is also noted that specimens were placed horizontally in the furnace so that heating coming from above was evenly distributed on the TRM strip, which faced upward (see Figure 4b). Four K-type thermocouples were used as heat sensors (Figure 4). One was placed at a close distance from the specimen's surface in order to record the air temperature (TKA). The furnace temperature was adjusted so that the air temperature recorded by TKA was equal with the target one. The rest of the sensors (insulated against radiation heating) were applied on the specimen as follows: two of them were attached on the textile projecting from each end of the TRM strip [one close to the loaded end (TKLE) and another close to the free end (TKFE)] while another sensor was placed on the strip's surface at a distance of 50 mm to 100 mm from the loaded end (TKM).

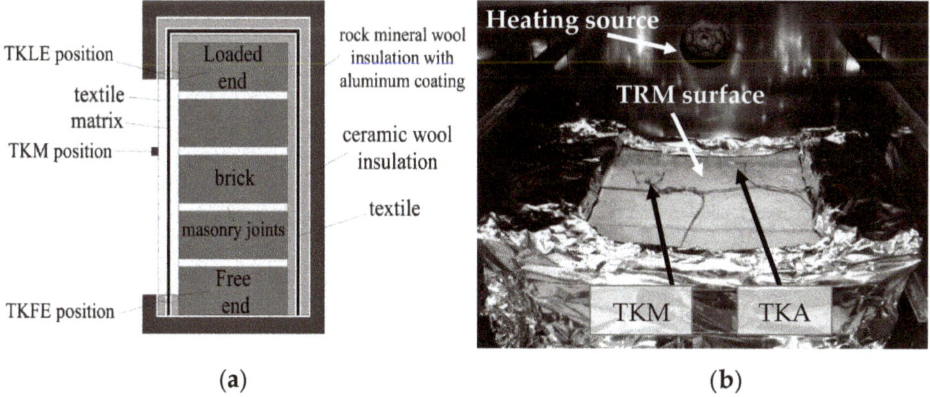

Figure 4. Insulated specimen (instrumented with thermocouples): (**a**) side view and (**b**) inside the electrical furnace; (K-type thermocouple: close to Loaded End, TKLE; close to Free End, TKFE; on the strip's surface, TKM; close to specimen's surface, TKA)

2.3.3. Shear Bond Tests Set-Up

For the execution of the single-lap/single-prism shear bond tests, each wallette was fitted in a steel frame. Then, the textile(s) projecting from the loaded end was (were) connected to the fixed part of the machine through a joint that provided full in-plane and partial out-of-plane rotation capacity (Figure 2—Section 2.2). The connection was realized by clamping the FRP tab between two bolted hinged steel plates. During testing, the projecting textile(s) was (were) being pulled away from the bonded TRM overlay. Tests were carried out at a piston displacement-controlled mode using a servohydraulic testing machine (load capacity 250 kN). The displacement rate was set to 0.005 mm/s. Each specimen was instrumented (see Figure 2—Section 2.2) with: (i) two digital dial gauges (DDG) firmly attached on the wall close to the overlay's loaded end acting against an aluminum plate glued on the first transversal yarn of the pulled textile in order to record the textile's relative displacement in respect to the wallette (under the assumption that the TRM-to-substrate slip was null), (ii) two DDG glued on the wall close to the overlay's free end acting against an aluminum plate glued on the second transversal yarn of the projecting textile in order to record the textile's slip (these sensors were applied only on the Tx_Sy_z_01 specimen of each group of identical specimens).

3. Results

3.1. Heating Effect

Visual inspection of specimens right after the conclusion of the heating-cooling regime revealed the formation of fine cracks on the top mortar layer of the TRM overlays. Cracks did not seem to propagate through the entire thickness of the TRM overlays (looking from the side of the TRM strips). In the case of single-layer TRM overlays, the number of cracks increased with increasing exposure temperature for both types of mortars whereas TRLMs were more prone to cracking than their TRNM counterparts. In the case of normal weight mortar, double-layer TRM overlays remained uncracked after specimens' exposure at 200 °C while, in the case of lightweight mortar, fewer cracks appeared on the double-layer TRM overlays than on the single-layer ones.

These fine cracks are attributed to differential shrinkage/swelling phenomena dissimilar between TRNM/masonry and TRLM/masonry joints. The 24-h-long drying at 40 °C was not sufficient for the complete drying of the TRM strips' matrices. The evaporation of the remaining moisture in the normal weight matrix (during the heating-cooling regime) is the cause for restrained shrinkage of the TRNM strip (while bed joints' mortar is also undergoing shrinkage but at a different pace). For lightweight matrices, restrained volume change phenomena are more complex. Initially, that is after the end of the (wet) curing period, the moisture content buffered in the lightweight (porous) aggregates (pumice, in this work, used in a fully saturated condition during mixing) is fed back into the paste, which causes swelling. The latter (depending on its magnitude) can even be perceived as a crack prevention mechanism due to differential volume changes. Nevertheless, during drying (especially during heating), a larger quantity of moisture escapes from the lightweight mortar in comparison to the normal weight one, which results in larger deformations of the TRLM strip due to differential shrinkage and in a lengthier TRLM shrinkage evolution period. Larger differential shrinkage deformations of the TRLM strip cause cracking provided that the bond strength of the TRM/substrate interface is larger than the tensile stresses developed in the TRLM strip due to shrinkage, which, in turn, should be larger than the tensile strength of the mortar comprising the strip. Lastly, as in all cementitious fiber-reinforced materials, shrinkage cracking was limited by increasing the fiber volume fraction (0.68% and 0.92% for single-layer and double-layer TRMs, respectively).

3.2. Failure Mode

All specimens failed due to textile slippage from within the mortar with simultaneous sleeve fibers' fracture along the textile projecting from the TRM overlay's loaded end (Figure 5). The pre-existing surface cracks in the matrix—which had been created during the heating-cooling regime —were not increased in number during shear bond testing.

Figure 5. Specimen at failure.

4. Discussion

The experimental results are presented in Table 2 in terms of: (i) the average—for each group of identical specimens—temperature recorded by the TKLE, TKFE, and TKM thermocouples at the end of 1 h exposure to the target furnace temperature, (ii) the maximum textile axial stress (σ_{max}) computed as the ratio of the maximum load carried by the TRM overlay to the cross sectional area of the longitudinal (load-aligned) fibers (equal to 6.545 mm^2 for the single-layer TRM overlay and 13.090 mm^2 for the double-layer one), (iii) the corresponding relative displacement of the textile with respect to the wall ($d_{r,max}$) being equal to the average of the readings from the DDG at the loaded end and (iv) the corresponding textile slip from within the mortar (s_{max}) being equal to the average of the readings from the DDG at the free end.

As far as the thermocouples' readings are concerned, temperature values do not vary significantly in relation to the sensors' position (per specimens' group), the number of textile layers (per mortar type), and the type of mortar (for the same number of textile layers). The TRM surface temperature and the temperature of the projected textile close to the loaded and free end for all types of TRM overlays investigated are found to be equal to roughly half of the ambient air temperature (for both mortar types: 50 °C for single-layer TRM overlays exposed to 120 °C; ~110 °C and ~120 °C for single-layer and double-layer TRM overlays exposed to 200 °C, respectively). In more detail, the rate of TRM surface and textile temperatures over the nominal exposure (air) temperature increase with increasing exposure temperature. The evolution of the thermocouples' values during the hour-long exposure is depicted in Figure 6 for three representative specimens furnished with TRLM overlays (the respective temperature records concerning specimens with TRNM overlays are almost identical).

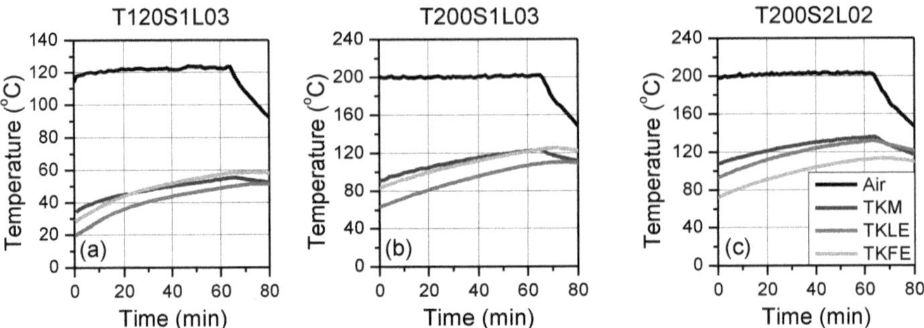

Figure 6. Thermocouples' temperature profiles for representative specimens: (**a**) T120S1L03, (**b**) T200S1L03, and (**c**) T200S2L02.

The maximum textile axial stress, σ_{max}, (average for each specimen group) versus the target furnace temperature plot is presented in Figure 7a for each type of mortar. It is observed that, at control conditions (non-heated specimens: 20 °C in Table 2) and after exposure at a nominal air temperature of 120 °C, both single-layer TRM systems (TRNM and TRLM) exhibit similar bond capacity differences between them being unimportant considering the statistical performance of each pair of compared specimen groups. The same does not apply for the nominal exposure temperature of 200 °C for which single-layer and double-layer TRNM overlays outperform the respective TRLM ones in terms of residual bond capacity.

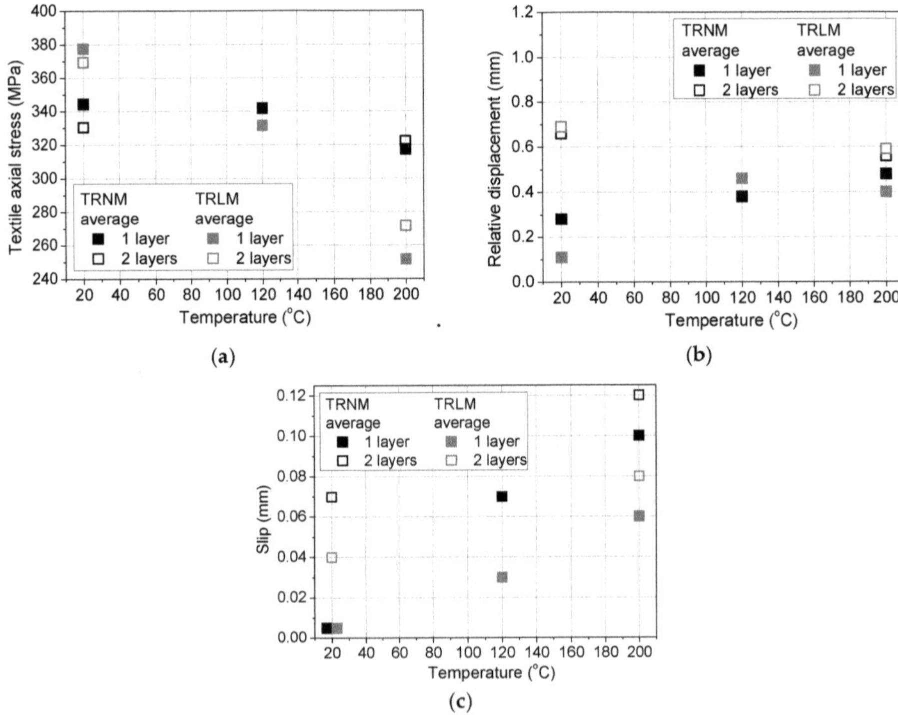

Figure 7. (**a**) Maximum textile axial stress, σ_{max}, and corresponding (**b**) relative displacement, $d_{r,max}$, and (**c**) slip, s_{max}, versus nominal exposure temperature.

Per type of matrix considered, the following trends are observed. For the same thermal treatment prior to testing (none, or exposure at either 120 °C or 200 °C for 1 h), σ_{max} stresses corresponding to double-layer overlays are comparable to those of single-layer ones (as also noticed by Askouni and Papanicolaou [6] for specimens tested without having been previously heated). The effect of exposure temperature increase prior to testing does not seem to substantially affect the residual bond capacity of single-layer and double-layer TRNM overlays. The maximum textile axial stress remains almost unchanged for specimens subjected to 120 °C and decreases by less than 10% for specimens subjected to 200 °C in comparison with the maximum textile axial stress of the reference specimens. The same does not apply for TRLM overlays. In this case, the maximum textile axial stress decreases slightly (by 14%) for specimens subjected to 120 °C (with single-layer TRMs) and by 50% and 36% for specimens subjected to 200 °C with single-layer and double-layer overlays, respectively, in comparison to the maximum textile axial stress of the reference specimens. An increase of mortar thickness (to accommodate higher fiber volume fractions, i.e., more textile layers) seems to be beneficial in terms of residual bond capacity.

In TRLM overlays, the maximum textile axial stress decrease for increasing exposure temperature is mainly attributed to the cracking that this exposure caused. Despite the fact that these cracks did not increase in number or in width during shear bond testing, it is believed that they comprised textile-to-matrix bond breaker points. Bond damage could also occur as the combined result of: (i) early-age swelling and (ii) stress buildup during moisture evaporation by heating. The latter should theoretically be minimal in lightweight mortars with highly porous aggregates. However, according to Chandra and Berntsson [20], lightweight cementitious matrices with a dense cement paste (like the one used herein) exhibit low vapor permeability and, hence, moisture transport results in the development of high internal stresses. The latter is expected to mainly affect the bond between the textile and the top (atmosphere-exposed) mortar layer. In double layer systems, the stress-relieving effect of higher fibers' content is combined with the preservation of the bond quality of the bottom textile layer with the surrounding (unexposed) matrix.

The plot of the textile relative displacement corresponding to the maximum textile axial stress, $d_{r,max}$, versus the target furnace temperature is presented in Figure 7b for each TRM configuration. These displacement values are less reliable in comparison to the stress ones due to their (inherently) higher CoV. Therefore, the corresponding data trends are only qualitatively commented. Average $d_{r,max}$ values: (i) depend more on the number of textile layers than on the type of matrix used and (ii) increase with increasing exposure temperature for single-layer configurations (for which d_r measurement is more straightforward compared to double-layer ones). The relative displacement is the result of two synergistic phenomena: the elongation and the slippage of the longitudinal yarns' fibers in the matrix (see Askouni and Papanicoloaou [5]). In the case of single-layer TRNM overlays, the increase of d_r is attributed mainly to the increase of the textile slippage from within the normal weight mortar with increasing exposure temperature (see slip values recorded by DDG at the TRNM overlays' free ends in Figure 5c). In the case of TRLM overlays, the textile-to-matrix bond is more severely damaged by volume change phenomena (especially those heat-induced), which lead also to sleeve fibers' rupture within the matrix (hence, the lower slip values compared to single-layer TRNMs in Figure 7c). Average $d_{r,max}$ values of double-layer TRM overlays are higher than the respective values of single layer ones. This is in agreement with observations done by Askouni and Papanicolaou [6] who also provide the relevant reasoning.

The response (textile axial stress versus relative displacement curves) of representative specimens of both TRM systems (TRNM and TRLM) is given in Figure 8. Most curves can be approximated as bilinear up to failure. In the first branch, both components (textile and mortar) behave in an elastic composite manner. The second branch extends between the change of the curve's inclination and the maximum load where the bond between fiber yarns and matrix is gradually deteriorated along a part of the bond length. It is highlighted that failure manifests the commencement of the load-aligned yarns' debonding from the matrix in combination with their sleeve fibers' progressive rupture. Post-peak stress decrease is significantly more rapid for the TRLM overlays due to the brittleness of the lightweight mortar.

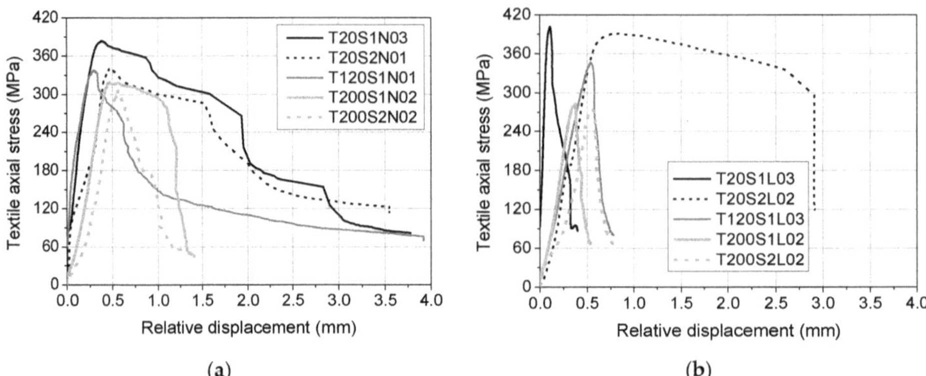

Figure 8. Response curves of representative specimens reinforced with: (**a**) TRNM and (**b**) TRLM overlays.

5. Conclusions

The main drive for use of lightweight matrices in TRMs is to provide heat shielding to the reinforcement (fibrous grid) since lightweight mortars are characterized by lower heat conductivity with respect to normal weight ones. Nevertheless, the design of lightweight aggregate (e.g., pumice) cement-based mortars suitable for TRMs (i.e., comparable to normal weight ones, in terms of—at least —strength) requires high strength pastes with low effective water-to-cementitious material ratios in order to compensate for the low crushing strength of the lightweight aggregates. The latter aggregates, which are often used in a fully saturated condition in the mixture, provide larger quantities of evaporable (and non-evaporable) moisture (in comparison to normal weight mortars) leading to: (i) early-age swelling (once wet curing is concluded), (ii) stress buildup during moisture evaporation by heating (dense paste causing high vapor pressure), and (iii) differential shrinkage cracking (along all TRM/masonry joints' interfaces). Hence, it is highlighted that a crucial parameter for the comparison of the post-heating residual shear bond response of different TRM systems on a reasonable basis is their moisture content. This must be kept at a constant (comparable) level, which is difficult to achieve and monitor. The main conclusions drawn from the experimental results presented in this work are summarized as follows. At control conditions (non-heated specimens) and after exposure at a nominal air temperature of 120 °C, both single-layer TRM systems (TRNM and TRLM) exhibit similar bond capacities. The same does not apply for the nominal exposure temperature of 200 °C for which single-layer and double-layer TRNM overlays outperform the respective TRLM ones in terms of residual bond capacity. For TRLM overlays, the maximum textile axial stress decreases by 50% and 36% for specimens subjected to 200 °C with single-layer and double- layer overlays, respectively, in comparison to the maximum textile axial stress of the reference specimens. Increase of mortar thickness (to accommodate higher fiber volume fractions, i.e., more textile layers) seems to be beneficial in terms of post-heating residual bond capacity.

The relevant knowledge curve is at the beginning of its ascending branch. There are still many open questions to be answered (apart from the issue of moisture control). The object of the current study is a multi-parametric problem, which is hard to draw generalized conclusions from. There is a multitude of constituent material combinations comprising different TRM systems, which, in turn, can be combined with a vast array of different masonry substrates (all—adjacent materials—possessing different shrinkage strain potentials, porosity, and transport properties to mention a few). Per TRM/substrate combination, it would be interesting to assess the effects of hygro-thermal fatigue scenarios.

Author Contributions: P.D.A. is the author of the draft manuscript. Additionally, she participated in the planning of the experimental program and was responsible for the construction/curing/handling of the specimens, the execution of the experimental program, the data treatment, and the derivation of preliminary conclusions. C.G.P. was in charge for the planning of the experimental program. She performed a critical evaluation of the experimental results and edited the manuscript. Lastly, M.I.K. was responsible for the mix design of the mortars. He also offered his help in the mechanical characterization of the materials used.

Funding: The research described in this paper has been co-financed by the European Union (European Social Fund—ESF) and Greek national funds through the Operational Program "Human resource development, education, and lifelong learning" of NSRF 2014-2020, Project "Supporting researchers with emphasis on young researchers—EDVM34".

Conflicts of Interest: The authors declare no conflict of interest.

References

1. Koutas, L.N.; Tetta, Z.; Bournas, D.A.; Triantafillou, T.C. Strengthening of Concrete Structures with Textile Reinforced Mortars: State-of-the-Art Review. *J. Compos. Constr.* **2019**, *23*, 03118001. [CrossRef]

2. Kouris, L.A.S.; Triantafillou, T.C. State-of-the-art on strengthening of masonry structures with textile reinforced mortar (TRM). *Constr. Build. Mater.* **2018**, *188*, 1221–1233. [CrossRef]

3. D' Antino, T.; Sneed, L.H.; Carloni, C.; Pellegrino, C. Influence of the substrate characteristics on the bond behavior of PBO FRCM-concrete joints. *Constr. Build. Mater.* **2015**, *101*, 838–850. [CrossRef]

4. Raoof, S.M.; Koutas, L.N.; Bournas, D.A. Bond between textile-reinforced mortar (TRM) and concrete substrates: Experimental investigation. *Compos. B Eng.* **2016**, *98*, 350–361. [CrossRef]

5. Askouni, P.D.; Papanicolaou, C.G. Experimental investigation of bond between glass textile reinforced mortar overlays and masonry: The effect of bond length. *Mater. Struct.* **2017**, *50*, 164. [CrossRef]

6. Askouni, P.D.; Papanicolaou, C.G. Textile Reinforced Mortar-to-masonry bond: Experimental investigation of bond-critical parameters. *Constr. Build. Mater.* **2019**, *207*, 535–547. [CrossRef]

7. Maroudas, S.R.; Papanicolaou, C.G. Effect of High Temperatures on the TRM-to-Masonry Bond. *Key Eng. Mater.* **2017**, *747*, 533–541. [CrossRef]

8. Ombres, L.; Iorfida, A.; Mazzuca, S.; Verre, S. Bond analysis of the thermally conditioned FRCM-masonry joints. *Measurements* **2018**, *125*, 509–515. [CrossRef]

9. Donnini, J.; De Caso y Basalo, F.; Corinaldesi, V.; Lancioni, G.; Nanni, A. Fabric-reinforced cementitious matrix behavior at high-temperature: Experimental and numerical results. *Compos. B Eng.* **2017**, *108*, 108–121. [CrossRef]

10. Ombres, L. Analysis of the bond between Fabric Reinforced Cementitious Mortar (FRCM) strengthening systems and concrete. *Compos. B Eng.* **2015**, *69*, 418–426. [CrossRef]

11. Raoof, S.M.; Bournas, D.A. Bond between TRM versus FRP composites and concrete at high temperatures. *Compos. B Eng.* **2017**, *127*, 150–165. [CrossRef]

12. Pancar, E.B.; Akpınar, M.V. Temperature Reduction of Concrete Pavement Using Glass Bead Materials. *Int. J. Concr. Struct. Mater.* **2016**, *10*, 39. [CrossRef]

13. Oxfall, M. Climatic Conditions Inside Nuclear Reactor Containments: Evaluation of Moisture Condition in the Concrete within Reactor Containments and Interaction with the Ambient Compartments. Ph.D. Thesis, Lund University, Lund, Sweden, 2016.

14. CEN. *EN 1996-1-1 Eurocode 6–Design of Masonry Structures–Part 1-1: General Rules for Reinforced and Unreinforced Masonry Structures*; European Committee for Standarization: Brussels, Belgium, 2005.

15. *RILEM TC 76: Technical Recommendations for Testing and Use of Constructions Materials: LUMB1-Compressive Strength of Small Walls and Prisms*; Chapman & Hall: London, UK, 1991.

16. CEN. *EN ISO 13934-1: Textiles-Tensile Properties of Fabrics–Part 1: Determination of Maximum Force and Elongation at Maximum Force Using the Strip Method*; British Standard: Brussels, Belgium, 1999.

17. CEN. *EN 1015-11: Methods of Test for Mortar for Masonry–Part 11: Determination of Flexural and Compressive Strength of Hardened Mortar*; European Committee for Standardization: Brussels, Belgium, 1993.

18. *AC434 ICC-ES: Masonry and Concrete Strengthening Using Fiber-Reinforced Cementitious Matrix (FRCM) Composite Systems*; ICC-Evaluation Service: Whittier, CA, USA, 2013.

19. De Felice, G.; Aiello, M.A.; Caggegi, C.; Ceroni, F.; De Santis, S.; Garbin, E.; Gattesco, N.; Hojdys, Ł.; Krajewski, P.; Kwiecień, A.; et al. Recommendation of RILEM Technical Committee 250-CSM: Test method for Textile Reinforced Mortar to substrate bond characterization. *Mater. Struct.* **2018**, *51*, 95. [CrossRef]

20. Chandra, S.; Berntsson, L. *Lightweight Aggregate Concrete*, 1st ed.; William Andrew Publishing: William, CA, USA, 2003; pp. 291–319.

 *applied*
sciences

Article

Thermomechanical Behavior of Textile Reinforced Cementitious Composites Subjected to Fire

Panagiotis Kapsalis [1,2,*], Michael El Kadi [1], Jolien Vervloet [1], Matthias De Munck [1], Jan Wastiels [1], Thanasis Triantafillou [2,3] and Tine Tysmans [1]

[1] Department Mechanics of Materials and Constructions, Vrije Universiteit Brussel (VUB), Pleinlaan 2, 1050 Brussels, Belgium; Michael.El.Kadi@vub.be (M.E.K.); Jolien.Vervloet@vub.be (J.V.); Matthias.De.Munck@vub.be (M.D.M.); Jan.Wastiels@vub.be (J.W.); Tine.Tysmans@vub.be (T.T.)
[2] Department of Civil Engineering, Structural Materials Laboratory, University of Patras, 26504 Rio, Greece; ttriant@upatras.gr
[3] Engineering Division, New York University Abu Dhabi, Saadiyat Island, PO Box 129188, UAE
[*] Correspondence: Panagiotis.Kapsalis@vub.be; Tel.: +32-489-180-273

Received: 31 January 2019; Accepted: 18 February 2019; Published: 21 February 2019

Abstract: The mechanical behavior of textile reinforced cementitious composites (TRC) has been a topic of wide investigation during the past 30 years. However, most of the investigation is focused on the behavior under ambient temperatures, while only a few studies about the behavior under high temperatures have been conducted thus far. This paper focused on the thermomechanical behavior of TRC after exposure to fire and the residual capacity was examined. The parameters that were considered were the fiber material, the thickness of the concrete cover, the moisture content and the temperature of exposure. The specimens were exposed to fire only from one side and the residual strength was measured by means of flexural capacity. The results showed that the critical factor that affects the residual strength was the coating of the textiles and the law of the coating mass loss with respect to temperature. The effect of the other parameters was not quantified. The degradation of the compressive strength of TRC was quantified with respect to temperature. It was also concluded that a highly asymmetrical design scheme might lead to premature failure.

Keywords: bending tests; fire; high temperature; textile coating; textile reinforced cementitious composites (TRC)

---

1. Introduction

Innovation in construction has always been a matter of great interest. In the past decades, the materials that play a leading role towards this have been the textile reinforced cementitious composites, usually referred to as textile reinforced mortars (TRM), textile reinforced concrete (TRC) or fabric reinforced cementitious matrix composites (FRCM). TRC is a material that combines the good compressive behavior of cementitious matrices with the good tensile properties of a proper reinforcement. The innovation lies in the very high tensile capacity of the thin fibers, but especially in the slenderness offered by the lightweight textiles with respect to the traditional bulky steel reinforcement. Additionally, the most common fiber materials (glass, carbon, basalt, aramid) are much less prone to corrosion than steel, which leads to lower needs in concrete cover, thus, thinner and more lightweight elements. Additionally, TRC presents a significant advantage with respect to the fiber reinforced polymer (FRP) composite materials, through the increased resistance of the cementitious matrices to high temperatures with respect to the polymer matrices of FRPs. Finally, another important advantage of TRC with respect to FRPs is the higher compatibility of the cementitious matrices with most substrates; thus, TRC can be used as a retrofitting material in more applications in construction.

A major concern with respect to TRC lies in the fact that the low thickness of TRC elements might end up being a drawback for their fire resistance, since the textile reinforcement is more exposed to high temperatures. At the same time, the most common failure mechanism of concrete due to high temperatures (spalling of the cover) is of small importance in thick elements, while in TRC elements with thickness of a few millimeters it can be critical. Therefore, even though TRC has been widely investigated and already practically used in the past few years for many applications (load bearing or non-load bearing elements in new constructions, strengthening, repairing and seismic retrofitting of existing concrete or masonry buildings, sandwich façade panels, bridge components, curved shell elements, etc.), there is still no high certainty about the materials' response in fire conditions.

The state-of-the-art in the literature includes several publications concerning the temperature effect on TRC. However, a significant amount on them is not focused on TRC alone, but on applications of TRC as a strengthening technique on existing concrete or masonry structures. Studies [1–6] investigate the effectiveness of a TRC layer as a means of strengthening concrete beams or slabs subjected to high temperatures. In studies [7,8] TRC has been used on existing concrete substrates; however, these studies have focused only on the effect of high temperatures on the bond between the two different materials. In publications [9–11], the structural capacity and the effectiveness of the bond between the TRC and masonry substrate has also been tested under exposure to high temperatures. Clearly, the results given by these publications cannot be used as data to work with in the design of TRC structures, since the high concrete (or masonry) mass of the existing building gives thermal inertia to the system which does not exist in slender TRC structures.

In studies [12–28], the structural capacity of TRC alone (not on a different substrate) under elevated or high temperature has been investigated. However, in [12–21], the maximum temperature that was tested was 650 °C, and only in publications [22–28], TRC was investigated under temperatures of 700 °C–1000 °C, which corresponds to the realistic temperatures developed in case of a cellulosic fire [29]. Moreover, out of the last group of publications, only in [26,28] tests have been performed on TRC specimens according to the standard fire curve proposed by EN1363 ([29]), while using glass or carbon fiber reinforcement, which are the most commonly used in structural applications.

In conclusion, publications that investigate TRC as a structural material under realistic fire conditions are scarce, and there is a large gap of knowledge on the behavior and design of this new material for the accidental load case of fire.

2. Materials and Methods

2.1. Matrix

The matrix that was used in this study is a commercially available cementitious mortar of ordinary Portland cement. It included quartz sand at a percentage of 25%–30% by weight as well as some additives, which were not disclosed by the manufacturer. The maximum grain size was 2.5 mm, and the mortar had a high flowability, which was necessary for casting properly through the textile reinforcement.

The compressive and flexural strength of the mortar were measured by conducting compression and three-point bending tests according to EN 12190 and EN 196-1, respectively. They were measured after 28 days of casting and with identical curing conditions as those of the TRC specimens that will be described below (cured at constant temperature of 20 °C and covered with constantly wet fabrics).

The compressive strength, measured by testing six specimens (cubes of 40 mm), was found equal to 61.45 MPa, with a variance of 4.16 MPa. The flexural strength, measured by testing five specimens (prisms of 40 × 40 × 160 mm), was found equal to 7.60 MPa with a variance of 0.93 MPa.

2.2. Textile Reinforcement

Three types of commercially available textiles have been used for this study, all consisting of coated glass or carbon fibers. A description of their properties is given next.

Two-dimensional (2D) glass styrene-butadiene (SBR) coated: Two-dimensional AR-glass textile with styrene-butadiene (SBR) coating (Figure 1a). The mesh size was equal to 12 mm in both directions and the weight of the textile before and after coating was equal to 568 g/m^2 and 653 g/m^2, respectively, according to the technical datasheet obtained from the provider.

Figure 1. (**a**) 2D styrene-butadiene (SBR) coated glass-fiber textile; (**b**) 3D SBR coated glass-fiber textile; (**c**) SBR coated carbon-fiber textile.

Three-dimensional (3D) glass SBR coated: Three-dimensional AR-glass textile with styrene-butadiene coating (Figure 1b). The mesh size differed among the two perpendicular directions and the two faces, being 10 mm for the front face (face 1) and either 9 or 18 mm at the back face (face 2). However, the cross- sectional area of the reinforcement was equal in both faces: 70.5 mm^2/m lengthways and 71.6 mm^2/m crossways. The weight of the textile before and after coating is 917 g/m^2 and 1055 g/m^2, respectively. The distance between the two faces is 12 mm. The distance holders were made of polyester and were randomly curved; thus, their purpose was to hold the two layers of glass textiles at the specified distance and not to provide extra mechanical performance.

2D carbon SBR coated: Two-dimensional carbon textile with styrene-butadiene coating (Figure 1c). The mesh size was equal to 12.7 mm in both directions and the weight of the textile before and after coating was equal to 516 g/m^2 and 578 g/m^2, respectively.

Useful technical information about the textiles is summarized in Table 1.

Table 1. Geometrical and mechanical data of the reinforcing textiles.

Type of Textile		Roving Distance (mm)		Weight before Finishing (gr/m^2)		Nominal Thickness (mm)		Yarn Failure Stress (MPa)	Yarn Stiffness (GPa)
		warp	weft	warp	weft	warp	weft		
Two-dimensional (2D) glass		12	12	284	284	0.106	0.106	526	67
Three-dimensional (3D) glass	Face 1	10	10	229.3	229.3	0.171	0.171	496	67
styrene-butadiene (SBR) coated	Face 2	18	9	229.3	229.3	0.171	0.171		
2D carbon SBR coated		12.7	12.7	258	258	0.143	0.143	814	93

2.3. Specimens

In total, six series of specimens were casted. Each series (consisting of six identical specimens) was made in order to investigate the influence of a different parameter. The parameters that were investigated are the following:

- Thickness of the concrete cover
- Time/temperature of exposure to fire
- Material of fibers
- Moisture content

The specimens were exposed to fire from only one side (referred to as "face 2"). The same side (face 2) was the one that was subjected to tension during the bending tests.

Each one of the series was casted as a plate of TRC of dimensions 500 mm × 435 mm (and a varying thickness according to the design of each case). After being cured at 20 °C and in a wet environment for 28 days, each plate was cut into six identical specimens of dimensions 500 mm × 70 mm, three of which were tested in bending without being subjected to a fire test, while the other three were tested in bending after being exposed to fire (the set-up of the fire tests is described in the next paragraph). In all cases, at least six yarns were present over the width of 70 mm, while these dimensions comply with the dimensional norm for testing TRC in tension, as per [30].

The differences between the geometry, the reinforcement and the duration of the fire test are provided in Figure 2.

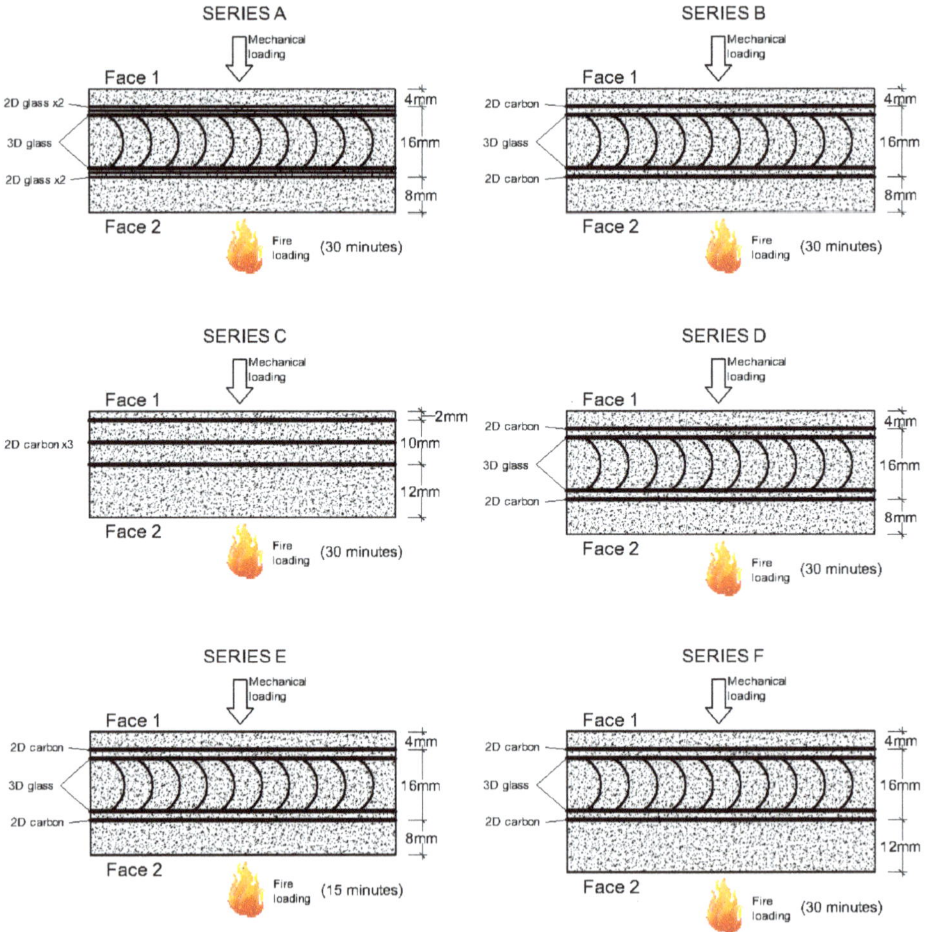

Figure 2. Cross sectional geometry of specimen Series A to F.

It is also noted that all the specimens were dried in a furnace before being subjected to the fire test. The drying process consisted of successive heating (up to 104 °C) and weighting of the specimens until the weight was constant. This corresponds to 0% of moisture content. Series D was the exception to this, as it was first submerged in water until it was saturated with water (also defined after successive weighting until constant weight), which corresponds to 100% of moisture content. Eventually, specimens in Series D were dried in the same oven and by the same process, until the moisture content reached 50%.

The most important geometrical data for all the specimen Series are summarized in Table 2.

Table 2. Geometrical data and testing parameters for Series A to F.

Fire Test	Series	Type of Reinforcement	Cover Thickness (mm)		Total Thickness (mm)	Effective Depth (mm)	Time of Exposure (min)	Moisture Saturation (%)	Fiber Volume Fraction (%)
			Face 1	Face 2					
TEST 1	A	Glass	4	8	28	20	30	0	2.17
TEST 2	B	Glass + carbon	4	8	28	20	30	0	1.47
	C	Carbon	2	12	24	12	30	0	1.82
	D	Glass + carbon	4	8	28	20	30	50	1.47
	F	Glass + carbon	4	12	32	20	30	0	1.29
TEST 3	E	Glass + carbon	4	8	28	20	15	0	1.47

It should be noted that:

- Face 2 is the face that was exposed to the fire during the fire test (also referred to as "surface" for the fire test). It is also the face that was subjected to tension during the bending test.
- Effective depth refers to the level of the most stressed fibers (starting from face 1) during the bending test. Thus, it is also the level of fibers which were closer to the exposed to fire surface, which are the fibers that were exposed to the highest temperature.
- Fiber volume fraction (V_f) was calculated only in the longitudinal direction of the specimens, in the direction of the tensile stresses during the bending test.

2.4. Fire Tests Set-up

For the fire tests, the standard fire curve given by EN 1363-1 was utilized. The fire curve was reproduced by the vertical furnace of the Fire Testing Facility at the University of Patras, Greece (see Figure 3).

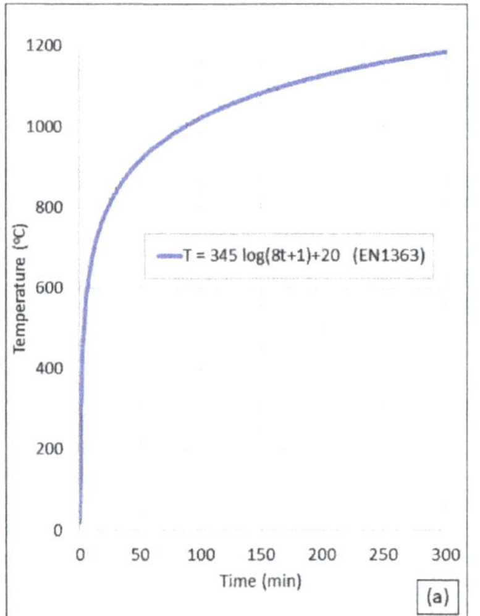

Figure 3. (**a**) Standard fire curve according to EN 1363-1; (**b**) Vertical furnace at the Fire Testing Facilities of the University of Patras (internal dimensions of 3 m × 3 m × 1.2 m).

As is apparent from Table 2, the specimens were not tested all at once but in several fire tests, for practical reasons such as different durations of exposure to fire, limited number of temperature sensors or due to the uncertainty of the expected residual strength and, thus, the possibility to re-evaluate and change the design of the specimens.

The sides of the specimens that were not directly exposed to fire were protected by using mineral wool, which is a fire-resistant insulating material. The insulation was tightened on the specimens using metallic wire (see Figure 4).

The temperature was measured with the use of thermocouples that were placed on the surface, in the middle and at the bottom of the specimens. The edge of the thermocouples placed at the surface was covered for a few millimeters with ceramic wool (which is thermally insulating and fire-resistant), in order to avoid being affected by the air temperature. The thermocouples in the middle were fixed inside a small cavity that was drilled a few days after unmolding the specimens. The thermocouples at the bottom were placed between the mineral wool and the specimens.

Figure 4. Set-up of specimens from Series B, C, D and F for the execution of the fire test. At the end of each fire test, the specimens were left to cool down naturally, without getting them out of the furnace before reaching the room temperature. The door of the furnace was only opened after the temperature had dropped to 180 °C, to avoid sudden cooling down which could harm the specimens and the furnace itself.

2.5. Mechanical Tests Set-up

The specimens were subjected to four-point bending according to Figure 5. The mechanical behavior (residual strength and stiffness) of the specimens that were subjected to the fire test was compared to the behavior of identical (control) specimens that were not subjected to fire. The results gave a good insight about the degradation of the specimens.

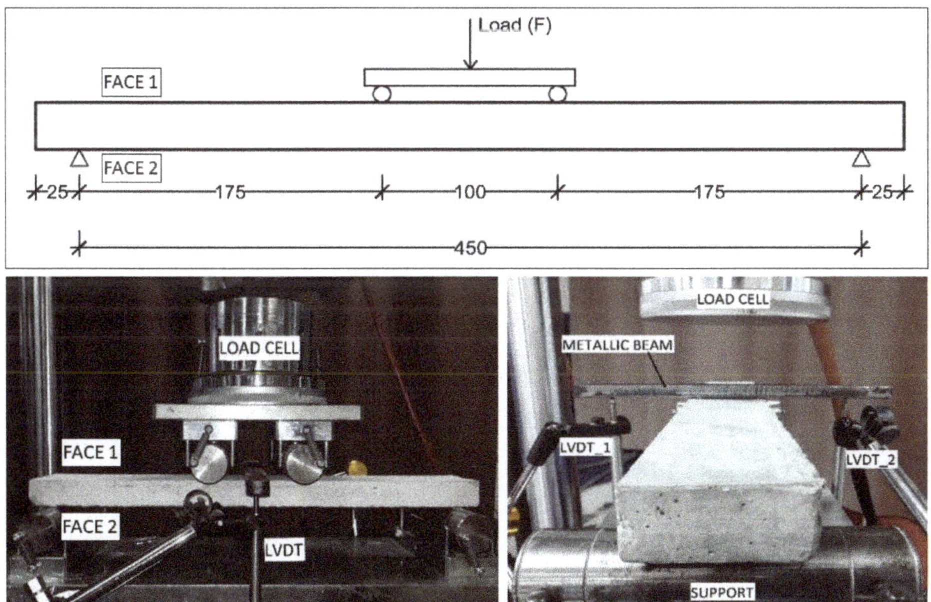

Figure 5. Experimental set-up details about the four-point bending tests.

The value of the applied load was measured directly from the load cell that was fixed on the testing machine. A spherical metallic insert with three degrees of rotational freedom was attached between the load cell and the specimen, to eliminate the influence of geometrical imperfections. The testing method was displacement controlled, with a rate of 1 mm/min. The deflection of the specimens was measured in the middle of their span using two Linear Variable Differential Transformers (LVDTs), one on each side, in order to take into account any displacements due to possible torsional rotation of the specimens. A stiff metallic beam was fixed at the top of the specimens, exactly at the middle section, and the LVDTs were taking measurements with respect to this beam.

3. Results and Discussion

3.1. Fire-Testing Results

As described in the previous paragraph, the specimens were not tested all at once, but rather in the following fire tests:

- Fire test 1: Series A was tested for a duration of 30 min.
- Fire test 2: Series B, C, D and F were tested for a duration of 30 min.
- Fire test 3: Series E was tested for a duration of 15 min.

In the previous paragraph, it was mentioned that nine thermocouples were used in each fire-test to monitor the temperature on the specimens (three sensors on the surface of the specimens, three in the middle and three at the bottom). However, due to failure of the specimens or the fixing of the sensors, not all measurements recorded could be trusted. Therefore, in the next figures and tables (Figure 6 and Table 3), only the most reliable measurements are presented.

Table 3. Temperature measurements from fire tests 1, 2 and 3.

Fire Test	Duration (min.)	Series	Temperature at the End of the Fire Test (°C)		
			surface	middle	bottom
1	30	A	638	525	437
2	30	B	648	526	343
		C	-	-	-
		D	**650**	**530**	343
		F	666	524	385
3	15	E	477	302	192

* Normal fonts: measured values; **Bold fonts**: estimated values.

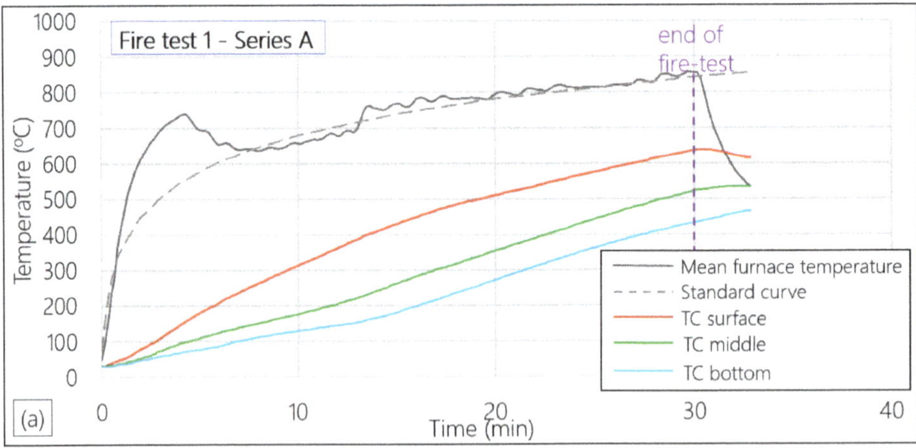

Figure 6. *Cont.*

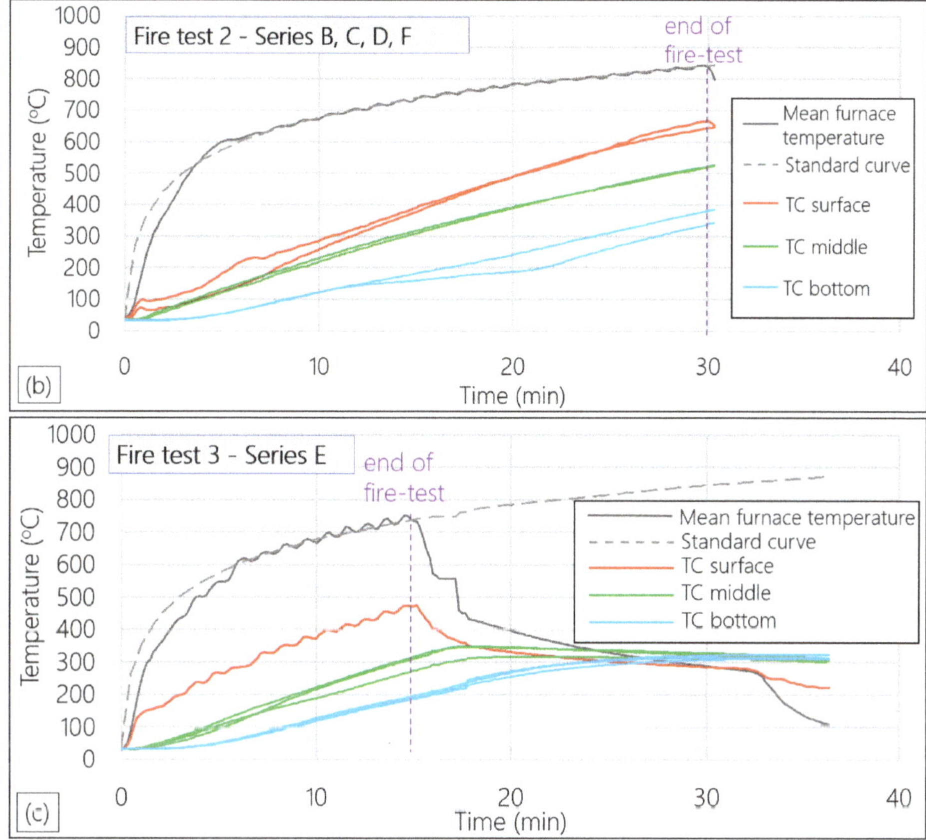

Figure 6. (**a**) Temperature measurements from fire test 1; (**b**) temperature measurements from fire test 2; (**c**) temperature measurements from fire test 3.

Additionally, it should be noted that since it was not possible to apply thermocouples in all specimens and all positions (also some of the applied thermocouples failed to give trustworthy measurements), some values of measured temperatures could be estimated based on the similar geometry of all the specimens. As a result, Table 3 is filled with values that were actually measured (normal fonts) or could be safely estimated (**bold fonts**).

3.2. Results from Coating Burn-off Tests

The most critical parameter seems to be the failure of the bond between the matrix and the reinforcement, which is caused by the coating burn-off. Thus, some additional tests were performed, where samples of textiles were exposed to several temperatures and their weight was measured before and after exposure. Thus, the mass loss of the coating could be calculated.

It is important to note that:

- The equipment that was used was a small electrical furnace with the capacity to reach 1000 °C.
- Apart from the temperature, the time of exposure also plays a significant role. The heating rate in the middle of the specimens (closest measurement to the level of the effective depth, thus, the fibers that are of interest) was almost the same in both the 15-min and the 30-min fire tests,

equal to 18–19 °C/min. Therefore, the heating time was decided each time according to the target temperature and a standard heating rate of 18 °C/min.

- The cooling down of the specimens, after reaching the maximum temperature, was performed with a rate of 1.5 °C/min until a temperature of 200 °C, which is also a good approach of the cooling down rate that was measured at the 15-min fire test.
- The initial mass of the coating was calculated based on the weight of the textiles before and after coating, as provided by the technical datasheets.

From the results which are presented in Figure 7, it was observed that the critical temperature after which the mass loss is becoming significant is close to 300 °C. It was also observed that after 500 °C, the coating was almost completely burnt-off.

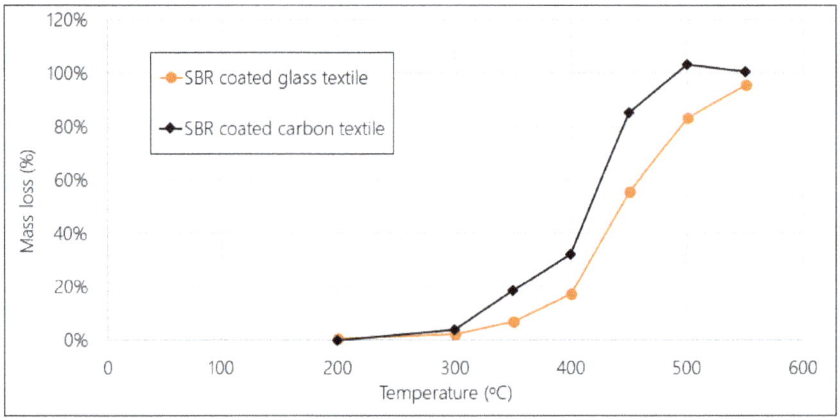

Figure 7. Mass loss of the textiles' coating after heating to different temperatures.

3.3. Results from TRC Heating and Compression Tests

Since the degradation of the matrix due to the exposure to high temperatures also plays a significant role in the specimens' mechanical response, some heating and compression tests were also performed on TRC specimens. This was chosen to be done on specimens of TRC rather than plain mortar, since the existence of the textiles might affect the compressive strength of the TRC sample even in ambient temperatures. This is because the interface between the concrete cover and the core of the element, where the textiles are placed, is a weak area in compression. Therefore, there is a chance that the failure will occur faster by spalling of the cover due to the weak connection of the cover to the core. This was actually observed, because the compressive strength of the TRC specimens was lower (by 20 MPa) than the compressive strength of the plain mortar specimens (see Paragraph 2.1). However, the shape and the dimensions of these specimens were different from the ones in the specimens of plain mortar (see Paragraph 2.1), which explains the difference in the measured strength.

The dimensions of these specimens were 70 × 110 × 28 mm. The loading direction was parallel to the height of 110 mm. Thus, the loaded cross section of 70 × 28 mm is the same as the cross section of the specimens subjected to bending. The applied load was given by the loading cell of the testing machine, while the deformation of the specimens was monitored by Digital Image Correlation, and thus, the strain and the elastic modulus were obtained.

Figure 8 gives the degradation of the TRC specimens due to heating at several temperatures, both in terms of strength and elastic modulus.

In Table 4, the reduction of the compressive strength and the elastic modulus at each elevated temperature is given in percentage, with the specimens in ambient temperature as reference.

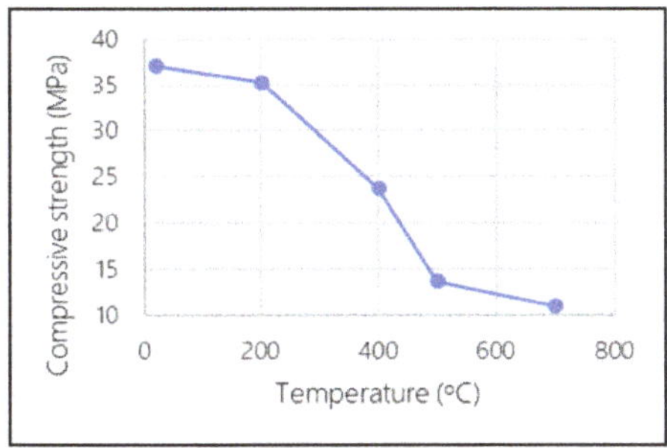
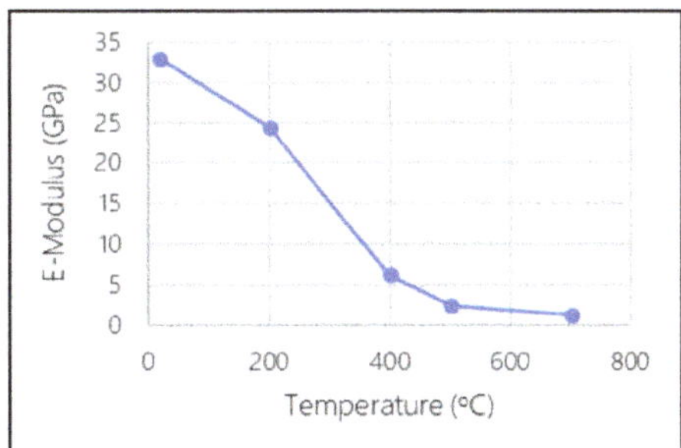

Figure 8. Reduction of compressive strength and elastic modulus of textile reinforced concrete (TRC) specimens after exposure to elevated temperatures.

Table 4. Reduction of compressive strength and elastic modulus of TRC after exposure to high temperatures.

Temperature (°C)	Reduction of Compressive Strength	Reduction of Elastic Modulus
20	-	-
200	5%	23%
400	36%	72%
500	63%	82%
700	71%	85%

3.4. Bending Tests Results

As it has been mentioned, three specimens of each series were tested in bending without being subjected to fire test and three specimens were tested after being exposed to fire. In this paragraph, the results from these tests are presented and discussed.

3.4.1. Results from 15-Minute Fire Tests

The only case where the specimens with SBR coated textiles presented a countable residual strength was the 15-min fire test. Even though the temperature at the level of the effective depth was not measured, by assuming a linear reduction of the temperature between the surface and the middle (where the temperatures are known), it was deduced that the temperature at the level of the effective depth was below 400 °C. However, as a precise calculation cannot be made, no numerical value is provided. Taking this estimation into account, it is not expected to have a severe degradation of the specimens of Series E, for the following reasons:

- The mass loss of the coating is in the order of 20% or lower (see Figure 7); therefore, since most of the coating is still in place, the bond between the textiles and the mortar will not be completely lost as in Series A, B, C, D and F.
- Even though it is well-known that glass fibers lose their strength after being exposed to temperatures higher than 300 °C, it is also well known that carbon fibers maintain their capacity to even higher temperatures if they are not in oxidizing atmosphere [31]. Therefore, even though the glass fibers within the specimens of Series E do not provide significant load bearing capacity, the carbon fibers do.
- The matrix was exposed to a maximum temperature of 477 °C at the surface (face 2), while at the bottom side (face 1, which is subjected in compression at the flexural test, thus, it is the most contributing part of the mortar) the maximum temperature reached 317 °C. According to Table 4, the degradation of the mortar is also not critical. The loss of compressive strength is close to 20% (interpolation between 5% and 36%), while the reduction of the elastic modulus is close to 48% (interpolation between 23% and 72%).

Figure 9 gives the comparison between the load-bearing capacity of the not subjected and the subjected to fire testing specimens in Series E. In the same figure the dashed lines correspond to the specimens that were tested "upside-down", which means that instead of having face 2 subjected to tension during bending, they had face 1. Additionally, for the specimen that was subjected to the fire test before the flexural test, it was again face 1 instead of face 2 that was directly exposed to fire. The reason why this happened was the difference in the concrete cover, since the specimens in Series E had a cover of 8 mm at face 2 and a cover of 4 mm at face 1. Of course, the effective depth during bending changed; thus, these two specimens cannot be compared to the others regarding their flexural strength (higher effective depth, and thus, higher flexural strength, as can be seen from the result of the not exposed specimen–the blue dashed line). However, it can be observed that the reduction of the load recorded for the exposed to fire specimen is greater than the respective reduction in the other specimens (not tested "upside down"–continuous blue lines). This is not surprising, because the concrete cover in the former case was smaller, thus, a heavier damage was done to the textiles at the level of the effective depth, as a result of the higher temperature reached. Therefore, it is logical that even though the not exposed, "upside-down" specimen (E3) has a higher strength than the other not exposed specimens (E1 and E2); this does not happen for the exposed-to-fire specimens (E6 is not stronger than E4 and E5).

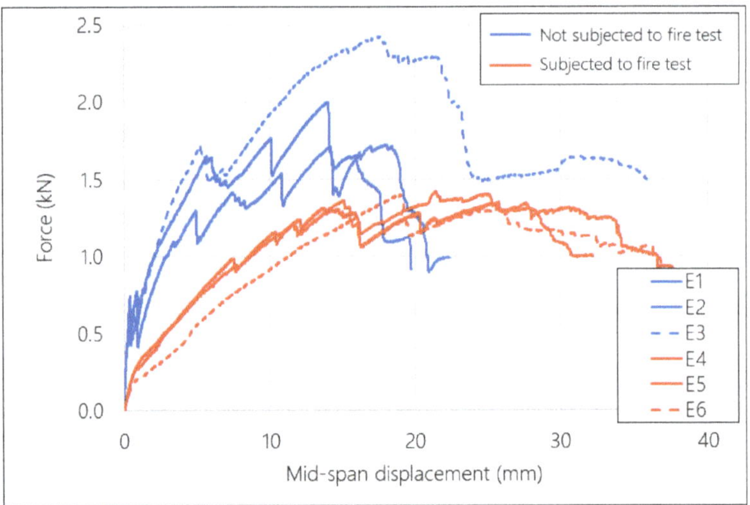

Figure 9. Force versus displacement curves for specimens in Series E.

Regarding Specimens E1, E2, E3 and E4, the initial and the post-cracking stiffness (k_1 and k_2, respectively) were calculated. Additionally, the maximum force (F_{max}) and the corresponding displacement δ_{max} were found. Eventually, the values of k_1, k_2, F_{max} and δ_{max} for the exposed and not exposed to fire specimens were compared and the degradation was calculated. The results can be seen in Table 5.

Table 5. Mechanical characteristics of specimens of Series E that were exposed or not exposed to fire.

Mechanical Properties	Not Exposed Specimens (E1, E2)	Exposed Specimens (E4, E5)	Difference (%)
k_1 (kN/m)	1.58	0.41	−74%
k_2 (kN/m)	0.20	0.10	−48%
F_{max} (kN)	1.86	1.39	−25%
δ_{max} (mm)	14.1	23.6	+68%

3.4.2. Results from 30-Minute Fire Tests

Regarding all specimens that were exposed to fire for 30 min, it can easily be derived from Table 3 that the temperature reached at the level of the effective depth (level of the most stressed fibers) was, in all cases, higher than 520 °C (since the temperature in the middle is around 520–530 °C and the effective depth was closer to the surface). During the subsequent bending tests, it was observed that the fibers started to pull-out at very low load levels, and that practically no residual strength was left (see Figure 10), even though no severe cracks or spalling of the cover was observed. It was obvious that the coating of the textiles, being a thermoplastic material, had been completely burnt-off, and thus, the bond between the textiles and the mortar had already failed before the test. As a matter of fact, the fibers could be pulled out of the specimens even bare-handedly (see Figure 11b). Coating burn-off tests showed that the SBR coating used was almost completely burnt-off after reaching 500 °C (see Figure 7).

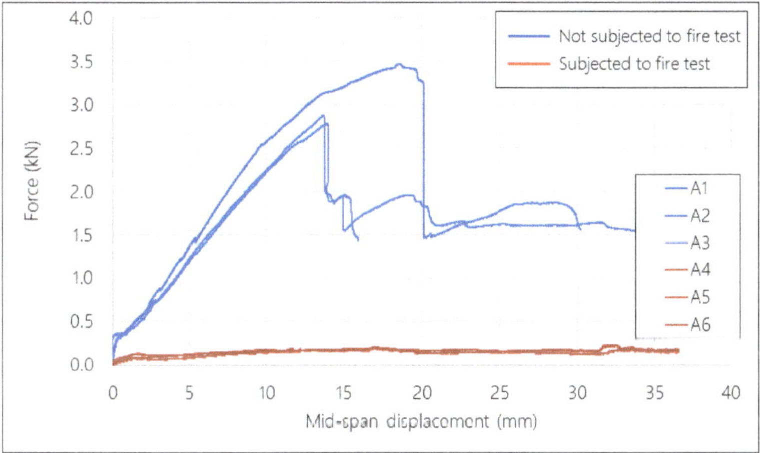

Figure 10. Force versus displacement curves for specimens in Series A.

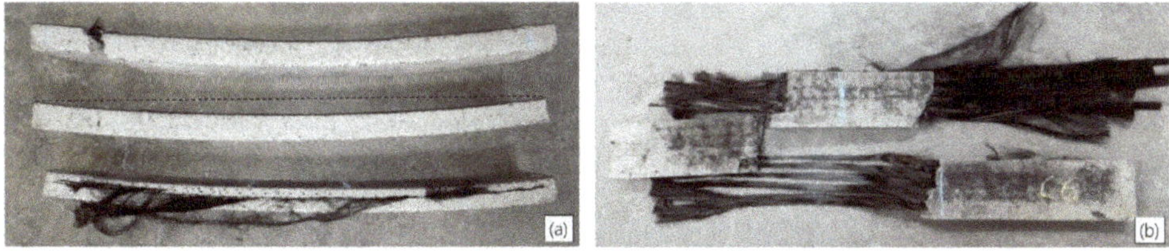

Figure 11. (**a**) Curvature of specimens from the one-sided fire loading and the asymmetrical placement of the reinforcement; (**b**) specimen in Series C that was subjected to fire testing. The fibers were easily pulled out by hand after the flexural test of the specimen.

Figure 10 gives the comparison of the load-bearing capacity of the not subjected and the subjected to fire testing specimens in Series A. Similar results were observed for Series B, C, D and F; therefore, the graphs for those series are not presented. No conclusions can be drawn regarding the effect of the nature of the fibers or the moisture content of the specimens.

Additionally, it is worth mentioning that specimens in Series C and F developed a residual curvature after the fire testing (Figure 11a). These two series were characterized by a high geometrical asymmetry, as can be seen from Figure 2. ue to this asymmetry, the top layer (concrete cover of 12 mm), which was exposed to the high temperature and had no reinforcement, suffered from an increased thermal expansion. On the other hand, the bottom layer was subjected to lower temperatures and the textile that was concentrated near the bottom offered an increased axial stiffness to the lower part of the specimens. As a result, the thermal expansion was much lower at the bottom with respect to the top, which led to the curving of the specimens due to the thermal loading. The curvature was so intense that the surface was severely damaged (visible cracks). Thus, the damage, and therefore, part of the deformation, were irreversible (the specimens remained curved even after cooling down). The residual strength of these specimens was so low that two of them broke while being transferred from the furnace to the bending test set-up; thus, they were not tested at all. In addition, since no numerical results were obtained from testing these series, the effect of the increased concrete cover could not be quantified.

The conclusions that can be drawn from these results, are:

- The temperature stability of the coating of the textiles seems to be the most decisive parameter regarding the residual strength of the TRC specimens, since it directly affects the bond between the matrix and the reinforcement. Thus, extra care must be given when thermoplastic coatings are used in applications with fire safety requirements.
- The increased concrete cover could potentially protect the reinforcement better than a thinner cover; however, it is suggested that the same cover be applied symmetrically, so that a high geometrical eccentricity is avoided.

4. Conclusions

This paper investigated the thermomechanical behavior of textile reinforced cementitious composites subjected to elevated temperatures. The specimens were made of a cementitious matrix with quartz sand and technical textiles coated with styrene-butadiene coating. The heating of the specimens was achieved by one-sided exposure, utilizing a standard fire curve (temperature versus time), which was followed for 15 and 30 min, after which the specimens were cooled down naturally. Flexural tests were performed to heated and not heated specimens, to determine their structural degradation due to the high temperatures. Additionally, compressive tests were performed to heated and unheated TRC specimens, for the same purpose. Moreover, coating burn-off tests were performed to the textiles, to determine the mass loss of coating as a function of temperature. The basic conclusions are the following:

- The most critical parameter that defines the residual strength of the TRC specimens after heating is the coating of the textiles. After the 15-min long fire test, where the temperature at the effective depth did not exceed 400 °C, the degradation was less severe, since the coating was not completely lost (less than 30%). The specimens in this case contained hybrid reinforcement of glass and carbon textiles and they suffered reductions of 74% and 48% in the initial and the post-cracking stiffness, respectively. The maximum force also dropped by 25%, while the corresponding maximum displacement increased by 68%.
- The textiles that were coated with a thermoplastic material retained a practically negligible residual strength after being subjected to a 30-min fire test, where the temperature at the level of the effective depth (most stressed fibers during the bending test) exceeded 500 °C. This corresponds to a mass loss of 90% or higher and is explained by the fact that the loss of the coating, which is an intermediate layer between the fibers and the matrix, leads to failure of the bond between the fibers and the matrix. The same result was observed regardless of the fiber material (glass or carbon), the thickness of the concrete cover (8 mm or 12 mm) and the moisture saturation of the specimens (0% or 50%).
- The degradation of the mortar due to the high temperature was also significant and could be another dominant parameter. Regarding the compressive strength and the elastic modulus, it was observed that the latter dropped faster with respect to the temperature of exposure. However,

in both cases the degradation was not severe until 200 °C (5% and 23%, respectively), while it became critical at 500 °C (63% and 82% loss, respectively).

- The temperature profile within the cross section of the one-sided exposed specimens of TRC was not uniform. Specifically, the temperature reduction through the top part of the specimens appeared to be higher due to the lower thermal conductivity of the top, hotter, layers. Thus, the concrete cover is also a potential critical parameter that could determine the residual strength of the heated specimens. The effect of the concrete cover, though, was not quantified in this study.
- Finally, it was concluded that a highly asymmetrical design scheme can be disastrous for the case of one-sided exposure to fire, since the double asymmetry (in heating and in axial stiffness) can lead to premature failure of the specimens solely due to thermal stresses.

Author Contributions: Conceptualization, P.K., T.T. (Thanasis Triantafillou) and T.T. (Tine Tysmans); Investigation, P.K. and M.E.K.; Methodology, J.W., T.T. (Thanasis Triantafillou) and T.T. (Tine Tysmans); Supervision, T.T. (Thanasis Triantafillou) and T.T. (Tine Tysmans); Writing—original draft, P.K.; Writing—review & editing, M.E.K., J.V., M.D.M., J.W., T.T. (Thanasis Triantafillou) and T.T. (Tine Tysmans).

Funding: This research was funded by the Agentschap voor Innovatie en Ondernemen (VLAIO), grant number IWT140070, and partially by the Structural Materials Laboratory at the University of Patras.

Acknowledgments: The authors gratefully all the members of the 'CeComStruct' project, of which this research is part of, for their advice. The authors would also like to thank the company 'Sika Hellas' for the donation of the cementitious material that was used in this research. Finally, the authors would like to acknowledge the priceless help of the laboratory technician Kyriakos Karlos for the laborious experimental work.

References

1. Hothan, S.; Ehlig, D. Reinforced concrete slabs strengthened with textile reinforced concrete subjected to fire. In Proceedings of the 2nd International RILEM Work Concrete Spalling Due to Fire Exposure, Delft, The Netherlands, 5–7 October 2011; RILEM Publications: Bagneux, France, 2011; pp. 419–426.

2. Hashemi, S.; Al-Mahaidi, R. Experimental and finite element analysis of flexural behavior of FRP-strengthened RC beams using cement-based adhesives. *Constr. Build. Mater.* **2012**, *26*, 268–273. [CrossRef]

3. Bisby, L. Design for fire of concrete elements strengthened or reinforced with fibre-reinforced polymer: State of the art and opportunities from performance-based approaches. *Can. J. Civ. Eng.* **2013**, *40*, 1034–1043. [CrossRef]

4. Michels, J.; Zwicky, D.; Scherer, J.; Harmanci, Y.E. Structural strengthening of concrete with fiber reinforced cementitious matrix (FRCM) at ambient and elevated temperature—Recent investigations in Switzerland. *Adv. Struct. Eng.* **2014**, *17*. [CrossRef]

5. Tetta, Z.C.; Bournas, D.A. TRM vs FRP jacketing in shear strengthening of concrete members subjected to high temperatures. *Compos. Part B* **2016**, *106*, 190–205. [CrossRef]

6. Raoof, S.M.; Bournas, D.A. TRM versus FRP in flexural strengthening of RC beams: Behaviour at high temperatures. *Constr. Build. Mater.* **2017**, *154*, 424–437. [CrossRef]

7. Raoof, S.M.; Bournas, D.A. Bond between TRM versus FRP composites and concrete at high temperatures. *Compos. Part B Eng.* **2017**, *127*, 150–165. [CrossRef]

8. Ombres, L. Analysis of the bond between Fabric Reinforced Cementitious Mortar (FRCM) strengthening systems and concrete. *Compos. Part B Eng.* **2015**, *69*, 418–426. [CrossRef]

9. Triantafillou, T.; Karlos, K.; Kefalou, K.; Argyropoulou, E. An innovative structural and energy retrofitting system for masonry walls using textile reinforced mortars combined with thermal insulation. *RILEM Bookser.* **2018**, *15*, 752–761.

10. Maroudas, S.R.; Papanicolaou, C.G. Effect of High Temperatures on the TRM-to-Masonry Bond. *Key Eng. Mater.* **2017**, *747*, 533–541. [CrossRef]

11. Ombres, L.; Iorfida, A.; Mazzuca, S.; Verre, S. Bond analysis of thermally conditioned FRCM-masonry joints. *Measurement* **2018**, *125*, 509–515. [CrossRef]

12. Ehlig, D.; Jesse, F.; Curbach, M. High temperature tests on textile reinforced concrete (TRC) strain specimens. In Proceedings of the International RILEM Conference on Material Science (MatSci), Aachen, Germany, 6–8 September 2010.

13. Hegger, J.; Horstmann, M.; Zell, M. Applications for TRC. In Proceedings of the 15th International Congress of the GRCA, Prague, Czech Republic, 20–23 April 2008.

14. Rambo, D.A.; Silva, F.D.A.; Filho, R.D.T.; Ukrainczyk, N.; Koenders, E. Tensile strength of a calcium-aluminate cementitious composite reinforced with basalt textile in a high-temperature environment. *Cem. Concr. Compos.* **2016**, *70*, 183–193. [CrossRef]

15. Silva, F.D.A.; Butler, M.; Hempel, S.; Filho, R.D.T.; Mechtcherine, V. Effects of elevated temperatures on the interface properties of carbon textile-reinforced concrete. *Cem. Concr. Compos.* **2014**, *48*, 26–34. [CrossRef]

16. Xu, S.; Shen, L.; Wang, J. The high-temperature resistance performance of TRC thin-plates with different cementitious materials: Experimental study. *Constr. Build. Mater.* **2016**, *115*, 506–519. [CrossRef]

17. Ward, M.; Bisby, L.; Stratford, T.; Roy, E. Fibre Reinforced Cementitious Matrix systems for fire-safe flexural strengthening of concrete: Pilot testing at ambient temperatures. In Proceedings of the 4th International Conference on Advanced Composites in Construction, Chesterfield, UK, 3–5 September 2009; NetComposites Ltd.: Chesterfield, UK, 2009; pp. 449–460.

18. Donnini, J.; Basalo, F.D.C.; Corinaldesi, V.; Lancioni, G.; Nanni, A. Fabric-reinforced cementitious matrix behavior at high-temperature: Experimental and numerical results. *Compos. Part B Eng.* **2017**, *108*, 108–121. [CrossRef]

19. Çavdar, A. A study on the effects of high temperature on mechanical properties of fiber reinforced cementitious composites. *Compos. Part B Eng.* **2012**, *43*, 2452–2463. [CrossRef]

20. Caverzan, A.; Colombo, M.; di Prisco, M.; Rivolta, B. High performance steel fibre reinforced concrete: Residual behaviour at high temperature. *Mater. Struct.* **2015**, *48*, 3317–3329. [CrossRef]

21. Colombo, I.; Colombo, M.; Magri, A.; Zani, G.; Di Prisco, M. Textile reinforced mortar at high temperatures. *Appl. Mech. Mater.* **2011**, *82*, 202–207. [CrossRef]

22. Rambo, D.A.S.; Silva, F.D.A.; Filho, R.D.T.; Da Gomes, O.F.M. Effect of elevated temperatures on the mechanical behavior of basalt textile reinforced refractory concrete. *Mater. Des.* **2015**, *65*, 24–33. [CrossRef]

23. Rambo, D.A.S.; Yao, Y.; Silva, F.D.A.; Filho, R.D.T.; Mobasher, B. Experimental investigation and modelling of the temperature effects on the tensile behavior of textile reinforced refractory concretes. *Cem. Concr. Compos.* **2017**, *75*, 51–61. [CrossRef]

24. Nguyen, T.H.; Vu, X.H.; Si Larbi, A.; Ferrier, E. Experimental study of the effect of simultaneous mechanical and high-temperature loadings on the behaviour of textile-reinforced concrete (TRC). *Constr. Build. Mater.* **2016**, *125*, 253–270. [CrossRef]

25. Keleştemur, O.; Arıcı, E.; Yıldız, S.; Gökçer, B. Performance evaluation of cement mortars containing marble dust and glass fiber exposed to high temperature by using Taguchi method. *Constr. Build. Mater.* **2014**, *60*, 17–24. [CrossRef]

26. Reinhardt, H.W.; Kruger, M.; Raupach, M. Behavior of textile-reinforced concrete in fire. *ACI Spec. Publ.* **2008**, *SP250*, 99–100.

27. Antons, U.; Hegger, J.; Kulas, C.; Raupach, M. High-temperature tests on concrete specimens reinforced with alkali-resistant glass rovings under bending loads. In Proceedings of the 6th International Conference on FRP Composites in Civil Engineering, Rome, Italy, 13–15 June 2012.

28. Buttner, R.M.; Orlowsky, J.; Raupach, M. *Fire Resistance Tests of Textile Reinforced Concrete under Static Loading—Results and Future Developments, Proceedings of the 5th International RILEM Workshop on High Performance Fiber Reinforced Cement Composites, Mainz, Germany, 10–13 July 2007*; Reinhardt, H.W., Naaman, A.E., Eds.; RILEM Publications: Bagneux, France, 2014.

29. BS EN 1363-1:1991. *Fire Resistance Tests—Part 1: General Requirements*; British Standards Institution (BSI): London, UK, August 1991.

30. Brameshuber, W. (RILEM Technical Committee). Recommendation of RILEM TC 232-TDT: test methodsand design of textile reinforced concrete. Uniaxial tensile test: test method to determine the load bearing behavior of tensile specimens made of textile reinforced concrete. *Mater. Struct.* **2016**, *49*, 4923–4927.

31. Triantafillou, T. *Textile Fibre Composites in Civil Engineering*; Woodhead Publishing: London, UK, 2016; pp. 173–174.

Article

Verification of the Structural Performance of Textile Reinforced Reactive Powder Concrete Sandwich Facade Elements

Mathias Flansbjer, Natalie Williams Portal * and Daniel Vennetti

Mechanics Research, RISE Research Institutes of Sweden, 50115 Borås, Sweden; mathias.flansbjer@ri.se (M.F.); daniel.vennetti@ri.se (D.V.)
* Correspondence: natalie.williamsportal@ri.se; Tel.: +46-10-516-6887

Received: 26 April 2019; Accepted: 11 June 2019; Published: 15 June 2019

Abstract: As a part of the SESBE (Smart Elements for Sustainable Building Envelopes) project, non-load bearing sandwich elements were developed with Textile Reinforced Reactive Powder Concrete (TRRPC) for outer and inner facings, Foam Concrete (FC) for the insulating core and Glass Fiber Reinforced Polymer (GFRP) continuous connectors. The structural performance of the developed elements was verified at various levels by means of a thorough experimental program coupled with numerical analysis. Experiments were conducted on individual materials (i.e., tensile and compressive tests), composites (i.e., uniaxial tensile, flexural and pull-out tests), as well as components (i.e., local anchorage failure, shear, flexural and wind loading tests). The experimentally yielded material properties were used as input for the developed models to verify the findings of various component tests and to allow for further material development. In this paper, the component tests related to local anchorage failure and wind loading are presented and coupled to a structural model of the sandwich element. The validated structural model provided a greater understanding of the physical mechanisms governing the element's structural behavior and its structural performance under various dead and wind load cases. Lastly, the performance of the sandwich elements, in terms of composite action, was shown to be greatly correlated to the properties of the GFRP connectors, such as stiffness and strength.

Keywords: reactive powder concrete (RPC); textile reinforced concrete (TRC); foam concrete (FC); sandwich elements; wind loading; finite element analysis (FEA)

---

1. Introduction

At the end of the 1950s, precast concrete elements emerged as a popular cladding solution for housing. Between the 1960s–70s, a renowned Swedish public housing project, entitled Million Program, made use of prefabricated modular concrete to construct residential buildings [1]. During this era, a number of realized European housing projects led to the extensive development of construction techniques related to precast concrete. During the 1960s–80s, the precast concrete industry, pertaining to the application of building envelopes, primarily made use of conventional steel reinforced concrete (RC). RC elements, however, pose certain disadvantages, such as the need for a thick concrete cover to protect the reinforcement. For instance, based on EN 206-1 [2], a recommended minimum concrete cover thickness can amount to 30–35 mm, considering XC3/XC4 exposure classes. Accordingly, the thickness of a facing can be around 80 mm, leading to not only a thick, but also a heavy, member. This issue was tackled in a project funded by the European Commission, SESBE (Smart Elements for Sustainable Building Envelopes). In SESBE, so-called smart facings were developed with several features: thin, lightweight, and adaptable via the inclusion of nanomaterials. A precast cladding solution taking the form of a sandwich element was developed using a combination of high-performance materials, such as Textile Reinforced Reactive Powder Concrete (TRRPC) for the facings, Foam Concrete (FC) for the insulating core, and glass fiber reinforced polymer (GFRP) continuous connectors.

The thickness and weight reduction of precast concrete has been successfully achieved by the development and application of new material alternatives. Conventional steel reinforcement has, for example, been replaced by textile reinforcement, while high-performance concrete, such as Ultra-High Performance Concrete (UHPC) or Reactive Powder Concrete (RPC), has replaced normal concrete. Lately, innovative façade elements have been produced using UHPC or Textile Reinforced Concrete (TRC), exemplified by ventilated façade cladding [3] and sandwich elements [4–6]. Progressively more UHPC (or RPC) has been applied in façade applications, as this composite material has revealed extraordinary features, such as durability and high strength [7–9]. By embedding textile reinforcement in this type of matrix, so-called Textile Reinforced Reactive Powder Concrete (TRRPC), a versatile precast product [10] which enhances the post-cracking behavior of high-strength concrete [5,11,12] can be assembled.

The design and verification of novel façade elements are typically realized by means of experiments combined with numerical modelling. Small-scale tests at the material or component levels can be initially conducted to gain knowledge related to flexural and composite behaviors of the developed elements. Full-scale testing can thereafter be performed to evaluate the structural performance according to e.g., service and ultimate loads. An example of this approach was presented in [13], wherein the structural performance of precast concrete sandwich facings developed with a system of FRP connectors was analyzed via small-scale and full-scale testing coupled with numerical analysis. Another study focused on the experimental testing of components, so-called small-scale, paired with the numerical analysis of the mechanical behavior of full-scale sandwich facings while using inverse analysis and relevant codes for parameter estimation [14–16]. The flexural behavior of TRC sandwich facings was investigated both experimentally and numerically in various works [17,18]. Moreover, multiscale mechanical modelling of TRC sandwich facings (i.e., micro, meso and macro) compared to macroscopic modelling in connection with experimental verification has also been shown to effectively predict the structural behavior of such elements [19].

This paper presents the validation of the structural performance of the developed TRRPC sandwich façade elements. Validation was established by means of a thorough experimental program coupled with finite element analysis (FEA). Within the SESBE project, experiments were conducted on individual materials (i.e., tensile and compressive tests) [11,20], composites (i.e., uniaxial tensile, flexural and pull-out tests) [21] and components (i.e., local anchorage failure, shear, flexural and wind loading tests) [22–24]. The experimentally yielded material properties were used as input for numerical models to better understand the findings of various component tests and allow for further material development. In this paper, a structural model of the element under wind loading was validated via experimental results. Lastly, this model was expanded to a full-size sandwich element with and without openings to further facilitate the prediction and analysis of its structural performance in relation to a given design scenario and SLS and ULS requirements.

2. Sandwich Façade Element Concept

2.1. Sandwich Element Details

The sandwich elements were designed as prefabricated concrete cladding with a surface area ranging from 7–10 × 2.7–3.0 m and weight of 2–5 ton, as per Figure 1. Conceptually, these elements cover the standard height of one storey and are attached to the main load-bearing structure via an anchorage system. Due to their immense size, these elements actively carry and transfer, e.g., self-weight and wind loads to the structure. Moreover, these elements consist of two facings made of 25 mm thick TRRPC, which are separated by a FC insulating layer of 150 mm. Connectors made of GFRP are embedded in the facings to ensure a certain level of composite action. Standard steel anchorage systems are installed to fasten the façade element to adjacent elements or structural members.

The individual components which make up the novel sandwich elements are further discussed in Section 3.

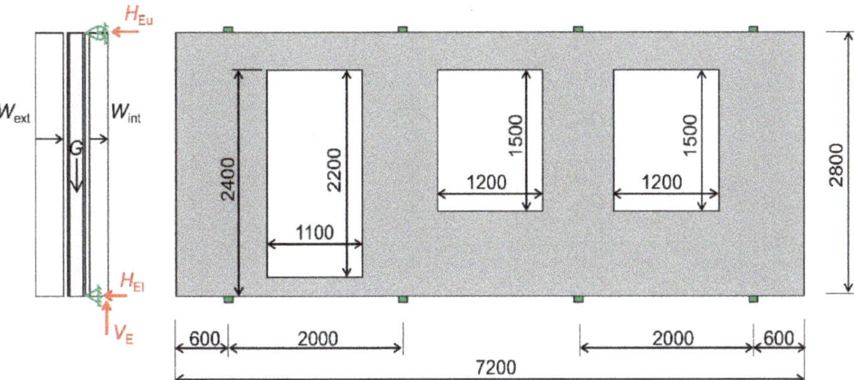

Figure 1. Illustration of sandwich element concept with structural loads (wind and self-weight, in black) and reaction forces (horizontal and vertical, in red).

Based on preliminary structural investigations in the conceptual phase, a thorough testing and modelling program was defined, as per Figure 2, to enable the verification of the structural performance of the elements at different phases, namely material development, component modelling and testing. The numerical analysis and experiments were conducted parallelly with the material development and characterization. Additionally, the evaluation was performed using an iterative process because of the underlying interaction between the materials and components. As emphasized in Figure 2, this paper focuses on presenting the methods and results pertaining to the local failure (anchorage) and wind load experimental tests, along with the verification of the overall behavior and detailed model of the sandwich element. The material development has been presented elsewhere for RPC [11,20], FC [25], GFRP [22] and TRRPC [21]. The component testing and modelling related to connector local failure and shear tests can be found in [24], and that concerning the four-point bending tests in [22].

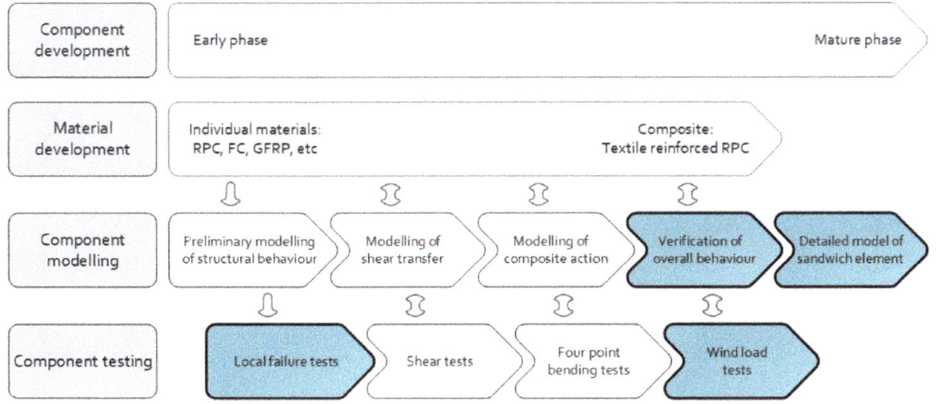

Figure 2. Workflow diagram referring to the component testing and modelling (highlighted boxes are principally covered in this paper). RPC: reactive powder concrete; FC: foam concrete; GFRP: glass fibre reinforced polymer.

2.2. Anchorage System

Based on the given design, the sandwich elements are subjected to two types of loads, namely vertical permanent loads, i.e., self-weight (G) of element and horizontal variable loads caused by wind (W_{ext} and W_{int}). The subjected loading and reaction forces are schematically illustrated in Figure 1. The self-weight of the element is assumed to be taken as a vertical reaction force (V_E) at the bottom anchors, and then transferred through the angle plate into the load bearing structure. The force will be taken as contact stress at the lower edge of the element and will be considerably lower than the compressive strength of the RPC. Alternatively, the vertical force can be transferred directly to the element below as a self-supporting façade system. In addition, the anchors need to withstand horizontal reaction forces (H_{Eu} and H_{El}) due to self-weight and both wind pressure and wind suction. At the upper anchorage point, the horizontal reaction force is transferred to the angle plate by two embedded bolt anchors (Figure 3a) and at the lower anchor details by one threaded stud inserted into the embedded bolt anchor (Figure 3b). Hence, the element anchors will mainly be subjected to shear load introduced at the bolt anchors. The shear load capacity of the anchors is more complicated to determine by calculations, and therefore, needs to be verified by tests. For the sake of obtaining design criteria, the shear capacity of the anchors was experimentally quantified in this project, as further explained in Section 4.1.1.

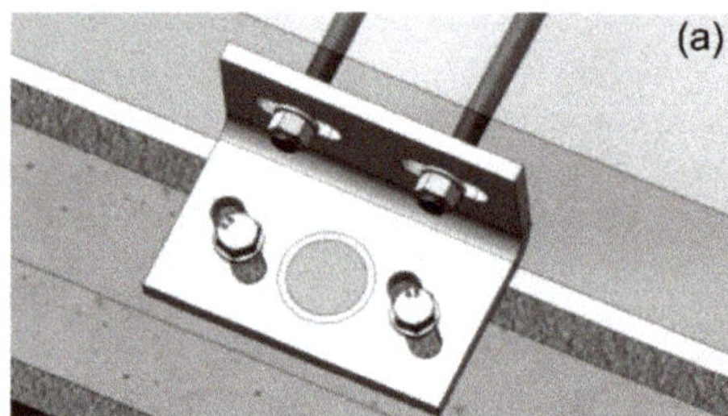

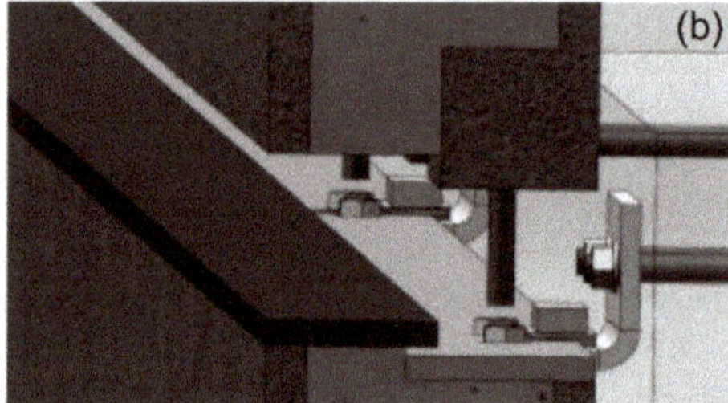

Figure 3. Details of upper (**a**) and lower (**b**) element anchors.

3. Materials

3.1. Textile Reinforced Reactive Powder Concrete (TRRPC)

TRRPC is composed of an RPC reinforced by a carbon-based textile grid coated by epoxy. Considering a precast concrete façade application, the RPC recipe includes large quantities of supplementary cementitious materials (SCMs). RPC is synonymous to UHPC such that it consists of six to eight different components and aggregate size of 2 mm or less. Table 1 presents the average strength values for RPC, while other details can be found in [11].

Table 1. Average strength properties (28 days) for RPC (standard deviation in parenthesis), source: [11].

Property	Average Values	Test Description
Compressive strength [MPa]	147.2 (2.3)	
E-modulus [GPa]	49.7 (1.7)	Compression tests
Ultimate strain [‰]	3.9 (0.2)	
Poisson's ratio [–]	0.22 (0.02)	
Tensile strength [MPa]	5.1 (0.5)	Uniaxial tensile tests

The textile grid applied in the TRRPC consists of carbon fibers with an epoxy coating. Superior bond properties between the concrete and textile are typically observed when epoxy is applied. Individual rovings were tested in tension as per [26], which indicated that the tensile strength in the warp and weft directions was 3433 MPa and 3878 MPa, respectively. The Young's modulus in the warp and weft directions was 233 GPa and 248 GPa, respectively. These average values are similar to those obtained by the producer.

The tensile behavior of thin TRRPC facings was quantified by means of uniaxial tensile tests performed according to RILEM TC 232-TDT [27], with the addition of Digital Image Correlation (DIC) measurements (refer to [21] for further details). The test specimens had dimensions of $700 \times 100 \times 25$ mm and were reinforced by two layers of carbon grid. During testing, a relatively stiff and linear behavior was noted prior to first cracking. First cracking thereafter occurred presumably when the tensile strength of the concrete was reached (3 MPa). This was followed by load jumps with minimal load increase, being indicative of multiple cracking along the specimen. Cracking typically initiated in proximity to the lateral rovings, which were observed to be a location prone to stress concentration.

3.2. Foam Concrete (FC)

FC, also known as cellular lightweight concrete (CLC), is applied as a thermally insulating layer in the developed sandwich element. It is made of a lightweight cementitious material with the following constituents: cement, sand, water and foam (water, air and surfactant). FC is typically made of a minimum of 20% by volume of mechanically entrained air in the fresh cement paste or mortar [28]. Based on project findings, FC has a minimal environmental impact compared to other insulation materials, such that it has ca. 70% lower embodied energy than expanded polystyrene (EPS) foam. Additionally, under fire exposure, neither smoke or toxic gases are released. FC was optimized in this project in terms of heat conductivity, density and compressive strength. Specifically, a thermal conductivity between 0.04–0.06 W/(m·K) and a wet density between 200–300 kg/m^3 were achieved. By adding Quartzene® (Svenska Aerogel AB, Gävle, Sweden) [25] to FC, the thermal conductivity can be reduced to 0.03–0.035 W/(m·K). The stiffness and compressive strength pertaining to FC, with a density of 200–400 kg/m^3, ranged between 5–37 MPa and 95–472 kPa, respectively. The addition of polypropylene fibers (length of 12 mm), with a dosage of 0.25%-vol., improved both the material's handleability and post-cracking behavior. These mentioned supplementary constituents were excluded in the FC incorporated in the sandwich elements.

3.3. Glass Fibre Reinforced Polymer (GFRP) Connectors

The composite action between the TRRPC facings of the element was enhanced by incorporating GFRP truss-like connectors. The connectors were fabricated using pultruded bars made of E-glass fibers impregnated with an epoxy resin. The bars, having a nominal diameter of ca. 6.1 mm, were reinforced by a bundle of E-glass fibers to form helical ribs on the bar's surface. In a half-cured state, the bars were bent into a zig-zag shape, followed by final curing. Two configurations were studied in this project, denoted as single (S) and double (D); see Figure 4a. A double connector is composed of two single connectors mirrored with respect to the longitudinal direction and fastened at intersecting points using plastic tie straps. The connector performance in an element was previously investigated via modelling and testing on a component level [22,24]. As a result, the diagonal bars were observed to be primarily loaded by axial tensile and compressive forces, as illustrated in Figure 4b.

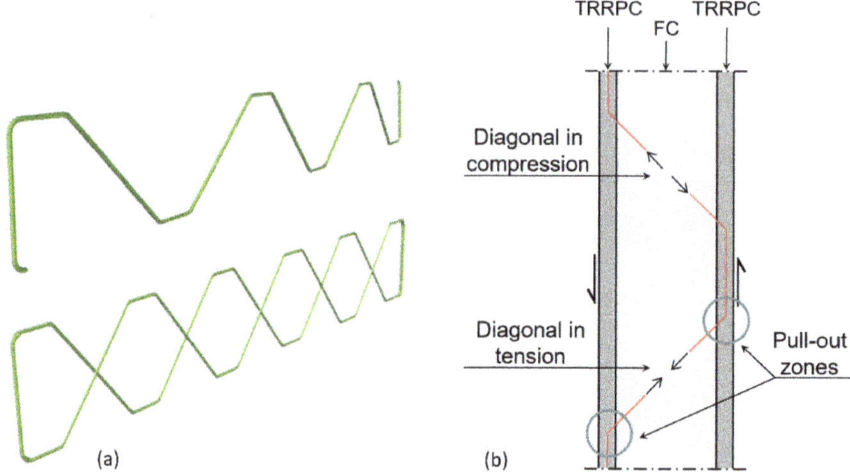

Figure 4. Photo of single (upper) and double connectors (lower) (**a**) and schematic of load transfer between the two TRRPC facings and single connector in a sandwich element (**b**).

The incorporation of connectors in thin facings is generally challenging, as it may prove difficult to enable load transfer without causing local pull-out failure at the connector. Accordingly, it was of key importance to further understand the properties on both material and component levels. Tensile, compression and pull-out tests were therefore performed; for details, refer to [22]. Table 2 provides the experimental results for the given GFRP connectors.

Table 2. Average properties for GFRP connectors (standard deviation in parenthesis), source: [22].

Property	Average Values	Test Description
Ultimate tensile capacity [MPa]	1012 (35)	
Ultimate strain [%]	2.5 (0.1)	Tensile test ISO 10406-1 [29]
Young's modulus [GPa]	40.3 (0.8)	
Critical buckling load [kN]	1.7 (0.1) [1]	Compression tests
Pull-out capacity [kN]	6.5 (0.5)	Connector pull-out test

[1] Critical buckling load for a buckling length of 212 mm, corresponding to TRRPC facing distance of 150 mm.

The critical buckling load in compression was experimentally quantified for different buckling lengths, based on the length of connector diagonals (inclination of 45°) in elements with different facing distances, i.e., dependent of the FC insulation thickness. In this study, the TRRPC facing distance was set to 150 mm which corresponds to a buckling length of approximately 212 mm.

The pull-out capacity of the connectors was determined from small-scale tests. Pull-out tests were conducted on connector segments cast in TRRPC panels (50 × 400 × 400 mm) with an embedment length of 10 mm. To simulate the actual loading of the connector in a facing (see Figure 4b), loading was applied axially along the connector at a 45° angle from the surface of the facing. The test parameters presented here were established based on a parametric study conducted in this project to initially evaluate the effect of embedment depths and connector types, see [24].

4. Methods

4.1. Experimental Methods

The material development and component testing phases consisted of an array of experimental investigations, as previously depicted in Figure 2. Most of the methods have been published elsewhere as specified in Section 2, apart from the tests performed on anchors embedded in a TRRPC facing and wind load testing on sandwich elements. Accordingly, the methods pertaining to these given component tests are described in detail.

4.1.1. Anchorage Testing

The element anchor system, previously discussed in Section 2.2, was experimentally investigated in this study. The main challenge associated with this system is such that the screw anchors should be embedded in a thin TRRPC facing (25 mm) all while being able to effectively transmit the horizontal forces from the sandwich element to the load bearing structure. The specimens (1360 × 1220 mm) were designed as small-scale façade elements, according to that shown in Figure 5. The elements consisted of two 25 mm TRRPC facings set apart by a 150 mm layer of FC. Both facings contained two carbon textile grid layers, placed symmetrically in the center of the facings. The two TRRPC facings were connected by two lines of GFRP connectors in each specimen. Each test specimen was provided with two upper element anchor details and two lower element anchor details. The inner facing was strengthened locally by increasing the facing thickness to 70 mm at the position of the two upper element anchoring details, each consisting of two bolt anchors (M16 × 140). The upper thickened sections were reinforced with one extra GFRP bar profile. The elements were also strengthened by a thicker section (70 mm) along the lower edge of the inner TRRPC facing, in which the two lower bolt anchors (M16 × 140) were incorporated. Six specimens were manufactured, and each element anchor detail was used for a given test configuration.

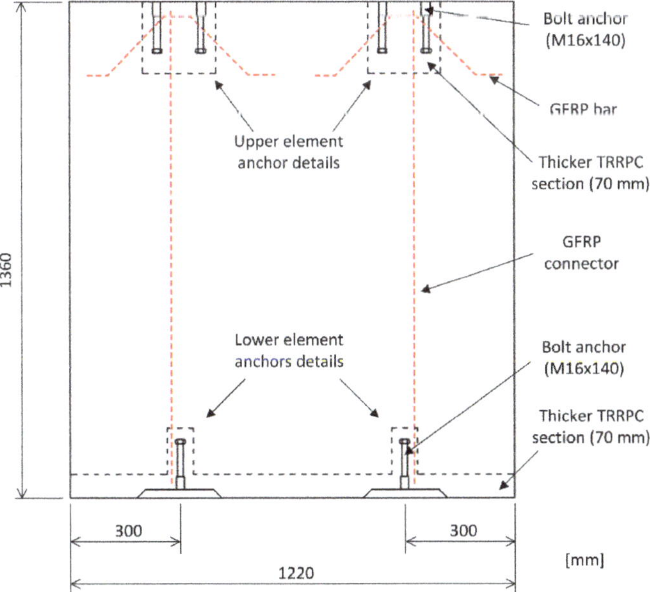

Figure 5. Illustration of the anchor tests specimens.

The shear load capacity tests of the anchor details were conducted using a servo-hydraulic testing machine. The shear load capacity related to the upper and lower element anchors was determined by four different test cases to account for positive and negative wind load: (a) positive shear load at upper anchors (H_{up}), (b) negative shear load at upper anchors (H_{un}), (c) positive shear load at lower anchors (H_{lp}) and (d) negative shear load at lower anchors (H_{ln}). Schematic illustrations of the set-ups for the four different test cases can be seen in Figure 6. In the upper element anchor tests (cases a and b), the shear load (H_{up} and H_{un}) was applied to a steel profile connected to the two bolt anchors. For the lower element anchors (cases c and d), the shear load (H_{lp} and H_{ln}) was applied directly to a threaded rod inserted in the bolt anchor. In test cases a and c, the specimens were placed directly on the laboratory floor, as for cases b and d, the specimens were placed on two supporting steel profiles. The load was applied using a displacement rate of 2.0 mm/min and was logged by a 100 kN rated load cell (accuracy greater than 1%).

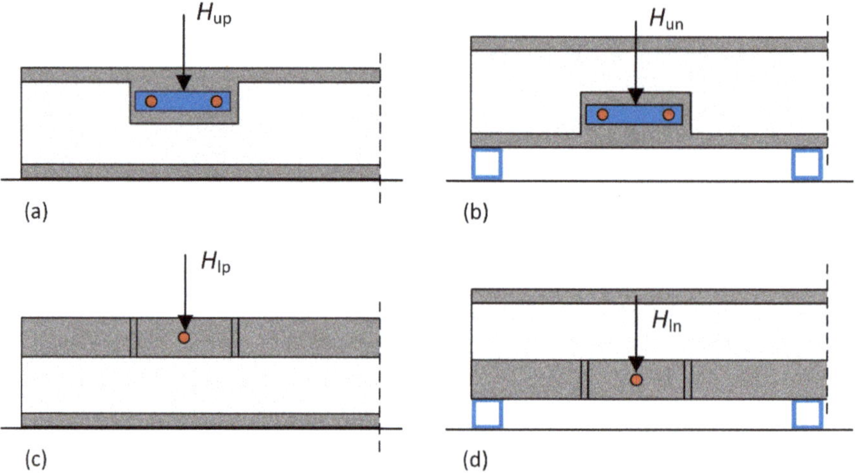

Figure 6. Schematic of the shear load capacity test set-ups; (**a**) positive shear load at upper anchors (H_{up}), (**b**) negative shear load at upper anchors (H_{un}), (**c**) positive shear load at lower anchors (H_{lp}) and (**d**) negative shear load at lower anchors (H_{ln}). Red filled circle indicates loaded bolt anchors.

4.1.2. Wind Load Testing

Wind load tests were conducted to verify the overall structural performance and validate the numerical model of the full sandwich element, all while considering the embedded connectors and anchorage details. The wind load was applied incrementally in pressure and suction on full-scale sandwich elements using a pressure chamber. The test specimens were produced to have surface dimensions of 3.0×2.8 m^2. The facings were made up of TRRPC with a nominal thickness of 25 mm and the core consisted of a 150 mm FC insulation layer. Both facings contained two layers of carbon textile grid.

Two element configurations underwent wind pressure tests, denoted as Single (S) and Double (D). In Figure 7, the first element comprised five rows of single connectors with a c/c distance of 0.5 m (dashed lines), while the second element had three rows of double connectors with a c/c distance of 1.0 m (dashed lines). The anchor details were designed similarly to that described for the anchor specimens above. However, in these specimens, the upper thickened sections were reinforced with one extra strip of carbon grid instead of a GFRP bar to simplify the production process of the elements.

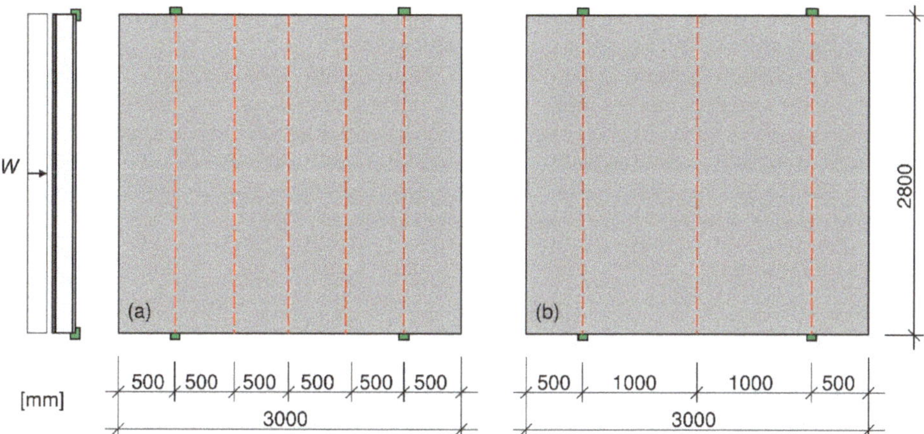

Figure 7. Illustration of wind load test on full sandwich elements: (**a**) single connector configuration (S) and (**b**) double connector configuration (D).

Testing was performed in a pressure chamber (capacity of ± 3 kPa) configured as a four-sided room with an opening on one side. The sandwich element was mounted in a steel frame, affixed at the prescribed four anchorage points, then placed in the chamber opening, as depicted in Figure 8a. This established connection allowed for the element to move freely during loading. To prevent air leakage and pressure drop during testing, the gap surrounding the frame was sealed with elastic sealing tape.

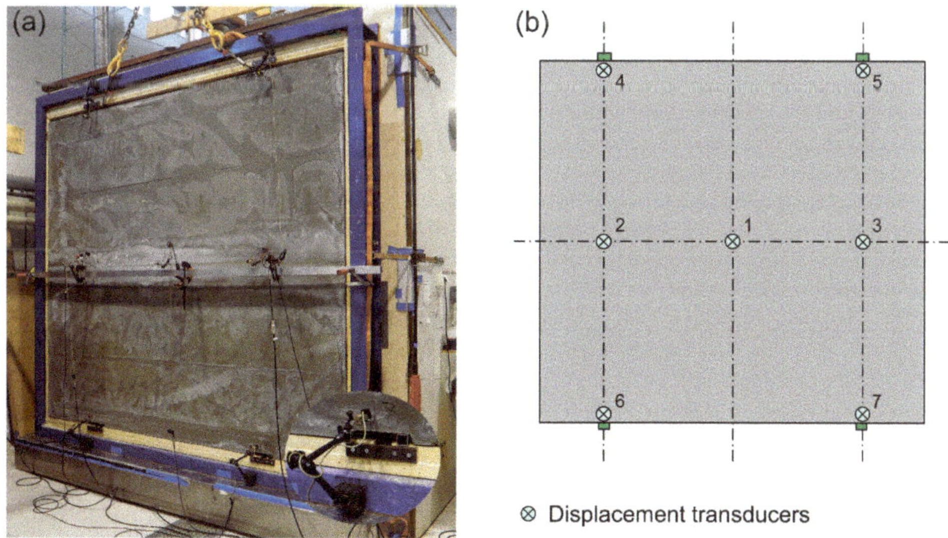

Figure 8. Photo of sandwich element mounted in the pressure test chamber (**a**) and location of displacement transducers on inner TRRPC facing (**b**).

The simulated wind load was applied to the TRRPC facing at the inside of the climate chamber as incremental sequences (0.5, −0.5, 1.0, −1.0, 1.5, −1.5, 2.0, −2.0 kPa) consisting of positive wind suction followed by negative wind pressure and so forth. Load increments were manually set to the given load level and then held constant for approximately 60 s before applying the following load level. The load was applied incrementally on each element as the capacities were initially unknown and it was of interest to apply both suction and pressure on the elements.

The out of plane deformation of the inner TRRPC facing (facing out of the chamber) was measured in relation to the steel frame of the chamber at seven points using displacement transducers with a measuring range of ± 25 mm. The locations of the seven measurement points are indicated in Figure 8b. Transducer positions 1, 2 and 3 measured displacements at mid-span, while positions 4, 5, 6 and 7 measured displacements near the top and bottom anchoring positions. During testing, the chamber pressure and displacements were measured at a sampling rate of 20 Hz.

4.2. Numerical Modelling

A conservative numerical modelling approach was chosen in this work to capture the element's overall structural behavior. The individual components, i.e., facings, insulation and connectors, were modelled as individual parts made up of structural elements, all while incorporating the interaction between the different parts. This type of detailed global structural model is limited in the sense that it is unable to capture local stress conditions fully accurately, e.g., locally at connectors and anchorage details. As such, mainly bending failures are reflected in the analysis, whereas pull-out failure and buckling of connectors or anchorage failures are not captured. These failure modes are thus verified by local resistance models and/or verified by experimental values. Moreover, as a first step, the chosen modelling concept was validated using the wind load test results, followed by a detailed analysis of a full-scale element. The presented models incorporate the same modelling parameters, but differing geometry, loading and boundary conditions.

4.2.1. General Parameters

To gain a deeper understanding of the performance of the developed façade elements, finite element (FE) calculations were conducted in Abaqus/CAE 6.14-1 (Dassault Systèmes, Vélizy-Villacoublay, France) [30]. The model consists of discrete parts describing TRRPC facings, FC insulation and GFRP connectors. The thicker section along the lower edge of the inner TRRPC facing and the local strengthening at the position of the two upper anchors were excluded in the model.

Based on the structural behavior observed in associated studies combining experimental and numerical results on a component level, as reported elsewhere [22,24], the shear transfer through the FC layer is assumed to be negligible. However, the FC takes on an important function of ensuring the transfer of normal compressive stress between facings, which stabilizes and maintains the spacing between the two facings. Accordingly, specific interaction conditions between the various layers were prescribed to replicate this observed behavior; see Figure 9. Tie constraints were defined at the interface between the inner facing and FC, which assumes full interaction between these layers. In contrast, a frictionless contact condition was defined at the interface of the outer facing and FC.

The FC core was modelled using linear continuum shell elements with 8-nodes. FC was modelled based on linear elastic material laws. A density of 300 kg/m^3 was defined along with an experimentally yielded value for the modulus of elasticity (10 MPa) and assumed Poisson's ratio (0.1).

The TRRPC facings were modelled using the same type of shell elements as that applied for FC. The mechanical behavior of RPC was incorporated by means of the Concrete damaged plasticity model available in Abaqus with default field variables (dilation angle, eccentricity, etc.), refer to [30]. This continuum damage model for concrete is based on plasticity and adopts two failure mechanisms: tensile cracking and compressive crushing of concrete. A linear elastic model was applied to describe the compressive behavior, since the compressive stresses in the facings were presumed to be minimal. As for uniaxial tension, the stress-strain response is linear elastic until reaching failure. A tensile strength of 3 MPa was defined, which corresponds to the experimentally measured tensile strength of a textile reinforced RPC facing, see Section 3.1. Moreover, in tension, a linear softening behavior was defined for the phase after reaching the failure stress, assuming a fracture energy of 70 Nm/m^2.

Experimental data, presented in Table 1, was applied such that the modulus of elasticity in tension and compression corresponded to 50 GPa, and Poisson's ratio to 0.22.

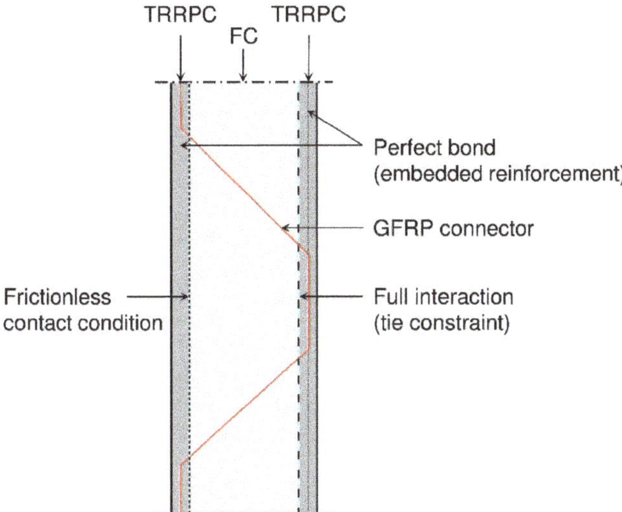

Figure 9. Schematic of the prescribed interaction conditions for the FE-model.

The carbon textile reinforcement grid was incorporated into the model as embedded reinforcement layers in the shell elements corresponding to the facings It is such that perfect bond between the reinforcement and the concrete is assumed. This interaction choice limits the crack formation from occurring within cracked regions according to the element size, as such individual localized cracking cannot be captured. The shell elements were however sized in accordance to observed crack distances of 40–50 mm, refer to [22]. The behavior of the reinforcement up to failure was modelled using linear elastic material models. Experimental values (refer to Section 3.1) were used for the nominal tensile strength (3433 MPa) and the modulus of elasticity (233 GPa). As a simplification, identical properties were assumed in the longitudinal and transversal directions of the textile grid. The cross-sectional area of each carbon grid layer was defined to be 85 mm^2/m.

Linear beam elements were used to model the GFRP connectors [30]. The connectors (nominal diameter of 6.1 mm), were attached to the center of the facings using tie constraints. On the conservative side, no interaction was defined at the connector-FC interface such that the connectors were free to deform, and a so-called initial connector imperfection was defined as 0.5 mm. Moreover, the GFRP bars were modelled according to linear elastic material behavior. The experimentally yielded material properties, i.e., modulus of elasticity (40.3 GPa) and nominal tensile strength (1012 MPa), applied in the model are listed in Table 2. Since Poisson's ratio was not tested, it was assumed to be 0.3 for the purpose of the analysis.

The Newton-Raphson iteration method was applied to find equilibrium within each load increment. Additionally, the feature named geometric nonlinear behavior was included in the analysis, i.e., second order theory related to large deformations. Accordingly, geometrical changes of, for instance the GFRP connectors, are included as a stiffening effect during the analysis (updating of stiffness matrix). Given this formulation, the GFRP bars can undergo large deformations in the model but the actual failure mode of the GFRP connectors is checked as a post-processing step with the experimentally yielded critical buckling load and pull-out capacity.

4.2.2. Wind Load Test Model

FE calculations were performed to simulate the wind load tests of the two sandwich elements, as aforementioned in Section 4.1.2. A schematic of the 3D models developed for the two test configurations are illustrated in Figure 10. The specimen was firstly subjected to the self-weight corresponding to the various components, followed by the wind load, applied as a distributed pressure on the outer surface of the facings. Subsequent steps alternated between positive wind suction and negative wind pressure according to the scheme used in the tests.

The lower anchor points were restrained in y- and z-directions, while the upper anchor points were only restrained in the z-direction; see Figure 10. However, the anchoring points were assumed free to move in the horizontal direction parallel to the element (x-direction).

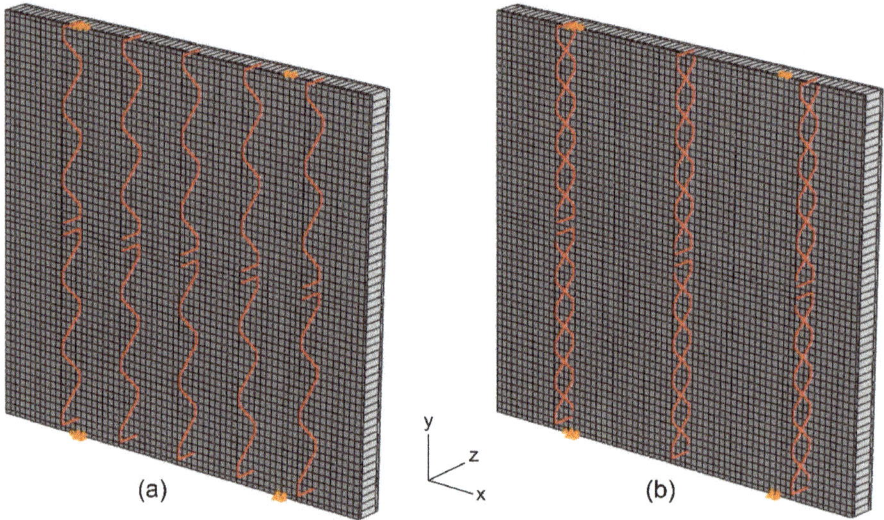

Figure 10. Schematic of FE model developed for the wind load testing with (**a**) single connector (S) configuration (**a**) and double connector (D) configuration (**b**).

4.2.3. Full-Size Sandwich Element Model

A concept building consisting of residential apartments was defined to calculate the loading schemes. The building is assumed to be situated on the west coast of Sweden. The building is prescribed dimensions of $20 \times 72 \times 12.4$ m (height × length × width). This scenario is limited to a typical element design, as illustrated in Figure 1, which consists of the materials and layer thicknesses, previously presented in Section 3. However, in this model, the TRRPC facings were limited to one layer of carbon grid reinforcement placed in the center of the facing, as this was found to be a sufficient amount of reinforcement for the applied design wind loads. Moreover, the study of the full-size sandwich element was limited to consider only the double connector configuration developed in this project.

The boundary conditions were the same as for the wind load test model described in Section 4.2.2, i.e., vertical (V) and horizontal (H_l) forces are transferred to the load bearing structure via the lower anchor points, while the horizontal (H_u) force is transferred by the upper anchor points.

The structural performance of the façade element was verified according to the limit state principle of EN 1990 [31]. Typically, two types of loads are included in normal design situations: vertical permanent loads from the facing's self-weight (G) and variable horizontal loads from wind (W); see Figure 1. The wind produces wind actions on the external (W_{ext}) and internal (W_{int}) surfaces according to EN 1991-1-4 [32]. Three load cases (LC) are thus investigated for the ultimate limit state (ULS) and the serviceability limit state (SLS). In *LC1*, the facing's external surface is under pressure, while the internal surface is under suction. In *LC2*, the external facing of the sandwich element is loaded in suction, while the internal layer is loaded in pressure. As for *LC3*, the external and internal facings are exposed to wind suction. The considered load combinations are stated in Equation (1) for SLS and in Equation (2) for ULS:

$$1.0G + 1.0 \, (W_{\text{ext}} + W_{\text{int}}) \tag{1}$$

$$1.35G + 1.5 \, (W_{\text{ext}} + W_{\text{int}}) \tag{2}$$

These load cases correspond to different wind directions to which the building could be exposed. The wind loads, given in Table 3, are calculated based on a concept building situated in Gothenburg (basic wind velocity $v_b = 25$ m/s) in terrain category IV, which is defined as an area in which at least 15% of the terrain surface is covered by buildings with an average height greater than 15 m. Furthermore, the presented numbers correspond to the most exposed parts of the building (worst case). The external and internal wind load acting on the doors and windows are assumed to act on the edges of the openings at the outer facing.

Table 3. Three wind load cases applied to sandwich element.

Load Case	SLS			ULS		
	W_{ext} [Pa]	W_{int} [Pa]	W_{SLS} [Pa]	W_{ext} [Pa]	W_{int} [Pa]	W_{ULS} [Pa]
LC1	514	192	706	771	288	1059
LC2	−771	−128	−899	−1157	−192	−1349
LC3	−771	192	−579	−1157	288	−869

In the developed model, the sandwich element was first loaded by the self-weight and the wind load actions up to the SLS. Thereafter, the additional self-weight and wind load actions corresponding to the ULS were applied. Finally, the wind load actions were increased further until failure of the element, if not reached at the ULS.

Verification at the ULS corresponds to the failure of the elements and related to human safety. For the sandwich element, this mainly concerns checking for connector failure, connector pull-out failure, textile grid failure and anchor failure. Concerning anchor failure, it should be verified that the horizontal reaction force, i.e., shear load at the anchor, at the different anchor positions and load combinations, are smaller or equal to the corresponding design shear load capacity according to Equation (3). The performance of the anchors was experimentally investigated according to that given in Section 4.1.1 and the design shear load capacities of the upper ($H_{\text{Rup,d}}$, $H_{\text{Run,d}}$) and lower ($H_{\text{Rlp,d}}$, $H_{\text{Rln,d}}$) anchors summarized in Table 4 in Section 5.1.

$$\text{If } H_{\text{Eu,d}} > 0 \text{ then } H_{\text{Eu,d}} \leq H_{\text{Rup,d}} \text{ otherwise } \left| H_{\text{Eu,d}} \right| \leq H_{\text{Run,d}}$$
$$\text{and} \tag{3}$$
$$\text{If } H_{\text{El,d}} > 0 \text{ then } H_{\text{El,d}} \leq H_{\text{Rlp,d}} \text{ otherwise } \left| H_{\text{El,d}} \right| \leq H_{\text{Rln,d}}$$

Verification at the SLS, representing a lower load level, usually relates to appearance, functioning and comfort of occupants, e.g., deflections and cracking. According to EN 1992-1-1 [33], the extent of cracking shall be limited in order to ensure the adequate functionality or durability of the structure, as well as to safeguard an aesthetically pleasing surface. The requirements of maximum crack width are normally only valid for steel reinforced concrete structures. When carbon textile reinforcement is used, corrosion is not a concerning issue because the grids are designed to be highly durable. By comparing between different codes given in fib bulletin 40 [34], the crack width limits are less restricted for FRP reinforced concrete. However, the knowledge is rather scarce, and it is stated that in absence of information the limitations for steel reinforced concrete could be adopted also for FRP reinforced concrete. For the lowest exposure classes given in [33], the crack width has no influence on durability and the given crack width limit of 0.4 mm is just set to guarantee acceptable appearance. However, crack widths can also be controlled to satisfy specific aesthetic requirements. As stated in ACI 533R-11 [35], the aesthetic effect of a crack in a facing is correlated to the surface's texture. For smooth surfaces, e.g., cast-in-place concrete, and coarse textured surfaces, e.g., exposed aggregate concrete, crack widths limited by structural demands are considered aesthetic. Concerning high quality smooth surfaces, it is recommended that cracking be limited to 0.13 mm for interior facings.

Table 4. Summary of anchor shear load capacity for the four load cases.

Test Case	A		B		C		D	
Anchor position Shear load direction Number of tests	upper positive 4		upper negative 4		lower positive 6		lower negative 6	
Average value	$H_{Rup,m}$ [kN]	12.4	$H_{Run,m}$ [kN]	12.1	$H_{Rlp,m}$ [kN]	8.7	$H_{Rln,m}$ [kN]	9.9
Standard deviation	σ [kN]	1.0	σ [kN]	2.5	σ [kN]	1.0	σ [kN]	1.5
Coefficient of variation	V_x [–]	0.08	V_x [–]	0.21	V_x [–]	0.12	V_x [–]	0.15
Characteristic fractile factor	k_n [–]	1.83	k_n [–]	1.83	k_n [–]	1.77	k_n [–]	1.77
Characteristic value	$H_{Rup,k}$ [kN]	10.5	$H_{Run,k}$ [kN]	7.5	$H_{Rlp,k}$ [kN]	6.9	$H_{Rln,k}$ [kN]	7.3
Design value	$H_{Rup,d}$ [kN]	7.0	$H_{Run,d}$ [kN]	5.0	$H_{Rlp,d}$ [kN]	4.6	$H_{Rln,d}$ [kN]	4.8

According to [33], the function or appearance of a member or structure should not be negatively impacted by deformation. In general, the limit of the design deflection is specified to either $L/250$ or $L/500$, where L is the effective span of the element. In [35], deflection limits are given specifically for non-load bearing precast wall elements, saying, deflection of any point on the facing measured from its original position should not exceed $L/480$. For the element in this study, the more restrictive limit according to [33] corresponds to a maximum deflection of 2800/500 = 5.6 mm.

5. Experimental Results of Element Tests

5.1. Anchor Shear Load Capacities

The anchor performance is studied based on the maximum shear load capacity yielded at failure. For the four different load cases mentioned in Section 4.1.1, the average shear load capacity is provided in Table 4, along with characteristic and design values. The design values of the shear load capacity $H_{R,d}$ were evaluated according to Equation (4) as per EN 1990 [31]:

$$H_{R,d} = \frac{H_{R,k}}{\gamma_M} = \frac{H_{R,m}}{\gamma_M}(1 - k_n V_x)$$ (4)

where $H_{R,k}$ and $H_{R,m}$ are the characteristic and average values of the shear load capacity, respectively, and γ_M is the partial factor for material properties. The partial factor for the RPC is assumed to be equal to that of concrete. Tensile failure in RPC was the governing failure mode observed during the anchor shear load tests, as shown in Figure 11. According to SIS-CEN 1992-4-1 [36], the partial factor for concrete related to tensile failure modes under shear loading of headed anchors can be set to $\gamma_M = 1.5$. Moreover, given that tensile failure is governing, the coefficient of variation V_x can be assumed to be known in the selection of the characteristic fractile factor k_n, based on the knowledge of the coefficient of variation related to RPC's tensile strength. The specified design values are only valid for this specific anchor design, and should be treated as indicative only.

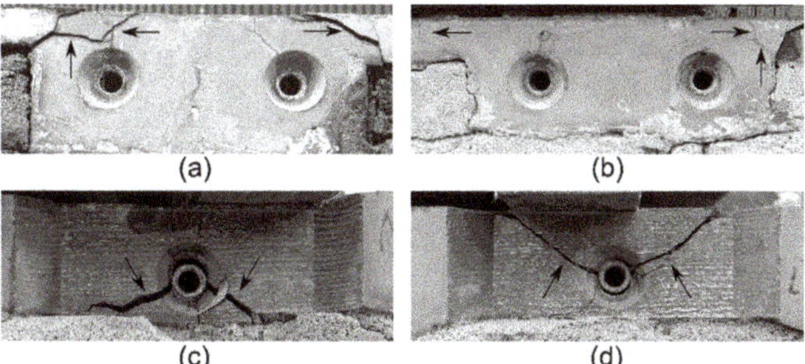

Figure 11. Observed failure modes after shear load testing for cases (**a–d**).

5.2. Wind Load Test Results

The element performance is analyzed according to the resulting mid-span deflection of the element in correlation with the applied wind load on the outer facing (i.e., facing interior of chamber). Wind suction and pressure at the facings are represented by positive and negative values, respectively. In Figure 12, the global behavior of the two tested elements, namely single (S) and double (D) connector configurations, is shown as the wind load versus mid-span deflection (at locations 1–3 in Figure 8). The deflections, $d1$–$d3$, were adjusted by deducting the average displacement at the position of the anchors, i.e., global displacements of the element with respect to the test rig. It should be noted that during the last wind load cycle, it was only possible to reach a pressure of approximately −1.9 kPa due to small air leakage from the test chamber. It can be observed that both tested elements performed similarly under cyclic loading. Only minor differences in element deflections at the position of the three displacement transducers were noted, which confirms that the elements mainly bend about the x-axis, as per Figure 10. Both elements exhibit a somewhat larger mid-span deflection during wind suction compared to wind pressure, with a maximum deflection of approximately 4 mm at maximum wind suction (2 kPa) which is less than $L/500$.

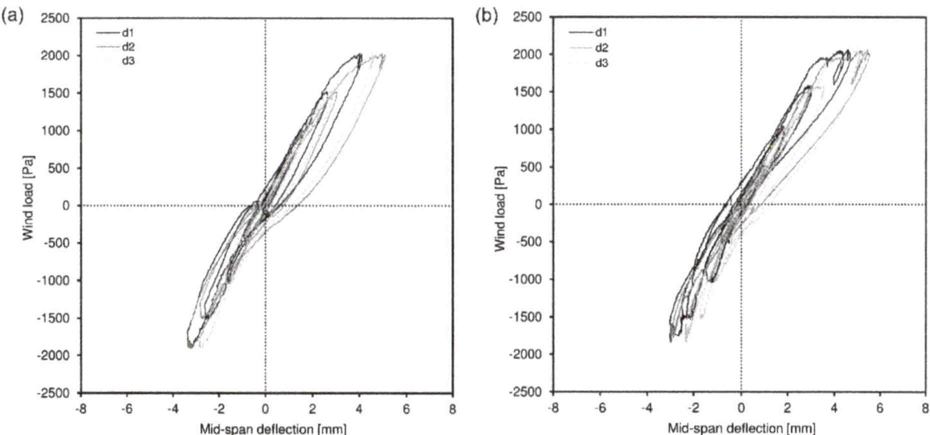

Figure 12. Wind load versus mid-span deflection ($d1$–$d3$) for single (S) connectors (**a**) and for double (D) connectors (**b**). The deflection values are adjusted with respect to displacements at the anchors.

Given that the total wind load is distributed evenly between the four anchors, it can be assumed that there is a linear relation between the horizontal reaction force at each of the anchors and the applied wind load. Accordingly, this amounts to a maximum shear load at the anchors during testing of approximately 4 kN at maximum wind suction and wind pressure. This shear load is well below the average measured capacities of the anchors reported in Section 5.1. Furthermore, after testing, some of the smaller pre-existing shrinkage cracks in the facings had propagated slightly, but only a few new minor cracks were detected. No cracks or damages were noted in the regions around the anchorages.

6. Numerical Results

FE calculations of the two wind load test configurations were performed to validate the model of the sandwich element configurations, refer to Section 4.2.2. The validated models were then used to analyze full-size sandwich elements with and without openings, refer to Section 4.2.3. It is worth noting, that the model pertaining to the wind loading is intended to capture the overall structural behavior of the element. As mentioned in Section 2, it is of key importance to validate that the developed model of the sandwich element can effectively describe the most important phenomena and failure modes governing the overall behavior of the elements.

6.1. Validation of Wind Load Test Model

Comparisons of the global behavior of the two sandwich elements, represented as wind load versus mid-span deflection at locations $d1$–$d3$, are shown in Figure 13 for both single and double connector configurations. It should be noted that the deflection related to FEA was measured in the middle of the inner facing. Compared to the experimental results, the stiffness during the loading sequences is captured rather well. The hysteresis effects in the unloading sequences are not fully captured in the model. The incremental damage in the facings due to cracking was included. However, in the experiments, a major part of the hysteresis effects can most likely be attributed to unforeseen movements in the anchoring positions, which was excluded in the developed FE model. Another factor which could influence the numerical results, is the fact that linear elastic material models were assigned for all materials, except for the RPC facings. As such, these materials recover perfectly after unloading in the model, which is not the case in the experiments. It is also important to note that pre-cracks existed in the facings which could have also likely influenced the presented experimental behavior. Despite these discrepancies, the outcome of the analysis is deemed suitable since the aim of the model was primarily to simulate the behavior under static wind loading (load increments).

Figure 14 depicts contour plots at a wind pressure of 2.0 kPa of the out-of-plane displacement for both single and double connector cases. From these figures, it can be seen that the elements mainly bend about the x-axis, as per Figure 10, which confirms the experimentally observed behavior.

At a wind pressure of 2.0 kPa, corresponding to the maximum wind pressure of the climate chamber, the compressive forces in the most stressed diagonal connectors were checked for both single and double connector cases. For both cases, it was observed that the maximum compressive force was −1.8 kN at this wind pressure. This compressive force very close to the experimentally yielded critical buckling load (1.7 kN), see Table 2. At a wind suction of 2.0 kPa, the maximum tensile force in the most stressed diagonal was 3.0 kN and 2.8 kN for single and double connector configurations, respectively. These loads are well below the average value of the connector pull-out resistance of 6.5 kN, reported in Table 2.

Contour plots of the maximum principle plastic strain, shown in Figure 15, indicate cracked regions of the facings around the attachment of the connectors. The cracks were found to be larger at the connector attachments at the location of the upper and lower anchors. For the element with single connectors, as per Figure 15a, cracks also propagate from the attachment of the connectors towards the vertical edges of the facing. It should be noted that all cracks can be defined as small if the strain values are translated into crack widths. Moreover, the tensile stresses in the carbon grid were found to be minimal at 2.0 kPa (for both pressure and suction), which is below the prescribed nominal tensile strength of the reinforcement. For this reason, subsequent analyses only incorporated one layer of carbon grid in each RPC facing.

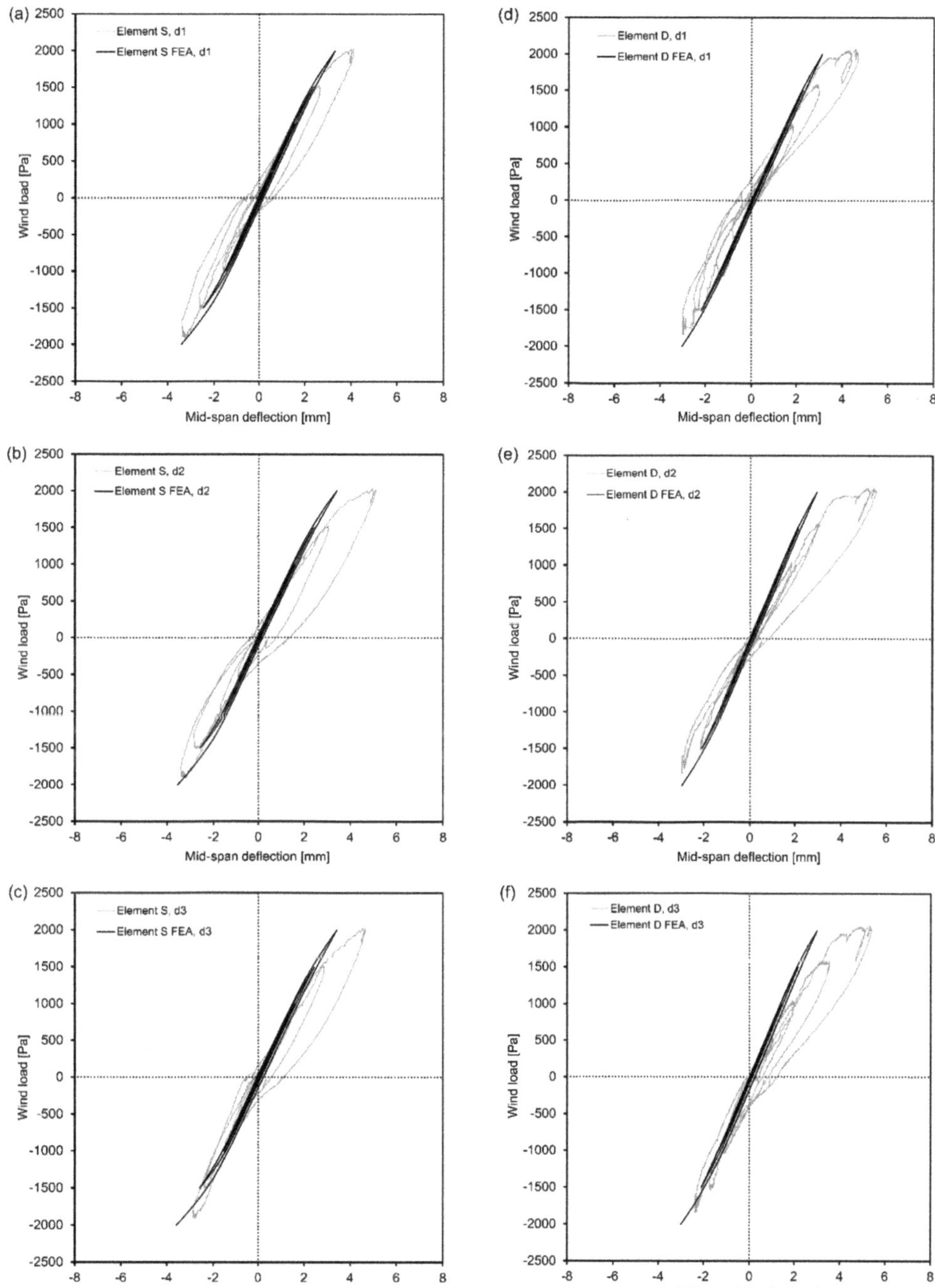

Figure 13. Comparison between experimental and finite element analysis (FEA) global behavior for the two sandwich element configurations with single (S) connectors (**a–c**) and double (D) connectors (**d–f**). Data pertaining to deflection locations *d*1–*d*3 are indicated in the figures.

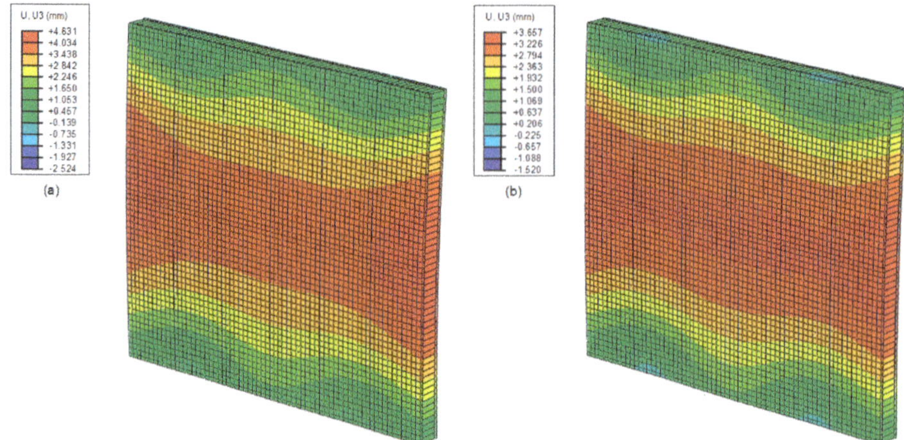

Figure 14. Contour plot of the out-of-plane displacement at a wind pressure of 2.0 kPa for element with single (S) connectors (**a**) and element with double (D) connectors (**b**).

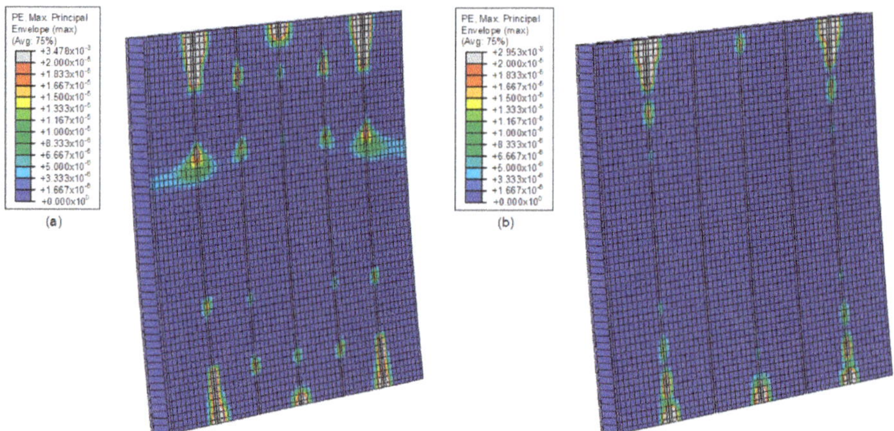

Figure 15. Contour plot of cracked regions, represented as maximum principle plastic strain, of the facing at a wind pressure of 2.0 kPa for element with single (S) connectors (**a**) and element with double (D) connectors (**b**).

6.2. Performance of Full-Size Sandwich Element

Analyses were performed on full sandwich elements having three different spacings between the connectors. The outer connector lines were placed 100 mm from the vertical edges and one connector line was placed at the position of each anchor line in all cases. Otherwise, the connector spacing for the three spacing options was set to 0.5, 1.0 and 2.0 m. All options were analyzed for the three load cases previously defined in Table 3.

The FEA results are summarized for the three different connector spacing options in Table 5. At the SLS, the maximum displacement, u_{max}, of the inner facing and the indication of cracking in the facings at W_{SLS} are given. At the ULS, the maximum horizontal reaction force at the upper anchors, $H_{El,d}$, and lower anchors, $H_{Eu,d}$, together with the maximum pull-out force, $F_{Epo,d}$, in the connectors are given at the design wind load W_{ULS}. Furthermore, the maximum wind load, W_{max}, is given, defined as the wind load when the first limiting design criteria was reached. The ratio W_{max}/W_{ULS} indicates how much wind load above ULS that the sandwich element can withstand before reaching failure, i.e., design criteria. The capacity of the anchors was the limiting factor in all cases, except in *LC2* and *LC3* for the configuration with the largest connector spacing (2.0 m), wherein the pull-out capacity of the connector was limiting.

Table 5. Summary of results for the different connector distance configurations.

Connector Spacing [m]	Load Case	SLS		ULS				Failure		
		U_{max} [mm]	Cracks	$H_{El,d}$ [kN]	$H_{Eu,d}$ [kN]	$F_{Epo,d}$ [kN]	W_{max} [Pa]	W_{max}/W_{ULS} [–]	Failure Mode	
	LC1	0.8	Minor	−3.5	−2.6	1.0	1420	1.3	Lower anchor	
0.5	*LC2*	−1.1	Minor	3.5	4.3	1.8	−1590	1.2	Upper anchor	
	LC3	−0.8	Minor	2.1	2.9	1.4	−1590	1.8	Upper anchor	
	LC1	1.1	Minor	−3.5	−2.7	1.5	1460	1.4	Lower anchor	
1.0	*LC2*	−1.6	Minor	3.5	4.3	2.7	−1610	1.2	Upper anchor	
	LC3	−1.0	Minor	2.1	2.9	1.9	−1620	1.9	Upper anchor	
	LC1	1.5	Minor	−3.4	−2.7	1.9	1470	1.4	Lower anchor	
2.0	*LC2*	−2.0	Minor	3.5	4.2	3.3	−1550	1.1	Conn. pull-out	
	LC3	−1.2	Minor	2.1	2.8	2.4	−1450	1.7	Conn. pull-out	

LC2 was found to be the worst load case for all three spacing options at the SLS with respect to maximum displacement, and at the ULS with respect to maximum possible design wind load. Put simply, the maximum wind load resistance implies that the concept building may also be situated in terrain category III or have a total height of approximately 30 m in terrain category IV.

As can be noted, the deformations are rather small at the SLS, but the sandwich elements behave slightly different depending on the connector spacing. In the element with the smallest connector spacing (0.5 m), both facings work together and have nearly the same deformed shape; see Figure 16a,b. However, in the element with the largest spacing (2.0 m) between the connectors, the facings work more independently, thereby making the deformations more related to local bending of the individual facings; see Figure 16c,d. Consequently, the behavior of the element with 1.0 m connector spacing is somewhere in between these two options. The outer facing separates from the FC at some positions because of wind suction. This effect is greatest for the facing with the largest connector spacing. However, from the analyses it can be concluded that the maximum separation between the outer facing and the FC is less than 0.6 mm at the SLS. At the SLS, only smaller cracked regions appear around the connector attachments at the locations of anchors similar to that observed in the analyses of the wind load tests. Accordingly, these regions increase in size at the ULS. For the element with a connector spacing of 2.0 m, vertical cracks in the two outer spans appear above the ULS load in the outer facing due to local bending.

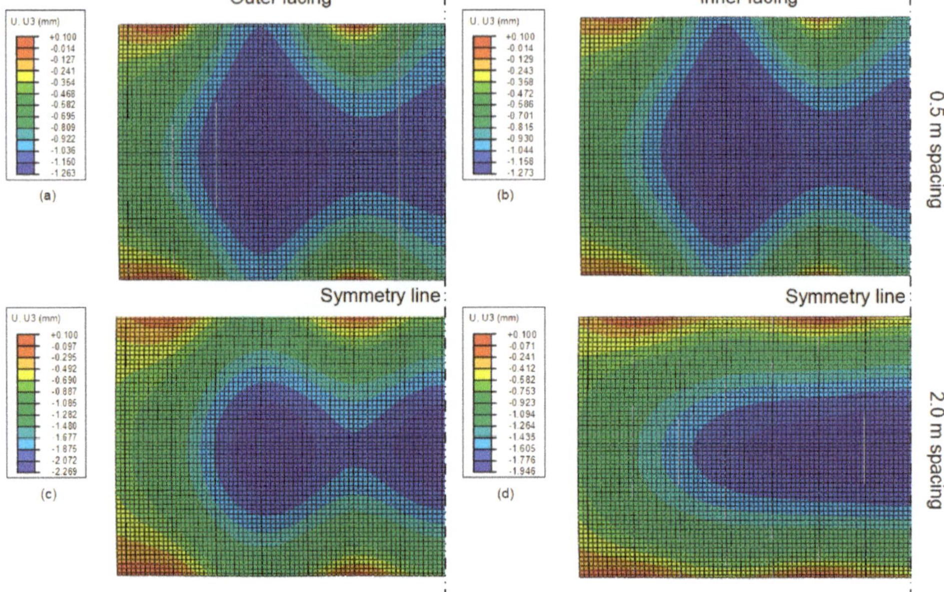

Figure 16. Out-of-plane deformation plots for *LC2* at SLS for outer facing (**a**) and inner facing (**b**) for element with 0.5 m connector spacing, and outer facing (**c**) and inner facing (**d**) for element with 2.0 m connector spacing. Only half of the element is shown due to symmetry.

6.3. Performance of Full-Size Sandwich Element with Openings

The concept described for the full-size element, as per Figure 1, was applied in this analysis. More specifically, the FE-model presented in Section 6.2 was modified to contain window and door openings. To analyze the appropriate placement of the connectors in this given element configuration, three cases were considered, namely Case I-III. The first case (Case I), i.e., reference case, consists of the connectors being placed at the outer vertical edges and at the positions of the anchors as was the case in full-sandwich element with a 2.0 m connector spacing. Furthermore, the analysis consisted of solely applying *LC2*, defined in Table 3, as this was found to be the most critical load case at both the SLS and the ULS. In *LC2*, the external facing of the sandwich element is under suction and the internal facing is under pressure.

The deformed shape together with the out of plane displacement and cracks, represented as maximum principle plastic strain, are shown for the SLS in Figure 17. As can be seen, the displacements of the element are rather small, with a maximum value of approximately 3 mm. However, the displacements due to local bending are rather pronounced at the openings, especially above and below the door opening. This also leads to severe cracking of the facing in these regions. Cracks were also observed around the connector attachments at the location of the upper and lower anchors and smaller cracks at the corners of the openings.

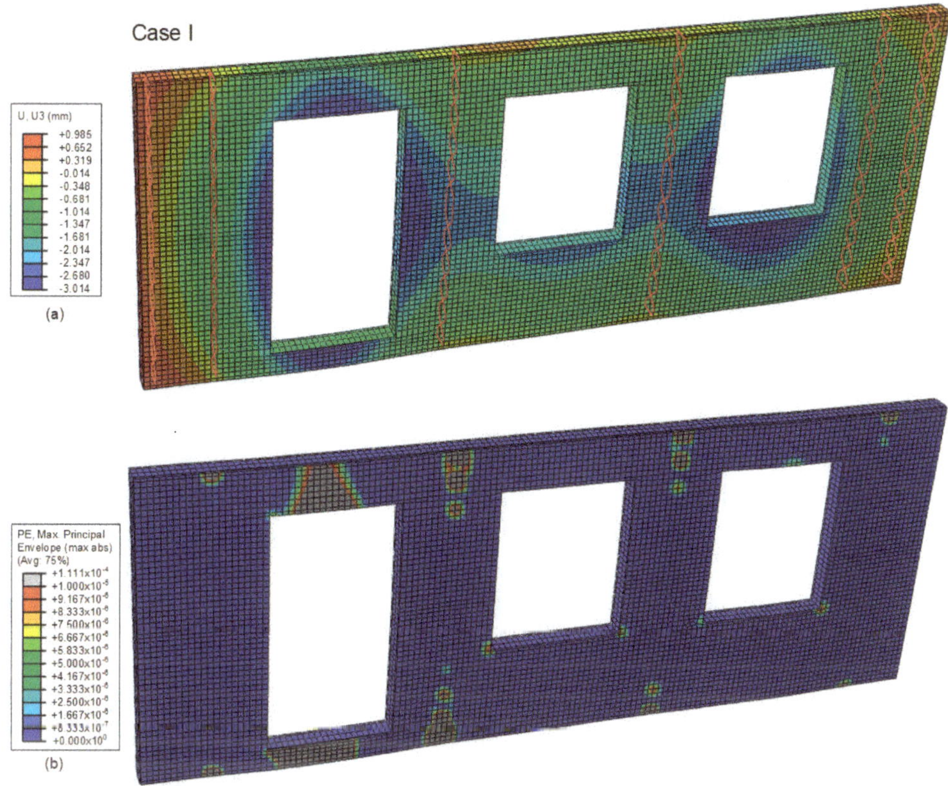

Figure 17. Case I: Contour plots of out-of-plane displacement (connectors in red) (**a**) and cracked regions at the SLS (**b**).

In the second case (Case II), additional connectors were placed between the outer and inner facings, horizontally along the upper and lower edges of the element. This stiffens the entire element and consequently reduces the displacements of the element, especially at the door opening; see Figure 18a. Furthermore, the amount of cracking was also drastically reduced, as depicted in Figure 18b. Nevertheless, the local bending of the outer facing was quite noticeable around the openings. To overcome this phenomenon, additional vertical connectors were placed on each side of the openings, together with one additional short connector below each window opening in the third case (Case III). These extra connectors reduced both the element's global bending about the x-axis, as per Figure 10, and the local affects around the openings, such that there were reduced displacements and cracking; see Figure 18c. The maximum displacement at the SLS was reduced to slightly above 1 mm and first cracking in the outer facing took place at load levels above the SLS. Accordingly, the cracking at the ULS is shown for Case III in Figure 18d.

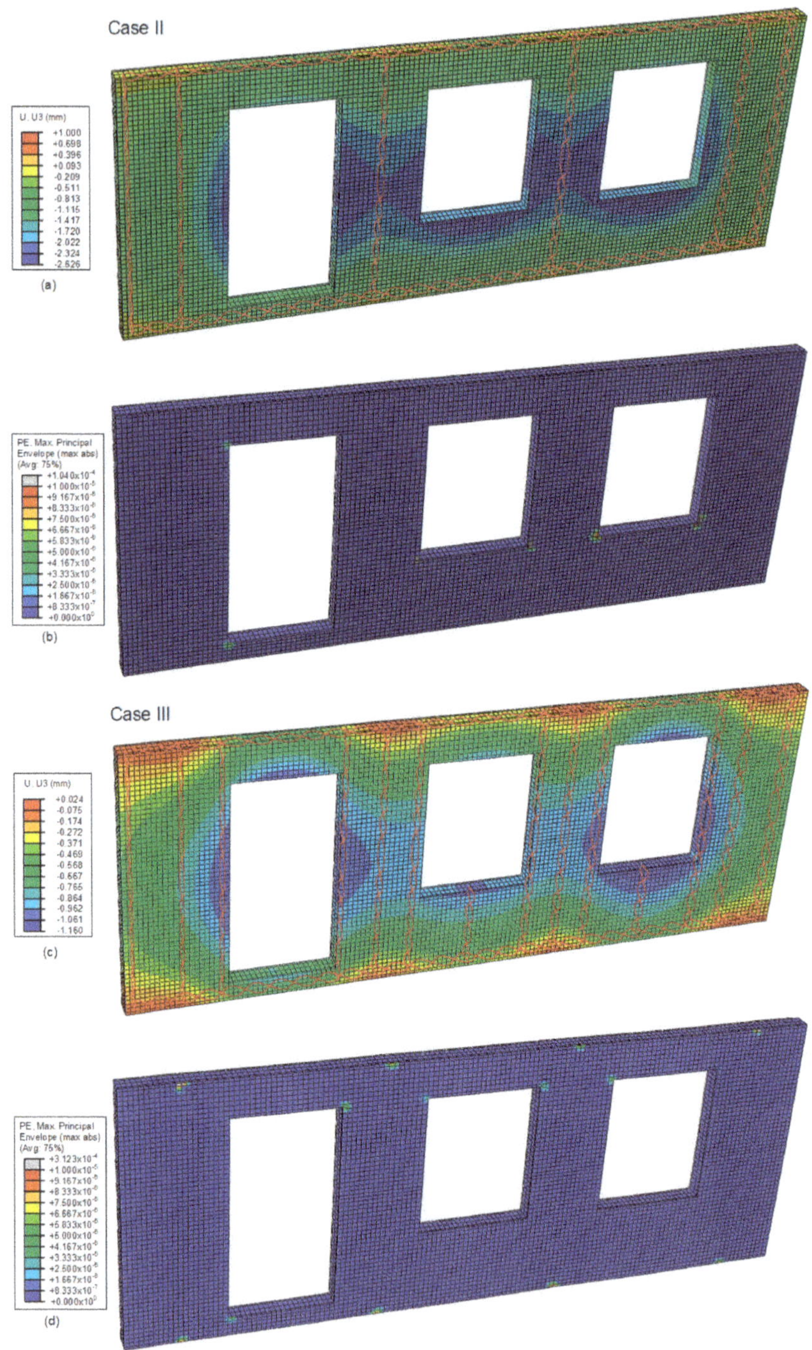

Figure 18. Case II: contour plots of out-of-plane displacement (**a**) and cracked regions at the SLS (**b**); and Case III: contour plots of out-of-plane displacement at the SLS (**c**) and cracked regions at the ULS (**d**).

A summary of the FEA results pertaining to the sandwich element with openings with different connector placement cases (Cases I-II) is provided in Table 6. For Cases I and II, the pull-out capacity of the connectors resulted in the limiting factor for the maximum wind load resistance, while the capacity of the anchors was governing for Case III.

Table 6. Summary of results for the different connector configurations subjected to LC2.

Case	Load Case	SLS		ULS			Failure		
		u_{max} [mm]	Cracks	$H_{El,d}$ [kN]	$H_{Eu,d}$ [kN]	$F_{EDO,d}$ [kN]	W_{max} [Pa]	W_{max}/W_{ULS} [−]	Failure Mode
I	LC2	−3.0	Major	3.8	4.3	3.7	−1350	1.0	Con. Pull-out
II	LC2	−2.6	Minor	3.7	4.1	3.3	−1530	1.1	Con. Pull-out
III	LC2	−1.2	Minor	3.6	4.1	1.6	−1680	1.2	Upper anchor

7. Discussion

Within the scope of the SESBE project, the structural performance of a developed TRRPC sandwich façade element was verified. Based on preliminary structural investigations in the conceptual phase, a thorough experimental and modelling program was established. Experiments were conducted on individual materials, composites and components. The experimentally yielded material properties were used as input for the FE models, and the model was validated by its ability to reproduce the findings in component tests. The modelling and testing have been performed in an iterative process, in parallel with the development and characterization of the materials.

This paper presents an overview of the sandwich element concept, together with a description of the different incorporated materials and components, e.g., TRRPC facings, FC insulation core, GFRP connectors and anchoring details. Moreover, the structural model of the element was validated via experimental data from wind load testing. Therefore, the overall behavior of the sandwich element could be modelled in a realistic way while being subjected to wind loads. The resulting deformations and cracking were also found to be within acceptable limits. Hence, within the project, it has been proven through experimental data validation at different investigational levels, i.e., composite and component, that the chosen modelling concept can describe the most important phenomena and failure modes governing the overall behavior of the TRRPC sandwich element. Local failure modes that are not directly captured by the structural model, such as connector pull out and anchorage failure, are accounted for by design criteria determined by tests. The validated model was expanded to a conceptual full-size sandwich element with and without openings to enable further prediction and analysis of its structural performance according to a design scenario, as well as the SLS and ULS requirements.

The sandwich element's composite action is mainly dependent on the mechanical properties of the GFRP connectors, i.e., strength and stiffness. The investigated connector solutions, single and double, were deemed to have sufficient load resistance for the studied load cases. Concerning the elements with a double connector configuration, the deflections were observed to be smaller as a result of superior composite action. The double connectors also present the advantage of being able to carry both wind suction and pressure. As demonstrated in the wind load tests and numerical modeling, composite action can be further improved by simply minimizing the spacing between the connectors. However, it was found in both practice and modelling that window or door openings in an element limit the ability to use tight connector spacing. As an alternative, a combination of single and double connectors can be applied in one element in accordance to the given design load.

For the sandwich element developed in this project, the FC core was shown to have an insignificant role concerning shear transfer between the TRRPC facings. Normal compressive stresses can however be transferred via the core, which in turn ensures a set distance between facings. The FC core also stabilizes the connector diagonals in compression to some extent, but since the level of restraint is uncertain, the connectors were assumed to be free to deform in the FE model. However, there is obvious potential in the further development of the mechanical performance of the FC core to increase its contribution to the composite action.

Overall, since the cracking in the facings has been shown to be minimal for the relevant load levels, it is acceptable to use only one layer of carbon grid reinforcement from a mechanical point of view. By doing so, the amount of reinforcement grid would be reduced, and the related physical labor simplified. Alternatively, additional reinforcement could be placed in specific regions where larger cracking is expected, e.g., around openings.

Author Contributions: Conceptualization and methodology, M.F. and N.W.P.; formal analysis and validation, D.V. and M.F.; Writing—Review and Editing, N.W.P., M.F. and D.V.

Funding: The SESBE (Smart Elements for Sustainable Building Envelopes) project was funded within the Framework Programme 7 under the Grant Agreement No. 608950. The authors would like to thank the European Commission for funding the project and making this work possible.

Acknowledgments: The GFRP connectors applied in this project were produced by Mostostal Warszawa S.A. and funding acquisition by Mueller U.

Conflicts of Interest: The authors of this paper declare no conflict of interest.

References

1. Stenberg, E. *Structural Systems of the Million Program Era*; KTH School of Architecture: Stockholm, Sweden, 2013.
2. EN 206-1. *Concrete-Specification, Performance, Production and Conformity*; European Standard: Brussels, Belgium, 2013.
3. Engberts, E. Large-size facade Elements of textile reinforced concrete. In *ICTRC'2006-1st International RILEM Conference on Textile Reinforced Concrete*; Hegger, J., Brameshuber, W., Will, N., Eds.; RILEM Publications SARL: Paris, France, 2006.
4. Hegger, J.; Horstmann, M.; Scholzen, A. Sandwich Panels with Thin-Walled Textile-Reinforced Concrete Facings. *Spec. Publ.* **2008**, *251*, 109–124.
5. Colombo, I.G.; Colombo, M.; di Prisco, M. Bending behaviour of Textile Reinforced Concrete sandwich beams. *Constr. Build. Mater.* **2015**, *95*, 675–685. [CrossRef]
6. Shams, A.; Stark, A.; Hoogen, F.; Hegger, J.; Schneider, H. Innovative sandwich structures made of high performance concrete and foamed polyurethane. *Compos. Struct.* **2015**, *121*, 271–279. [CrossRef]
7. Miccoli, L.; Fontana, P.; Silva, N.; Klinge, A.; Cederqvist, C.; Kreft, O.; Sjöström, C. Composite UHPC-AAC/CLC facade elements with modified interior plaster for new buildings and refurbishment. *J. Facade Des. Eng.* **2015**, *3*, 91–102.
8. Ghoneim, G.; El-Hacha, R.; Carson, G.; Zakariasen, D. Precast Ultra High Perfromance Fibre Reinforced Concrete Replaces Stone And Granite On Building Façade. In Proceedings of the 3rd International Fib Congress incorporating the PCI Annual Convention and Bridge Conference, Washington, DC, USA, 29 May–2 June 2010; pp. 1–15.
9. Rebentrost, M.; Wight, G.; Fehling, E. Experience and applications of ultra-high performance concrete in Asia. In Proceedings of the 2nd International Symposium on Ultra High Performance Concrete, Kassel, Germany, 5–7 March 2008; pp. 19–30.
10. Papanicolaou, C. Applications of textile-reinforced concrete in the precast industry. In *Textile Fibre Composites in Civil Engineering*; Thanasis, T., Ed.; Elsevier: Amsterdam, The Netherlands, 2016; pp. 227–244.
11. Mueller, U.; Williams Portal, N.; Chozas, V.; Flansbjer, M.; Larazza, I.; da Silva, N.; Malaga, K. Reactive powder concrete for facade elements—A sustainable approach. *J. Facade Des. Eng.* **2016**, *4*, 53–66. [CrossRef]
12. Hegger, J.; Kulas, C.; Horstmann, M. Spatial textile reinforcement structures for ventilated and sandwich façade elements. *Adv. Struct. Eng.* **2012**, *15*, 665–675. [CrossRef]
13. Salmon, D.C.; Einea, A.; Tadros, M.K.; Culp, T.D. Full scale testing of precast concrete sandwich panels. *ACI Struct. J.* **1997**, *94*, 239–247.
14. Lameiras, R.; Barros, J.; Valente, I.B.; Azenha, M. Development of sandwich panels combining fibre reinforced concrete layers and fibre reinforced polymer connectors. Part I: Conception and pull-out tests. *Compos. Struct.* **2013**, *105*, 446–459. [CrossRef]

15. Lameiras, R.; Barros, J.; Azenha, M.; Valente, I.B. Development of sandwich panels combining fibre reinforced concrete layers and fibre reinforced polymer connectors. Part II: Evaluation of mechanical behaviour. *Compos. Struct.* **2013**, *105*, 460–470. [CrossRef]
16. Assaad, J.; Chakar, E.; Zéhil, G.-P. Testing and modeling the behavior of sandwich lightweight panels against wind and seismic loads. *Eng. Struct.* **2018**, *175*, 457–466. [CrossRef]
17. Cuypers, H.; Wastiels, J. Analysis and verification of the performance of sandwich panels with textile reinforced concrete faces. *J. Sandw. Struct. Mater.* **2011**, *13*, 589–603. [CrossRef]
18. De Munck, M.; Vervloet, J.; Kadi, M.E.; Verbruggen, S.; Wastiels, J.; Remy, O.; Tysmans, T. Modelling and experimental verification of flexural behaviour of textile reinforced cementitious composite sandwich renovation panels. In Proceedings of the 12th Fib International PhD Symposium in Civil Engineering, Prague, Czech Republic, 29–31 August 2018; pp. 179–186.
19. Djamai, Z.I.; Bahrar, M.; Salvatore, F.; Si Larbi, A.; El Mankibi, M. Textile reinforced concrete multiscale mechanical modelling: Application to TRC sandwich panels. *Finite Elem. Anal. Des.* **2017**, *135*, 22–35. [CrossRef]
20. Mueller, U.; Williams Portal, N.; Flansbjer, M.; Malaga, K. Textile Reinforced Reactive Powder Concrete and its Application for Facades. In Proceedings of the Eleventh High Performance Concrete (11th HPC) and the Second Concrete Innovation Conference (2nd CIC), Tromsø, Norway, 6–8 March 2017.
21. Williams Portal, N.; Flansbjer, M.; Mueller, U. Experimental Study on Anchorage in Textile Reinforced Reactive Powder Concrete. *Nord. Concr. Res.* **2017**, *2*, 33.
22. Flansbjer, M.; Williams Portal, N.; Vennetti, D.; Mueller, U. Composite Behaviour of Textile Reinforced Reactive Powder Concrete Sandwich Façade Elements. *Int. J. Concr. Struct. Mater.* **2018**, *12*, 71. [CrossRef]
23. Flansbjer, M.; Honfi, D.; Mueller, U.; Wlasak, L.; Williams Portal, N.; Edgar, J.-O.; Larraza, I. Structural behaviour of RPC sandwich facade elements with GFRP connectors. In Proceedings of the 7th International Congress on Architectural Envelopes, San Sebastian-Donostia, Spain, 27–29 May 2015.
24. Flansbjer, M.; Honfi, D.; Vennetti, D.; Mueller, U.; Williams Portal, N.; Wlasak, L. Structural performance of GFRP connectors in composite sandwich façade elements. *J. Facade Des. Eng.* **2016**, *4*, 35–52. [CrossRef]
25. Silva, N.; Mueller, U.; Malaga, K.; Hallingberg, P.; Cederqvist, C. Foam concrete-aerogel composite for thermal insulation in lightweight sandwich facade elements. In Proceedings of the 27th Biennial National Conference of the Concrete Institute of Australia in Conjunction with the 69th RILEM Week, Melbourne, Australia, 30 August–2 September 2018.
26. Williams Portal, N.; Flansbjer, M.; Johanesson, P.; Malaga, K.; Lundgren, K. Tensile behaviour of textile reinforcement under accelerated ageing conditions. *J. Build. Eng.* **2015**, *5*, 57–66. [CrossRef]
27. RILEM TC 232-TDT. *Recommendation of RILEM TC 232-TDT: Test Methods and Design of Textile Reinforced Concrete-Uniaxial Tensile Test: Test Method to Determine the Load Bearing Behavior of Tensile Specimens Made of Textile Reinforced Concrete*; RILEM TC 232-TDT: Paris, France, 2016; pp. 4923–4927.
28. Van Deijk, S. Foam concrete. *Concrete* **1991**, *25*, 49–54.
29. ISO 10406-1. *Fibre-Reinforced Polymer (FRP) Reinforcement of Concrete-Test Methods. Part 1: FRP Bars and Grids*; International Organization for Standardization: Geneva, Switzerland, 2008.
30. Dassault Systèmes Abaqus/CAE User's Guide. *ABAQUS Version 6.14*; Groupe Dassault: Paris, France, 2014.
31. EN 1990. *Eurocode-Basis of Structural Design European Standard*; EN 1990: Brussels, Belgium, 2005.
32. EN 1991-1-4. *Eurocode 1: Actions on Structures-Part 1-4: General Actions-Wind Actions European Standard*; EN 1991-1-4: Brussels, Belgium, 2005.
33. EN 1992-1-1. *Eurocode 2: Design of Concrete Structures-Part 1-1: General Rules and Rules for Buildings. European Standard*; EN 1992-1-1: Brussels, Belgium, 2014.
34. Fib. Fib Bulletin No. 40-FRP Reinforcement in RC Structures. 2007. Available online: https://www.fib-international.org/publications/fib-bulletins/frp-reinforcement-in-rc-structures-115-detail.html (accessed on 3 December 2018).

35. ACI Commitee 533. 533R-11 Guide for Precast Concrete Wall Panels. 2012. Available online: https://www.concrete.org/publications/internationalconcreteabstractsportal/m/details/id/51683674 (accessed on 3 December 2018).
36. SIS-CEN/TS-1992-4-1. *Design of Fastenings for Use in Concrete-Part 4-1: General European Standard*; SIS-CEN/TS-1992-4-1: Brussels, Belgium, 2010.

Article

Validation of a Numerical Bending Model for Sandwich Beams with Textile-Reinforced Cement Faces by Means of Digital Image Correlation

Jolien Vervloet [1,*], Tine Tysmans [1], Michael El Kadi [1], Matthias De Munck [1], Panagiotis Kapsalis [1], Petra Van Itterbeeck [2], Jan Wastiels [1] and Danny Van Hemelrijck [1]

[1] Department Mechanics of Materials and Constructions, Vrije Universiteit Brussel (VUB), Pleinlaan 2, 1050 Brussels, Belgium; tine.tysmans@vub.be (T.T.); michael.el.kadi@vub.be (M.E.K.); matthias.de.munck@vub.be (M.D.M.); panagiotis.kapsalis@vub.be (P.K.); jan.wastiels@vub.be (J.W.); danny.van.hemelrijck@vub.be (D.V.H.)

[2] Department of Structures, Belgian Building Research Institute (BBRI), Avenue P. Holoffe 21, 1342 Limelette, Belgium; petra.van.itterbeeck@bbri.be

[*] Correspondence: jolien.vervloet@vub.be; Tel.: +32-(0)2-629-29-24

Received: 31 January 2019; Accepted: 19 March 2019; Published: 25 March 2019

Abstract: Sandwich panels with textile-reinforced cement (TRC) faces merge both structural and insulating performance into one lightweight construction element. To design with sandwich panels, predictive numerical models need to be thoroughly validated, in order to use them with high confidence and reliability. Numerical bending models established in literature have been validated by means of local displacement measurements, but are missing a full surface strain validation. Therefore, four-point bending tests monitored by a digital image correlation system were compared with a numerical bending model, leading to a thorough validation of that numerical model. Monitoring with a digital image correlation (DIC) system gave a highly detailed image of behaviour during bending and the strains in the different materials of the sandwich panel. The measured strains validated the numerical model predictions of, amongst others, the multiple cracking of the TRC tensile face and the shear deformation of the core.

Keywords: finite element model; real scale bending experiments; shear; structural insulating sandwich panel

---

1. Introduction

Structural insulating sandwich panels combine a lightweight insulating core with two thin stiff faces, hence they can improve the energy efficiency of the building and provide the necessary structural capacity. Due to this dual function, these panels are gaining more interest from the building industry, as they are very suitable for nearly zero-energy buildings and contribute to reach the energy efficiency objectives of the European Union.

Nowadays, pre-cast concrete sandwich panels are frequently used for walls in residential and commercial buildings, since their energy efficiency and structural capacity are well-known [1–4]. The weight of these concrete sandwich panels can be drastically reduced by replacing the steel-reinforced concrete faces by textile-reinforced cement (TRC) faces. Due to the use of textiles instead of steel, the thick concrete cover (needed for durability reasons in case of steel) can be avoided. This reduces the thickness of the faces, and therefore the weight as well.

The research groups of Hegger et al., Colombo et al., and Cuypers et al. investigated the behaviour of sandwich panels with TRC faces by bending experiments [5–7]. Hegger et al. also added connectors between the two faces to enhance the composite action of the panel [8]. The behaviour of TRC sandwich panels in compression, as well as their durability, has been recently explored by Tysmans et al. [9–11]. These studies represent the first step towards the application of TRC sandwich panels in residential, public and industrial buildings, as cladding or wall panels [12,13].

In order to accurately and safely design TRC sandwich panels for their application, the prediction of their behaviour under different loading conditions is indispensable. A few analytical models have been already established in [14–16], while numerical models can be found in [17–19]. The established numerical models were validated by experiments measuring the force-displacement behaviour or local strains of the sandwich panels. Accurate full-field strains of the bending behaviour of TRC sandwich panels to validate the existing models are, however, still missing in the current state of the art. Therefore, four-point bending tests monitored by digital image correlation (DIC) were performed in the scope of this paper, and were compared to results of a numerical model.

This paper shows a thorough validation of numerical bending models of TRC sandwich beams, available in literature [6,19], by full-field DIC results on four-point bending experiments. While in previous literature, the validation of the numerical model has been limited by local displacement measurements, this paper shows a detailed comparison of the strains in the faces and the core. The full-field analysis of the DIC measurements reveals four stages in the bending behaviour of the sandwich beams, and shows the behaviour of each component material (faces and core) during the experiments. This provides a more in-depth comparison and shows a good agreement between the numerical prediction and experimental results. As a conclusion, it can be stated that the established numerical model was validated and was able to simulate the behaviour in bending of TRC sandwich panels with high confidence.

2. Materials and Methods

2.1. Textile-Reinforced Cement

The faces of the used sandwich panels were made of TRC plates consisting of a cement matrix cast onto fibre textiles. The cement matrix was a self-compacting ordinary Portland cement (OPC) composed of CEM I 52.5 R cement, fly ash, silica fume, silica flour, superplasticizer, and a water/cement ratio of 0.15. The cement was commercially available as SikaGrout 217 [20], and had a maximum grain size of 1.6 mm and a density of 2000 kg/m^3. The compressive strength and compressive E-modulus were 58 MPa and 26 Gpa, respectively. The compressive strength of the cement was experimentally determined by calculating the average of seventeen cubes with dimensions of 150 mm × 150 mm × 150 mm, in accordance with NBN EN 12390-3 [21]. The E-modulus was measured by applying strain gauges on three cylindrical specimens (VUB, Brussels, Belgium) with a height of 230 mm and a diameter of 104 mm, which were subjected to compression test according to [22].

The textile reinforcement used for the TRC faces was a combination of three-dimensional (3D) and two-dimensional (2D) textiles. The 3D textile was a spacer textile, composed of two layers of 2D textiles kept at a distance of 11 mm by polyester PET fibres. The 3D textile was combined with two 2D textiles, one placed on the top and one on the bottom of the 3D textile, to increase the fibre volume a fraction above the critical one (0.73%). The critical fibre volume fraction has to be exceeded in order to create the strain hardening behaviour of the TRC. The critical fibre volume fraction was calculated by the matrix tensile stress σ_{mu}, the E-modulus of the matrix E_m, the fibre tensile failure stress σ_{fu}, and the E-modulus of the fibres E_f:

$$V_f > \frac{\sigma_{mu}}{-\frac{E_f}{E_m}\sigma_{mu} + \sigma_{mu} + \sigma_{fu}} \tag{1}$$

Both textiles consisted of alkali-resistant (AR) glass fibres placed in an orthogonal grid structure, and are presented in Figure 1. With a thickness of the faces at 22 mm, the total fibre volume fraction was 2.98%, and the effective fibre volume fraction in the loading direction was 1.49%. The specifications of both textiles are given in Table 1.

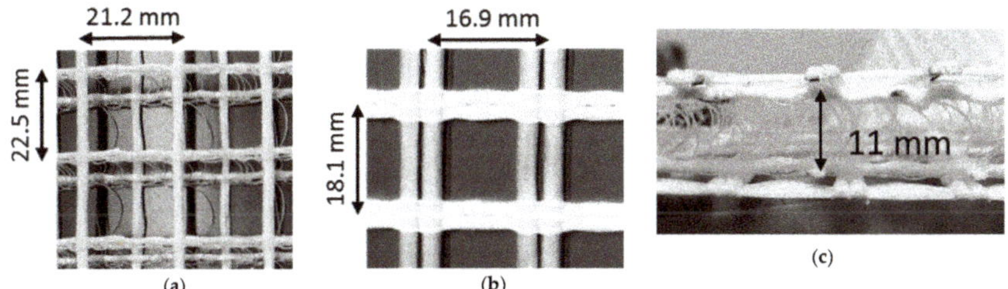

Figure 1. Textiles used in textile-reinforced cement (TRC): (**a**) three-dimensional (3D) textile, (**b**) two-dimensional (2D) textile, and (**c**) a combination of 2D and 3D textiles.

Table 1. Specifications of both 2D and 3D textiles.

	Fibre Material	Density (g/m^2)	Spacer Distance (mm)
3D textile [23]	AR-glass	536	11
2D textile [24]	AR-glass	568	-

The textile-reinforced cement faces, made for mechanical characterization, were cast in wooden moulds with dimensions as follows: height = 450 mm, width = 500 mm, and thickness = 22 mm. Before placing the 2D and 3D textiles, a layer of 5 mm cement was cast in the moulds. When the textiles were placed, the mould was filled with cement until a thickness of 22 mm was reached. Due to the self-compacting nature and small grain size of the cement, it could easily flow through the textiles and fill the mould. The moulds were covered with plastic foil, and the textile reinforcement cement plates were left to harden for 28 days.

2.1.1. Tensile Behaviour of Textile-Reinforced Cement

The tensile behaviour of TRC faces was investigated in detail in [25]. A brief description is given hereafter. The TRC faces of the sandwich beams were tested by a tensile test based on the recommendation of RILEM TC 232-TDT [26]. A schematic presentation of the test is given in Figure 2a. The dimensions of the specimens were as follows: length = 450 mm, width = 59 mm, and thickness = 22 mm. A total of 15 specimens were tested in tension, with a rate of 1 mm/min. The obtained stress–strain behaviour is presented in Figure 2b, which clearly shows three stages (indicated by I, II and III). In the first stage, the matrix and textiles showed composite action resulting in an E-modulus of 10.7 GPa. After reaching the matrix cracking stress of 2.28 MPa the second stage of multiple cracking occurred until crack saturation was reached (at a strain of 0.0033), and resulted in an E-modulus of 0.34 MPa. The third stage—post-cracking—was mainly determined by the textiles, and resulted in an E-modulus of 0.75 GPa and an ultimate stress of 7.43 MPa. The previously mentioned properties were average values of 15 specimens, and were used to establish the average tri-linear tensile stress–strain curve shown in Figure 2b.

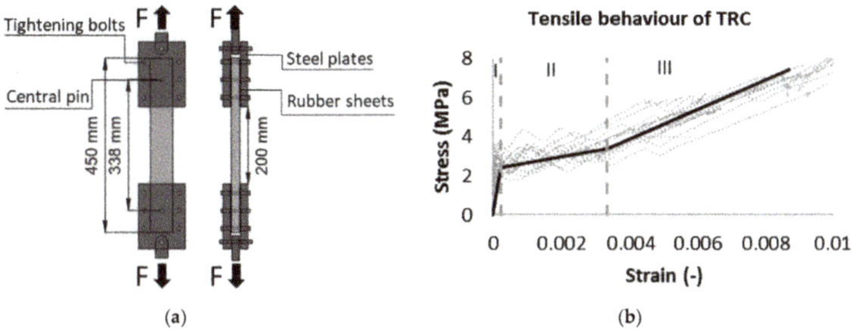

Figure 2. (**a**) Tensile test set-up, and (**b**) tensile behaviour of TRC faces.

2.2. Extruded Polystyrene Foam

The thermal insulation used for the sandwich beams was extruded polystyrene foam (XPS), in the form of rigid plates with a density of 33.5 kg/m^3, experimentally determined from six specimens. The top and bottom surfaces of the rigid insulation plates were imprinted with a rhombus pattern to provide mechanical interlocking and a better stress transfer between the TRC faces and the core. The finishing of the surfaces is shown in Figure 3a. The thickness of the foam blocks was 160 mm.

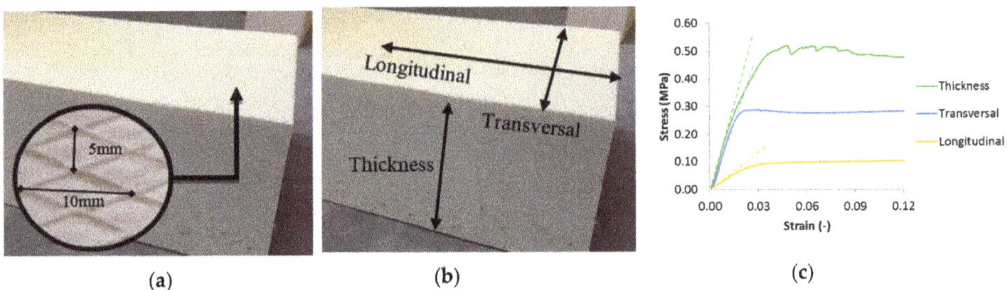

(a)　　　　(b)　　　　(c)

Figure 3. (**a**) Rhombus pattern on the faces of the rigid insulating extruded polystyrene foam (XPS) plates, (**b**) directions of the foam, and (**c**) compression test results in the different directions of the XPS foam.

2.2.1. Compressive Behaviour of XPS

Due to the extrusion production technique, the foam behaved differently in all three directions. Four compression tests on XPS cubes in every direction of 160 mm × 160 mm × 160 mm were performed in accordance with ASTM C165 [27], in order to determine the E-modulus and ultimate compressive strength of the foam. The best performance, in terms of stiffness and strength, was found in the thickness direction (see Figure 3c). The ultimate strength for the longitudinal, transversal, and thickness direction equalled 0.09MPa (σ_{cl}), 0.29 Pa (σ_{ctr}), and 0.52 MPa (σ_{cth}), respectively. The E-modulus was equal to 3.61 MPa (E_{cl}), 17.04 MPa (E_{cp}), and 20.6 MPa (E_{ct}) for the longitudinal, transversal, and thickness directions, respectively.

The shear strength and modulus of the XPS foam were determined by bending tests on four sandwich beams, with a span of 1 m and a width of 400 mm each, as described by the standard NBN EN 14,509 (2013) [28] (see Figure 4). This led to a shear modulus (G_c) of 4.7 MPa, and an ultimate shear strength (τ_c) of 0.24 MPa.

(a)　　　　(b)

Figure 4. (**a**) Force-displacement curves of bending tests performed to determine the shear strength of the core, and (**b**) the set-up of the bending test.

2.3. Production of Sandwich Beams

The construction of the sandwich beams was done in multiple phases. First, the XPS foam was placed into the mould so that the transversal direction of the foam (see Figure 3b) was aligned with the span of the beam. A thin cement layer of 5 mm was cast onto the XPS foam block, on which the 2D and 3D textiles were placed. The advantage of using a 3D textile was that the textile layers were kept directly at the right distance from each other. Afterwards, fresh cement paste was cast on the textile, until the face thickness of 22 mm was achieved. The self-compacting properties of the cement paste made it flow easily through the textile reinforcement and spread over the whole surface. In the next step, the surface was covered to reduce evaporation. After 24 h of hardening, the beam was turned over, and the second face was cast onto the XPS foam in the same way as the first face.

2.4. Four-Point Bending Set-Up

The load-deformation behaviour of the sandwich sections was investigated by means of a four-point bending set-up. This set-up allows for an area with a constant moment, where tensile stresses in the lower TRC face dominate. Furthermore, the set-up provoked shear stresses in the core between the supports and loading beams.

Four sandwich beams, with a span of 2.2 m between the supports, were tested in four-point bending with a displacement rate of 10 mm/min. The distance between the applied forces was 0.5 m, while the width and thickness of the beam were 0.4 m and 0.204 m (see Figure 5). The production process of the sandwich panels was explained in Section 2.3. During the test, the specimens were monitored with two DIC systems. DIC is a non-destructive measurement technique that records displacements of the entire observed specimen surface (by means of a speckle pattern), from which strains can be calculated. The displacements are related to a reference image taken at a zero-load step [29]. This measurement technique has proven to provide detailed information on textile and fibre reinforced cement application, as explained in [30] and [31]. The field of view of each system is captured a length of 600 mm along the length of the sandwich beam, starting from the middle of the beam (see Figure 5b).

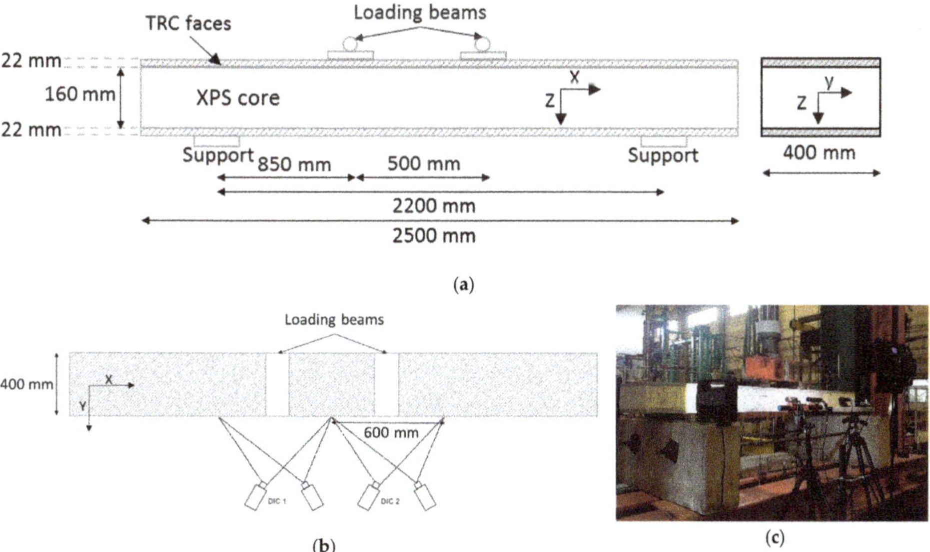

Figure 5. (a) front view of four-point bending test, (b) top view of four-point bending test, and (c) picture of test set-up.

3. Numerical Model Definition

3.1. Material Definition

The numerical modelling was performed in the finite element software ABAQUS/Explicit [32]. Non-linear material behaviour is applied by means of different prescribed material models in the program. The tensile and compressive behaviour of the TRC faces was implemented by combining the elastic and concrete damaged plasticity (CDP) material model. The compressive behaviour of the TRC faces was implemented in the elastic material model. Hence, the linear elastic behaviour was described by the compressive E-modulus of 26 GPa and the Poisson ratio of 0.21 [33] of the cement. For the CDP model, the requested input parameters were the dilation angle ($\psi = 36$), the potential flow eccentricity ($\varepsilon = 0.1$), the proportion of the ultimate compressive stress in a biaxial test to the uniaxial compressive stress ($f_{b0}/f_c = 1.0$), the shape of the deviatoric cross section ($K_c = 0.667$) and the numerical viscosity parameter ($\mu = 10^{-7}$). The values of these parameters were based on the ones described in [34]. Besides previously mentioned parameters, the non-linear tensile behaviour and ultimate compressive strength of the TRC were implemented in the uniaxial tensile stress-strain input of the CDP model. The compressive strength was limited to 58 MPa, while the complete tensile behaviour of the TRC faces, including the linear elastic part, was inserted. The used characteristic values are shown in Table 2.

Table 2. Characteristic values for the average tensile TRC curve.

Matrix Cracking Stress	Matrix Cracking Strain	End of Multiple Cracking Stress	End of Multiple Cracking Strain	Ultimate Failure Stress	Ultimate Failure Strain
2.28 MPa	0.00022	3.38 MPa	0.0033	7.43 MPa	0.0087

The initial linear elastic behaviour of the XPS foam was implemented by the elastic material model, and defined by the E-modulus and the Poisson ratio. The E-modulus, inputted into the numerical model, was based on the previously determined shear characteristics. Linear elastic analytical bending models for sandwich panels, described in [35,36], show that the deflection due to shear (80%) was significantly larger than the deflection due to bending (20%). Therefore, the applied E-modulus was calculated from the shear modulus (see Section 2.2) and the Poisson ratio (0.5) [19] of the XPS core.

The non-linear behaviour of the foam was modelled by the crushable foam–volumetric foam hardening material model. This model took into account the increased deformation of the foam in compression due to buckling of the cell walls, but required an isotropic material [32]. The crushable foam model requires the following parameters: the ratio between the initial yield stress in uniaxial compression and the initial yield stress in hydrostatic compression σ_c^0/p_c^0 (compression yield stress ratio), as well as the ratio between the hydrostatic tension and the initial yield stress in hydrostatic compression p_t/p_c^0 (hydrostatic yield stress ratio). The hydrostatic tension and initial yield stress in hydrostatic compression were set to 0.15 MPa and 0.14 MPa, respectively, as described in [19]. The initial yield stress in uniaxial compression was determined experimentally and set to 0.21 MPa (see Figure 3c). These values led to a compression yield stress ratio of 1.5 and a hydrostatic yield stress ratio of 1.07. The nonlinear behaviour of the foam was implemented through the yield stress and uniaxial plastic strain obtained from the average stress–strain curve of the thickness direction (Figure 3c). This non-linear material model implies the use of a dynamic explicit analysis, which is implemented with a time period of 10 and a mass scaling factor of 0.000001, in order to improve the speed of the analysis.

3.2. Cross Section and Boundary Conditions Modelling

A numerical model was established to simulate the four-point bending behaviour of sandwich beams with TRC faces, in order to compare it with experimental results of the same sandwich beams. In this way, more confidence in the numerical model was gained.

Both the faces and the core were defined as solid homogeneous sections in the numerical model and, were discretised by eight-node linear brick elements (C3D8R). The elements size was 35.7 mm $\times$ 3.6 mm $\times$ 200 mm (w $\times$ h $\times$ d) for the faces and 35.7 mm $\times$ 13.3 mm $\times$ 200 mm (w $\times$ h $\times$ d) for the core. Six elements were stacked over the thickness of the faces, and twelve over the thickness of the core of the sandwich beam. The mesh size was the result of a convergence study on the force-displacement behaviour of the sandwich beam, as well as on the stress and strain output. Multiple elements were necessary throughout the thickness of the faces to evaluate the stress/strain distribution over the thickness. Only one element was assumed over the width of the beam, since the load distribution, and therefore also the beam response, was assumed to be uniform. The mesh is shown in Figure 6.

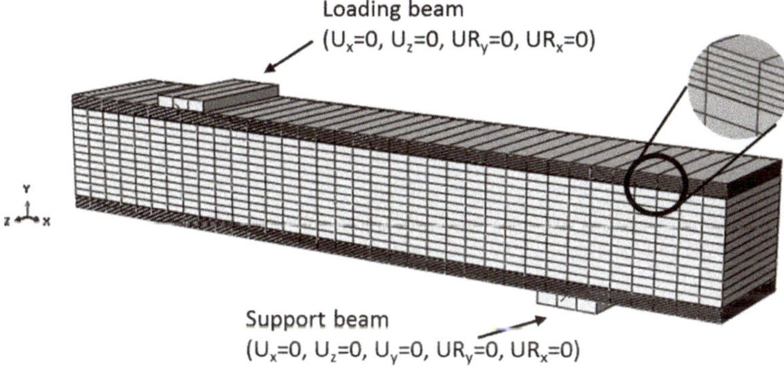

Figure 6. Numerical bending setup.

The contact between the rigid bodies (loading beams and supports) and the sandwich panels was established by a frictionless and hard contact interaction. The bond between the core and faces, however, was considered perfect, since no debonding was encountered during the experiments. Hence, the surfaces of the core and faces were modelled by a cohesive surface interaction, without defining damage interaction, defining a perfect bond. The default contact enforcement method was implemented, meaning that the elastic properties of the bond are based on the underlying element stiffness [32].

To simulate the bending behaviour of the sandwich beams, two rigid bodies and symmetry planes were used. In order to limit the number of elements, and therefore also the calculation time, symmetry boundary conditions were used in the *XY* and *YZ* planes. The results, however, can be plotted for the whole beam. One of the rigid bodies represents the support, while the other represents one of the loading beams. The support was restricted in the *X*, *Y* and *Z* directions, and could only rotate around the *Z*-axis. The loading beam was restricted in the *X* and *Z* directions, and could rotate only around the *Z*-axis.

For convergence reasons, the imposed displacement was performed with a smooth amplitude, so that the increments were smaller.

4. Results and Discussion

Figure 7a shows the force-displacement graph of a sandwich beam under four-point bending (as described in Section 2.4. Four-Point Bending Set-Up), where the displacement is measured at the tensile face of the beam underneath the loading pins by Linear Variable Differential Transformers (LVDTs). The orange curve shows the prediction by the numerical model, and the blue curve gives the average of the experimental results. The grey area shows the scatter on the experimental results. During the experimental campaign, four sandwich beams were tested. All sandwich beams failed by shear failure of the core, as shown in Figure 7b.

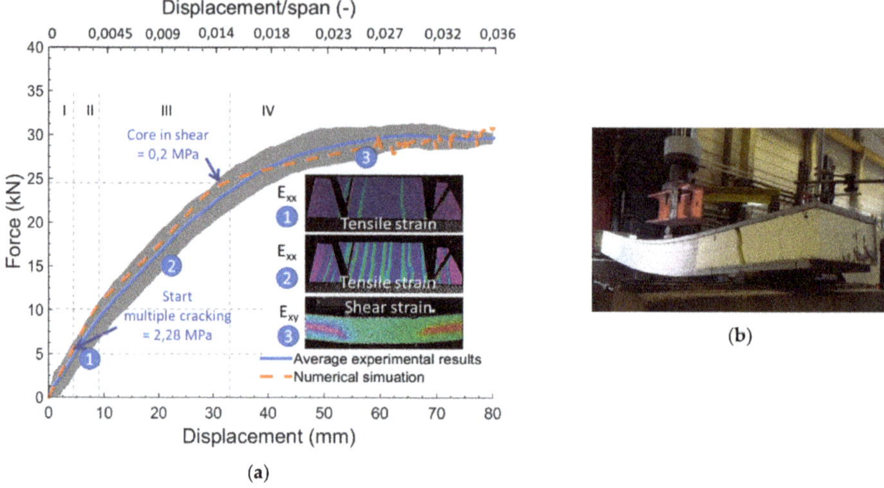

Figure 7. (**a**) force-displacement curve of the four-point bending tests on sandwich beams with TRC faces and the numerical prediction, and (**b**) failure in the shear of the core of a representative sandwich beam.

4.1. Numerical Model

The established numerical model revealed multiple stages in the bending behaviour of the TRC sandwich beams, based on the stress and strain development in the different materials of the sandwich beam. Four stages were distinguished, and indicated with I, II, III, and IV in Figure 7a. The first stage showed linear elastic behaviour of the beam. At a load of 5 kN (start of stage II), the matrix cracking stress of 2.28 MPa was reached at the surface of the tensile face, in the area with the constant moment (see Figure 8), which physically corresponds to the initiation of the first crack. The first crack initiation and the development of multiple cracks in the tensile face of the beam are specified for the second stage in Figure 7. Once the matrix cracking stress reaches through the complete thickness of the face (at a load of 10 kN in Figure 7, and illustrated in Figure 9), a clear reduction in stiffness was noticed, leading to the start of the third stage. Starting from a load of 25.5 kN, the core no longer deformed linearly and elastically but plasticly (see Figure 10), which led to another reduction in stiffness and the start of the fourth stage. The part of the plastic shear strain and total shear strain strain are presented in Figures 10 and 11, respectively. The maximum displacement in Figure 7 was a result of the applied maximum displacement of 100 mm during the analysis. Failure of the TRC sandwich beam was obtained when the ultimate shear stress (0.24 MPa) of the core was reached, which happened at a displacement of 91 mm as illustrated in Figure 12.

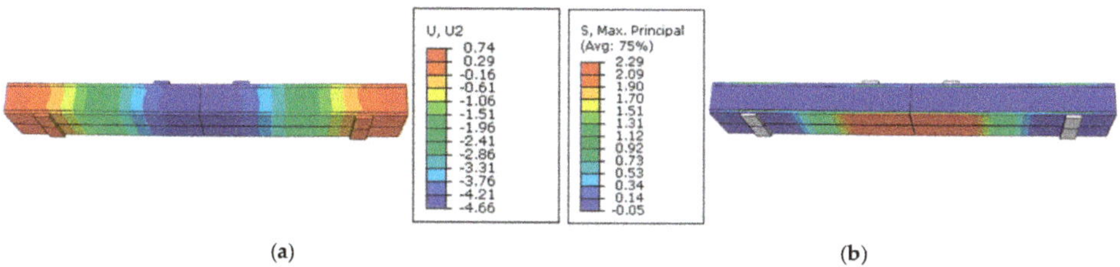

(a)   (b)

Figure 8. Start of the second stage, where the matrix cracking stress is reached at a vertical load of 5 kN. (**a**) Vertical displacement (U, in mm) and (**b**) horizontal stress (S, in MPa) in the tensile TRC face.

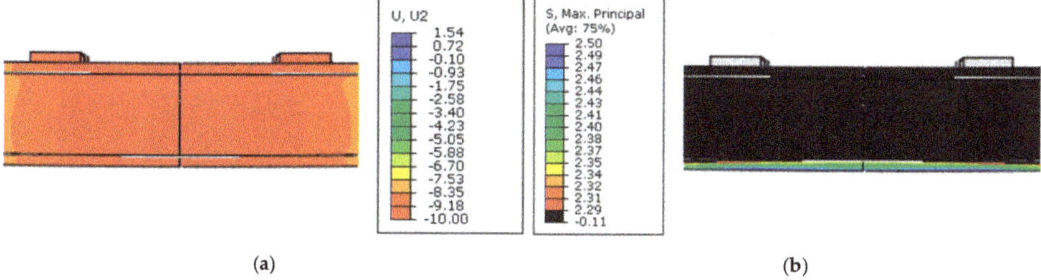

(a)   (b)

Figure 9. Start of the third stage at a load of 10 kN. (**a**) Vertical displacement (U, mm) in the middle of the beam; (**b**) the matrix cracking stress (S, in MPa) reaches through the entire cross-section of the tensile face in the constant moment area.

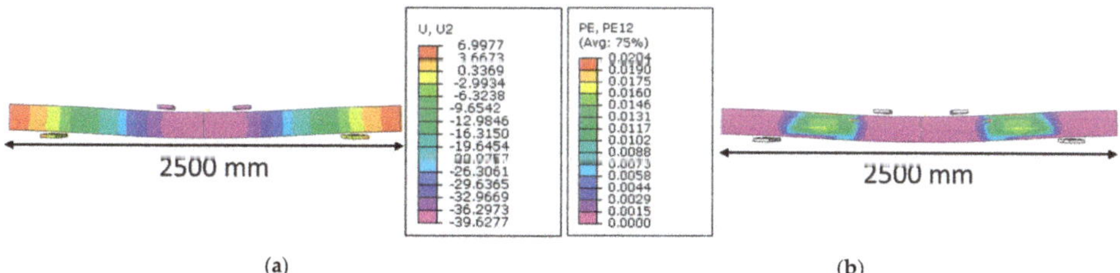

(a)   (b)

Figure 10. (**a**) Vertical displacement (U, in mm) at the start of the fourth stage (25.5 kN), and (**b**) plastic shear strain (PE [-]) of the core at a load of 25.5 kN.

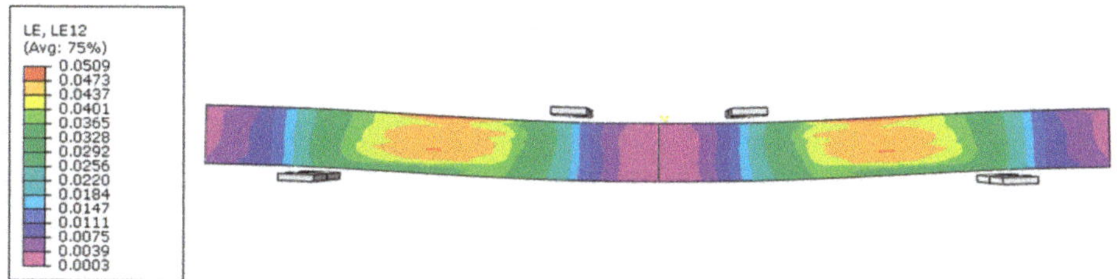

Figure 11. Total shear strain at a load of 25.5 kN.

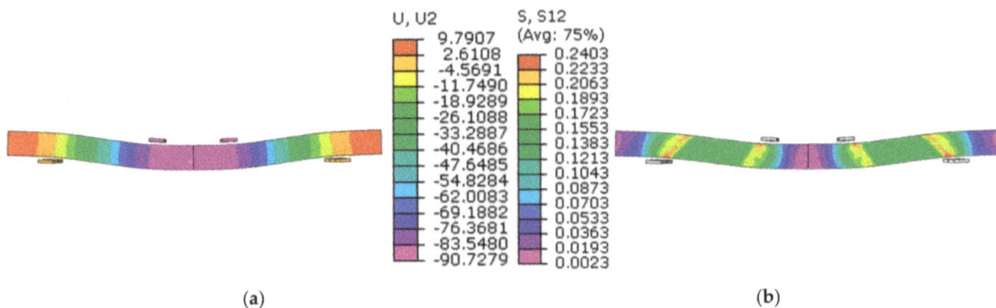

(a)   (b)

Figure 12. (**a**) Vertical displacement (U, in mm) of 91 mm, and (**b**) shear stress in the core at a displacement of 91 mm.

4.2. Experimental Results

A good correspondence between the numerical prediction and the experimental results was obtained; however, only three of the four stages were clearly visible in the experimental results.

In the first stage, both the faces and the core behaved linearly elastically. After reaching the matrix cracking stress in the area of the constant moment, the bottom face started to crack, indicating the start of stage II in Figure 7a. Figure 13 shows the longitudinal strains in the sandwich beams measured by both DIC systems, and therefore shows the appearance of the first crack after reaching the ultimate matrix cracking strength. These strain plots must be interpreted carefully. Strain results of the DIC technique were calculated from the average displacements, meaning that the displacements in the neighbourhood of cracks were responsible for apparent high strains at the location of the cracks. In reality, however, the strain in a crack is zero, so strain colormaps, as in Figure 13, can only be used to identify crack patterns; no significance should be attributed to the value of the strain in the vicinity of a crack. Simultaneously, the core showed a linear elastic shear strain, as shown in Figure 14.

Figure 13. Longitudinal strain at the start of stage II, when the first crack appears at a bending load of 5 kN.

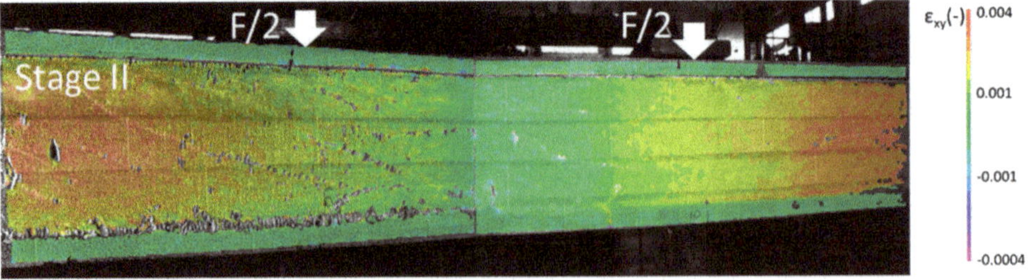

Figure 14. Shear strain of the core in stage II, at a bending load of 5 kN.

During stage III, cracking and propagation of the cracks occurred as shown in Figure 15. Since the highest tensile stress occurred in the area with the constant moment, most of the developed cracks are located between the loading beams.

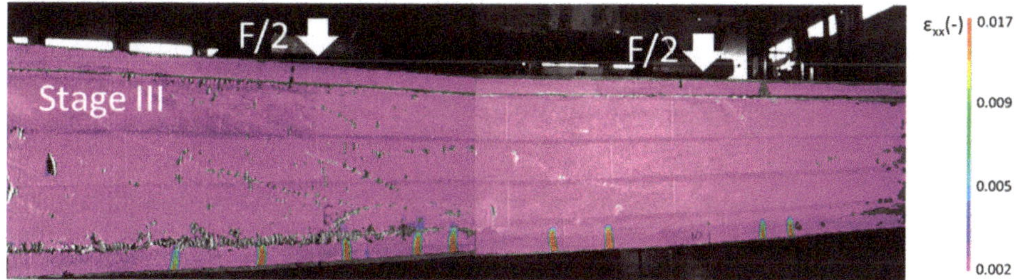

Figure 15. Longitudinal strain ε_{xx} at a load of 16 kN.

At a load of 26 kN (stage IV), the tensile face showed multiple cracking and the saturation of cracks between the loading beams (see Section 2.1.1), while the core reached a shear strain of 0.019, which is equal to the plastic shear strain of XPS. Both the tensile strain of the tensile face and the shear strain of the core are shown in Figures 16 and 17. The numerical model, however, predicted the plastic shear deformation of the core at a load of 25.5 kN. The observed phenomenon, plastic shear deformation of the core, was the same for the experiments and the numerical prediction. Also, the degradation of the stiffness corresponded well. The core continued deforming plastically, until its ultimate shear stress was reached and failure of the core occurred.

Figure 16. Longitudinal strain ε_{xx} at a load of 26 kN.

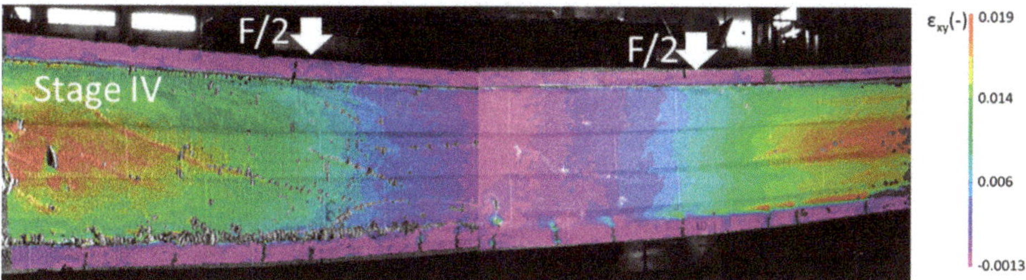

Figure 17. Shear strain ε_{xy} at a load of 26 kN.

4.3. Strain Comparison with the Numerical Model

The numerical prediction of the bending behaviour of sandwich beams, indicated by the orange dotted line in Figure 7, was established as explained in Section 3. In terms of force-displacement behaviour, a good agreement between the experimental and numerical results was obtained. Due to the full-surface DIC analysis, a more detailed comparison between the experimental results and numerical model could be performed in terms of strains, leading to a thorough validation of the model.

The strains in the TRC faces during the experiments were derived from the DIC results by artificially adding an extensometer between the loading beam and the middle of the beam (in the area of the constant moment). These artificial extensometers calculated the strain between two points, by dividing the measured displacements during loading by the initial calculated distance. In this way, an average shear strain in the area with the constant moment was obtained and quantified during the experiment. Since two DIC systems were used to monitor the full beam, the average of both systems was calculated, so that the complete area of the constant moment was covered (see Figure 18b). The strains of the numerical model were determined by calculating the ratio of the difference in longitudinal displacement between the middle of the beam and a point below the loading pin at the tensile face, and the initial distance between the same point (250 mm).

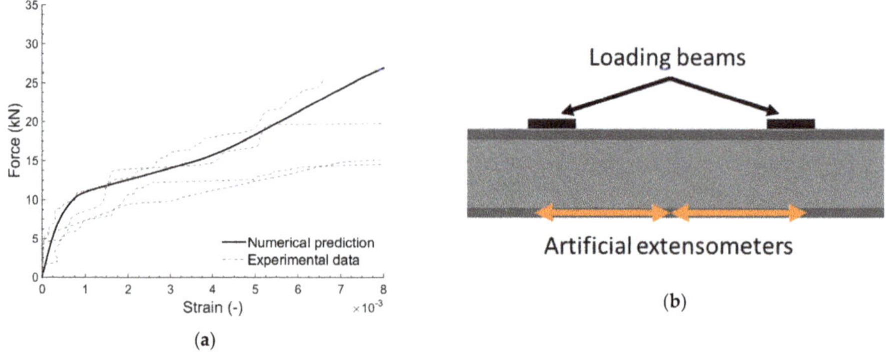

Figure 18. (**a**) Longitudinal strain in the tensile face of sandwich beam, obtained from the experimental results and numerical results; (**b**) schematic presentation of the artificial extensometers added on the DIC images.

The comparison of the experimental and numerical strains in the tensile TRC face are shown in Figure 18a. Both experimental and numerical longitudinal strains in the tensile face of the sandwich beam showed non-linear behaviour, as depicted in Figure 18a. The numerically implemented tri-linear tensile behaviour of the TRC is clearly visible in the tensile face of the sandwich beam in the bending of the numerical model. The number and place of the cracks in the tensile face cannot be predicted, which resulted in scattered experimental strain results in the tensile face. Due to this scatter, the experimental results showed a less pronounced tri-linear behaviour, but still follow the numerical tendency and showed the non-linear behaviour as predicted by the numerical model.

The core behaved plastically between the loading beams and the supports in the fourth stage, meaning that the plastic shear strain was reached at the beginning of this stage. Figure 19 shows the shear strain of the numerical prediction at a load of 26 kN, which was the start of the plastic shear behaviour in the experiments. These strains were compared with the experimental shear strains shown in Figure 19 and gave an identical strain distribution, with the maximal shear strain at the middle of the beam's height. Nonetheless, in the numerical analysis shear strain concentrations were noticed at the interface between the face and core, causing an overestimation of the shear strain. These concentrations were assumed to be due to the perfect bond, as simulated in the numerical model.

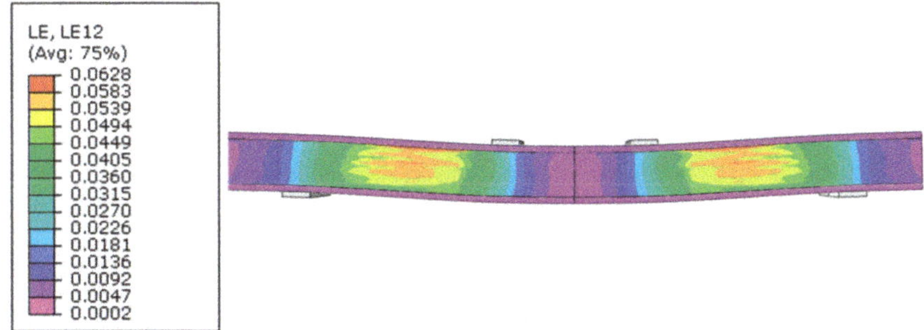

Figure 19. Shear strain in the core predicted by the numerical model, at a load of 26 kN.

Figure 20 shows the shear strains in the core at ultimate failure for both the experimental and numerical results of the sandwich beam. The same tendency was noticed for the numerical and experimental shear strain, which was limited in the area with the constant moment and increased outside the loading beams, where the highest shear forces were expected. Due to the perfect bond hypothesis in the numerical model, the shear strains were slightly overestimated.

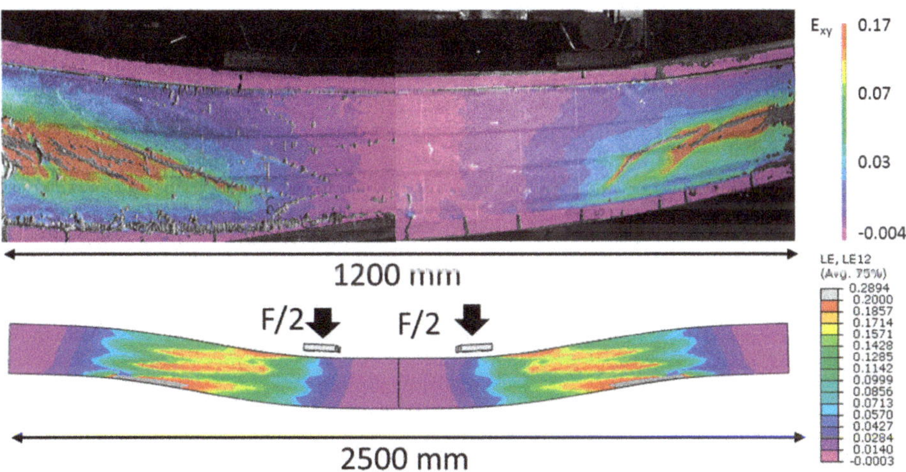

Figure 20. Comparison of the shear strain at failure load.

5. Conclusions

This paper presents a detailed comparison between a numerical prediction and experimental results of TRC sandwich beams under four-point bending by means of DIC. The numerical model considered the non-linear behaviour of both the TRC faces and the XPS foam core. A first comparison was made based on the force-displacement behaviour, which gave a good correspondence between the numerical prediction and the experimental results.

The stress and strain predictions of the numerical model identified multiple stages in the bending behaviour of the sandwich beams, which were confirmed by the experimental results. In the first stage, both the TRC faces and the XPS core behaved linearly elastically. Once the TRC tensile face started to crack, the second stage started. The four-point bending tests ended when the sandwich beams failed by shear failure in the core.

Thirdly, the tensile and shear strains obtained from the experiments and numerical simulation were compared. The TRC tensile strain was taken in the lowermost layer of the tensile face in the area of the constant moment, both for the numerical model and for the experiments. Both the tensile strain in the face and the shear strain distribution in the core corresponded well, indicating that the numerical model can reliably predict the experimental strains.

In conclusion, this paper showed how the use of the digital image correlation measurement technique allows for a full-field displacement and strain measurement of TRC sandwich beams, as well as the monitoring of the evolution of the crack pattern in the TRC faces. This detailed validation of the established finite element model contributes to the state of the art on the behaviour of TRC sandwich panels.

Author Contributions: Conceptualization, J.V., T.T. and P.V.I.; methodology, J.V. and T.T.; formal analysis, J.V.; investigation, J.V.; writing—original draft preparation, J.V.; writing—review and editing, T.T., J.W., M.D.M.; M.E.K.; P.K.; visualization, J.V.; supervision, T.T., J.W.; project administration, T.T.; funding acquisition, D.V.H.

Funding: This research was funded by Agentschap voor Innovatie en Ondernemen (VLAIO) grant number IWT140070.

Acknowledgments: The authors gratefully acknowledge Agentschap voor Innovatie en Ondernemen (VLAIO) for funding the research, and the Belgian Building Research Institute (BBRI) for the collaboration in the tests performed in this research.

Conflicts of Interest: The authors declare no conflict of interest.

References

1. Benayoune, A.; Samad, A.; Abang Ali, A.; Trikha, D.N. Response of pre-cast reinforced composite sandwich panels to axial loading. *Constr. Build. Mater.* **2007**, *21*, 677–685. [CrossRef]
2. Benayoune, A.; Samad, A.; Trikha, D.N.; Abdullah Abang Ali, A.; Ashrabov, A.A. Structural behaviour of eccentrically loaded precast sandwich panels. *Constr. Build. Mater.* **2006**, *20*, 713–724. [CrossRef]
3. Losch, E.D.; Hynes, P.W.; Andrews, R., Jr. State of the Art of Precast/Prestressed Concrete Sandwich wall Panels. *PCI J.* **2011**, *56*, 131–176.
4. Precast/Prestressed Concrete Institute. *Architectural Precast Concrete*; Precast/Prestressed Concrete Institute: Chicago, IL, USA, 2007.
5. Shams, A.; Horstmann, M.; Hegger, J. Experimental investigations on Textile-Reinforced Concrete (TRC) sandwich sections. *Compos. Struct.* **2014**, *118*, 643–653. [CrossRef]
6. Colombo, I.G.; Colombo, M.; Prisco, M. Bending behaviour of Textile Reinforced Concrete sandwich beams. *Constr. Build. Mater.* **2015**, *95*, 675–685. [CrossRef]
7. Cuypers, H.; Wastiels, J. Analysis and verification of the performance of sandwich panels with textile reinforced concrete faces. *J. Sandw. Struct. Mater.* **2011**, *13*, 589–603. [CrossRef]
8. Hegger, J.; Horstmann, M.; Feldmann, M.; Pyschny, D.; Raupach, M.; Feger, C. Sandwich panels made of TRC and discrete and continuous connectors. In Proceedings of the International RILEM Conference on Material Science, Aachen, Germany, 6–8 September 2010; Volume I, pp. 381–392.
9. Vervloet, J.; Van Itterbeeck, P.; Verbruggen, S.; El Kadi, M.; De Munck, M.; Wastiels, J.; Tysmans, T. Buckling Behaviour of Structural Insulating Sandwich Walls with Textile Reinforced Cement Faces. In *Strain-Hardening Cement-Based Composites*; RILEM Bookseries: Dordrecht, The Netherlands, 2018; pp. 535–543.
10. Vervloet, J.; Van Itterbeeck, P.; Verbruggen, S.; El Kadi, M.; De Munck, M.; Wastiels, J.; Tysmans, T. Sandwich panels with Textile Reinforced Cementitious skins as new insulating wall system: A case study. In Proceedings of the IASS Annual Symposium 2017, Hamburg, Germany, 25–28 September 2017; p. 10.
11. De Munck, M.; Tysmans, T.; Verbruggen, S.; Vervloet, J.; El Kadi, M.; Wastiels, J.; Remy, O. Influence of Weathering Conditions on TRC Sandwich Renovation Panels. In *Strain-Hardening Cement-Based Composites*; RILEM Bookseries: Dordrecht, The Netherlands, 2018; Volume 15, pp. 659–667.
12. Hegger, J.; Zell, M.; Horstmann, M. Textile Reinforced Concrete—Realization in applications. *Tailor Made Concr. Struct.* **2008**, 357–362. [CrossRef]
13. Hegger, J.; Will, N.; Horstmann, M. *Summary of Results for the Project INSUSHELL*; RWTH Aachen: Aachen, Germany, 2009.

14. Junes, A.; Larbi, A.S.; Claude, U.; Lyon, B.; Bohr, N. An experimental and theoretical study of sandwich panels with TRC facings: Use of metallic connectors and TRC stiffeners. *Eng. Struct.* **2016**, *113*, 174–185. [CrossRef]

15. Junes, A.; Si Larbi, A. An indirect non-linear approach for the analysis of sandwich panels with TRC facings. *Constr. Build. Mater.* **2016**, *112*, 406–415. [CrossRef]

16. Shams, A.; Hegger, J.; Horstmann, M. An analytical model for sandwich panels made of textile-reinforced concrete. *Constr. Build. Mater.* **2014**, *64*, 451–459. [CrossRef]

17. Colombo, I.G.; Colombo, M.; di Prisco, M.; Pouyaei, F. Analytical and numerical prediction of the bending behaviour of textile reinforced concrete sandwich beams. *J. Build. Eng.* **2018**, *17*, 183–195. [CrossRef]

18. Ilyes Djamai, Z.; Bahrar, M.; Salvatore, F.; Si Larbi, A.; El Mankibi, M. Textile reinforced concrete multiscale mechanical modelling: Application to TRC sandwich panels. *Finite Elem. Anal. Des.* **2017**, *135*, 22–35. [CrossRef]

19. Horstmann, M. Zum Tragverhalten von Sandwichkonstruktionen aus Textilbewehrtem Beton, RWTH Aachen. Ph.D. Thesis, RWTH Aachen University, Aachen, Germany, 2010.

20. *Sika Sikagrout-217 Fine Concrete Data Sheet*; Sika AG: Baar, Switzerland, 2016; p. 3.

21. Bureau voor Normalisatie. *Beproeving van Verhard Beton—Deel 3: Druksterkte van Proefstukken*; Wetenschappelijk en Technisch Centrum voor het Bouwbedrijf (WTCB). Available online: https://www.wtcb. be/homepage/index.cfm?cat=services&sub=standards_regulations&pag=norm_concrete&art=standards (accessed on 22 March 2019).

22. ASTM International. *ASTM C469M-14, Static Modulus of Elasticity and Poisson's Ratio of Concrete in Compression.* 2014. Available online: www.astm.org (accessed on 22 March 2019).

23. SitGrid 701. Available online: https://solutions-in-textile.com/produkte/fertigteilbau (accessed on 2 March 2019).

24. SitGrid 200. Available online: https://solutions-in-textile.com/produkte/fertigteilbau (accessed on 2 March 2019).

25. El Kadi, M.; Verbruggen, S.; Vervloet, J.; De Munck, M.; Wastiels, J.; Van Hemelrijck, D.; Tysmans, T. Experimental Investigation and Benchmarking of 3D Textile Reinforced Cementitious Composites. In *Strain-Hardening Cement-Based Composites*; RILEM Bookseries: Dordrecht, The Netherlands, 2018; pp. 400–408.

26. Brameshuber, W. Recommandation of RILEM TC 232-TDT: Test methods and design of textile reinforced concrete. *Mater. Struct.* **2016**, *49*, 4923–4927.

27. ASTM International. *ASTM C165-00—Standard Test Method for Measuring Compressive Properties of Thermal Insulations*; ASTM International: West Conshohocken, PA, USA, 2000.

28. Bureau voor Normalisatie. *EN 14509: Self-Supporting Double Skin Metal Faced Insulating Panels—Factory Made Products—Specifications*; Bureau voor Normalisatie: Brussel, Belgium, 2013; p. 177.

29. Sutton, M.A.; Orteu, J.J.; Schreier, H. *Image Correlation for Shape, Motion and Deformation Measurements*; Springer US: New York, NY, USA, 2009; ISBN 978-0-387-78747-3.

30. Verbruggen, S.; Remy, O.; Wastiels, J.; Tysmans, T. Stay-in-Place Formwork of TRC Designed as Shear Reinforcement for Concrete Beams. *Adv. Mater. Sci. Eng.* **2013**, *2013*, 1–9. [CrossRef]

31. Bilotta, A.; Ceroni, F.; Lignola, G.P.; Prota, A. Use of DIC technique for investigating the behaviour of FRCM materials for strengthening masonry elements. *Compos. Part B Eng.* **2017**, *129*, 251–270. [CrossRef]

32. *Dassault Systèmes Simulia Abaqus CAE User's Manual.* Abaqus 6.12. Available online: http://130.149.89.49: 2080/v6.12/books/usi/default.htm?startat=pt06ch60s01.html (accessed on 22 March 2019).

33. Brockmann, T. Mechanical and Fracture Mechanical Properties of Fine Grained Concrete for TRC Structures. In *Advances in Construction Materials 2007*; Grosse, C.U., Ed.; Springer: Berlin/Heidelberg, Germany, 2007; pp. 119–130. ISBN 9783540724476.

34. Tysmans, T.; Wozniak, M.; Remy, O.; Vantomme, J. Finite element modelling of the biaxial behaviour of high-performance fibre-reinforced cement composites (HPFRCC) using Concrete Damaged Plasticity. *Finite Elem. Anal. Des.* **2015**, *100*, 47–53. [CrossRef]

35. Allen, H.G. *Analysis and Design of Structural Sandwich Panels*, 2nd ed.; Elsevier Ltd.: Amsterdam, The Netherlands, 1969; p. 269.
36. Zenkert, D. *The Handbook of Sandwich Construction*, 2nd ed.; Engineering Materials Advisory Service Ltd.: London, UK, 1997; p. 432.

 *applied sciences*

Article

Shear Capacity of Textile-Reinforced Concrete Slabs without Shear Reinforcement

Jan Bielak *, Viviane Adam, Josef Hegger and Martin Classen

Institute of Structural Concrete, RWTH Aachen University, 52074 Aachen, Germany;
vadam@imb.rwth-aachen.de (V.A.); jhegger@imb.rwth-aachen.de (J.H.); mclassen@imb.rwth-aachen.de (M.C.)
* Correspondence: jbielak@imb.rwth-aachen.de; Tel.: +49-241-80-26830

Received: 13 March 2019; Accepted: 27 March 2019; Published: 1 April 2019

Abstract: A reliable and economic utilization of textile-reinforced concrete in construction requires appropriate design concepts. Unlike designs for bending, the development of models for shear is still the subject of current research. Especially for thin slabs, systematic experimental investigations are lacking. In this paper, the results of an experimental campaign on 27 carbon-textile reinforced slab segments tested in three-point bending are presented. The shear-span to depth ratio and member size were key variation parameters in this study. Increasing the structural depth of members led to a reduction in relative shear strength, while variation of shear slenderness controlled the efficiency of direct stress fields between load introduction and support. Interestingly, direct load transfer was activated up to a shear slenderness ratio of 4, which is significantly higher than in reinforced concrete ($a/d < 2.5$–3) and may result from the bond characteristics of the textile reinforcement. The experimental shear strengths were compared to predictions from existing models for shear of fiber-reinforced polymer (FRP)-reinforced concrete. The study shows that these FRP calculation models also predict the ultimate shear force for textile-reinforced concrete (TRC) tests presented in this paper with sufficient accuracy. Existing approaches for the size effect seem transferable as well. In order to validate the models for general use in TRC shear design, a compilation and comparison with larger experimental databases is required in future works.

Keywords: shear; textile-reinforced concrete; carbon concrete composite; design provisions; size effect; shear span

1. Introduction

Textile-reinforced concrete (TRC) combines high-performance non-metallic textile grids as aligned internal reinforcement with state-of-the art concrete technology. The resulting composite material makes a re-thinking of established construction methods possible [1–4]. The resistance to corrosion of the textiles permits reduced concrete covers and structural depths and supersedes additional protective polymeric layers (e.g., [5–7]). The higher tensile strength of reinforcement fibers, such as carbon, compared to typical reinforcement steel allows for further optimization of cross-sectional designs. With smart use of these materials, large resource savings can be realized in specific areas of concrete construction [8,9]. However, successful dissemination of TRC in practice depends on the availability of accurate and reliable, yet easy-to-use, design models [10].

In contrast to design for bending, both engineers and researchers are still confronted with fundamental questions regarding shear design of textile reinforced concrete for new constructions. While numerous applications [11–14] and first general approvals for thin façade panels exist [15], there is no design model for thicker TRC slabs between 5 cm and 20 cm with substantial shear loads, e.g., due to concentrated loads near supports. Slabs with such dimensions have high potential, both for bridges and in high-rise construction. Recent projects in Germany show the application for pedestrian bridges [16–19] as well as for small road bridges [20], especially in the transversal structural system. TRC slabs do not require additional protective layers (e.g., epoxy coating or bitumen) and thus the concrete can be driven or walked on directly. In high-rise construction, a promising application for TRC slabs are multi-storey car parks [21], where the question of corrosion-resistance because of exposition to deicing-salt as well as the maximization of clear floor height without increasing the building height dominate the design—both are strong arguments for the use of TRC.

The ongoing research on fundamental shear design models [22–27] and the numerous current research projects on shear in Europe [28–37] indicate that this topic is far from being solved for steel reinforced concrete. This foreshadows the long and tedious way toward an adequate level of knowledge on shear design of TRC.

TRC distinctly differs from fiber reinforced polymer (FRP) bar reinforced concrete, which is much more comparable to conventional concrete in component size, reinforcement diameter, stiffness in transversal direction, and shape. For FRP, extensive research exists on elements without (e.g., [38–46] and with (e.g., [47–51]) transversal reinforcement. Due to the great experience in research and practice, there are design provisions in several international codes (e.g., [52–54]). However, in contrast to FRP-reinforced concrete, research on TRC is still in its infancy. The existing models for FRP are an excellent starting point, but research on TRC should check unbiasedly fundamental assumptions on load-bearing mechanisms in order to avoid fallacies in design. The "riddle of shear" (Kani's famous dictum in [55]) for TRC is one of those fundamental topics targeted in a large-scale coordinated research program on TRC and carbon reinforced concrete in Germany named "Carbon Concrete Composites (C^3)-Project" [56]. In the subproject C^3-B3 [57], experimental and theoretical investigations on shear were performed by the Institute of Structural Concrete at RWTH Aachen University. Meanwhile, other researchers in Europe are investigating similar issues on shear capacity of filigree TRC beams [3,58] or the capacity of 3D textile reinforced elements [59].

The aim of the present article is to give insight on fundamental questions for shear design of TRC without shear reinforcement. Using the results of a systematic experimental investigation on slab segments, the influence of the component's height and the effect of shear slenderness are discussed. The comparison of shear capacity predictions from selected existing models to the test results indicates that TRC with epoxy-impregnated carbon textiles as longitudinal reinforcement exhibits a similar shear behavior compared to steel- or FRP-reinforced concrete components. This is the first step toward the transfer or adaption of existing shear design models to TRC.

2. Experimental Investigation on Shear Capacity

2.1. Test Setup and Instrumentation

For the experimental study of the shear capacity, single-span slab segments with single loading in mid-span were tested. The test setup was variable and scalable, which allowed a systematic parameter study. Figure 1 shows the test setup. The load was introduced under displacement control by means of a steel roll along the width of the specimen. An elastomer strip prevented local stress concentration on the upper surfaces of the specimens. The load rate was chosen to 1 mm/min. One fixed and one free roll were used to guarantee vertical support without horizontal constraints.

The vertical displacements were measured by two linear variable displacement transducers (LVDT) at mid-span. The longitudinal strain was tracked by a concrete strain gauge on top of the specimen in mid-span between two steel load-introduction plates and by an LVDT with a measuring length of 25 cm fixed to the bottom of the specimen. The load was measured continuously by a built-in load cell of the electric testing machine (100 kN maximum load capacity).

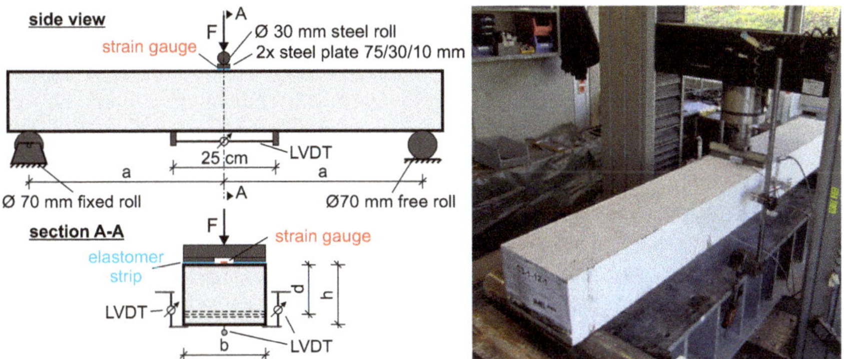

Figure 1. Test setup and instrumentation for three-point bending tests.

2.2. Variation of Parameters

The experimental program focused on the investigation on the variation of two parameters, the shear slenderness a/d and the effective depth d (Table 1). By varying the shear slenderness, one could investigate the increase in shear capacity due to concentrated loading near the supports and the formation of direct compression stress fields. The variation of the effective depth allowed for analysis of the size effect. All other parameters with an assumed influence on the shear capacity were kept constant. The geometrical ratio of longitudinal reinforcement ρ (cross-sectional area of reinforcement divided by effective depth and width) was chosen as ~0.24%, aiming at avoiding bending failure. Note that this reinforcement ratio is still typical for TRC plates in this depth range. This is relevant, because over-reinforced cross-sections may show a disproportionate amount of dowel action as shear transfer mechanism. The concrete compressive (and tensile) strength chosen for this study resulted from the idea of matching suitable high-performance materials, i.e., to be able to fully use the high tensile and bond strength of the non-metallic reinforcement. Details are given in Section 2.4. The width of 20 cm was chosen for all specimens considering the maximum specimen height, the maximum grain size, the number of reinforcement elements per layer resulting from the reinforcement grid spacing, and the maximum test load and dimensions of the test machine.

Table 1. Parameter variation for the experimental study.

Shear Slenderness a/d	Effective Depth
4	4 cm
5	8 cm
6	12 cm

2.3. Reinforcement

Non-metallic textile reinforcement can be categorized in different ways, regarding its fiber material, its impregnation material, or its geometry. As fiber materials, carbon, alkali resistant (AR-) glass, aramid, and basalt are usually utilized. Carbon and AR-Glass are the most common and there are several commercially available products world-wide. As impregnation materials, elastic rubber-based systems such as styrene-butadiene-rubber, as well as stiffer types based on epoxy resin or polyacrylate, are widespread. Nowadays, non-impregnated textiles (e.g., used in the construction of shell structures in [5] or [60]) are less common for new constructions due to the low efficiency and the low stiffness, which complicates the handling during manufacturing of TRC members. However, they are still used for repair and external retrofitting, especially for masonry walls. Regarding the geometry, one can distinguish the following types: Planar 2D textiles (biaxial or multiaxial), preformed 3D elements made from 2D textiles, and full 3D textiles (see e.g., [59]). Due to the variety of available products and combinations, a detailed characterization of the material properties of the reinforcement is indispensable for an experimental campaign on textile-reinforced concrete. It is interesting to note that the mechanical properties of FRP and textile reinforcement are related to different geometrical bases. For textile reinforcement, typically only the filament area without impregnation resin is counted. Reasons for this approach are the non-uniform geometrical cross-section along the axis of individual yarns (Figure 2) and the rather difficult determination of the cross-sectional area due to the small size. As the titer and the filament count of the individual non-impregnated rovings used in production of the textiles are known, the determination of the total filament area per meter is very simple. This is a significant difference to FRP bars or stirrups, for which the geometrical area given by the manufacturer includes the polymeric matrix. In consequence, area-related material characteristics such as stress or modulus of elasticity of impregnated textiles appear higher than for FRP bars, despite similar basis materials and similar compactness.

In this study, an epoxy-impregnated biaxial carbon grid was utilized as longitudinal reinforcement. The epoxy-resin was applied and hardened during production by the reinforcement manufacturer. Due to the high stiffness of the mesh, it could not be rolled on regular-sized reels. Hence, in this case, it was delivered in 5 m × 1.2 m planar panels. Figure 2a shows the biaxial open grid structure with a 38 mm axial spacing of the yarns, both in longitudinal (warp) and transversal (weft) direction of the fabric. The cross-sectional area of this reinforcement was symmetrical, with 95 mm^2/m in both directions. The reinforcement layout for this study aimed toward achieving a similar reinforcement ratio for different effective depths. In consequence, the specimens had one to three layers of reinforcement mesh strips of 5 yarns each (Figure 2c). During production, the bottom layer passed beyond the formwork in order to apply a slight prestressing to avoid sagging of the reinforcement (Figure 2b). All layers were aligned so each layers' weft and warp yarns stacked directly on top of each other. This resembles the typical production in practice, as the alignment of openings allows the concrete to pass. In mid-span, an additional fixation for one transversal yarn guaranteed the necessary concrete cover. Furthermore, due to the lower density of the impregnated carbon compared to fresh concrete, the buoyancy could be effectively controlled. Conventional spacers made from fiber-reinforced mortar or plastic could function as a suitable alternative but were avoided here to minimize the risk of floating up of the reinforcement and to eliminate a possible influence on the cracking process.

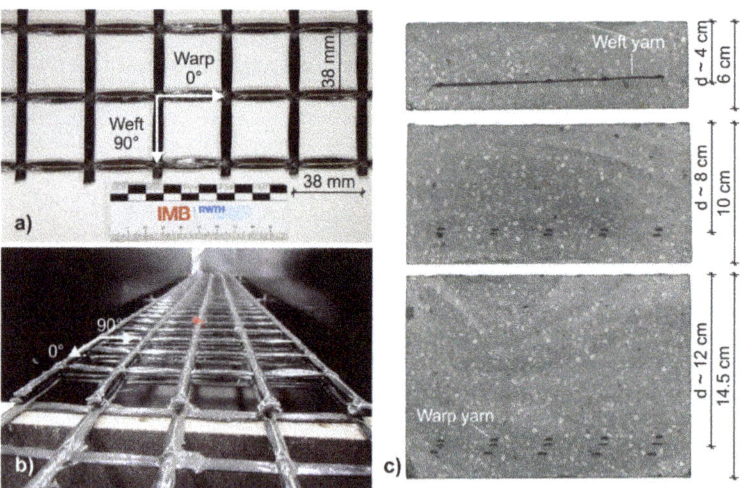

Figure 2. Reinforcement layout for test series: (**a**) Planar biaxial grid made of carbon with epoxy impregnation; (**b**) formwork with bottom reinforcement layer projecting beyond the end; (**c**) cross-section of the three specimen types.

The material characteristics of non-impregnated textiles and textiles with partial impregnation (surface coating) needed to be determined with composite specimens via uniaxial tensile tests [10,61–63]. For fully impregnated textiles with a homogenous stress distribution over the yarn area and simultaneous activation of all filaments, the properties could be determined on the textile without surrounding concrete [64]. Table 2 lists the material properties. The ultimate stress and modulus of elasticity were analyzed in uniaxial single yarn tests according to the setup described in [65], where individual yarns were extracted from the hardened grid and connected to the testing machine with an variable pressure along the clamping length. Note that, due to the well-known statistical effects for a bundle of linear-elastic yarns with brittle failure, the strength of a single yarn does not equal the strength of the fabric [66,67]. This can be taken into account by a reduction factor which depends on the number n of yarns [68]. For the reinforcement used in this study, a reduction factor of 0.85 for $n = \infty$ was proposed by Rempel [64].

Table 2. Reinforcement characteristics for solidian Grid Q95/95-CCE-38 (properties of one individual yarn, from [64] with test setup according to [65]).

Characteristic	Unit	Warp Direction (0°)	Weft Direction (90°)
Modulus of elasticity	[MPa]	244,835	243,828
Mean ultimate stress	[MPa]	3221 (n = 204 tests)	3334 (n = 218 tests)
5% quantile ultimate stress	[MPa]	2737	2762
Mean ultimate strain	[‰]	13.2	13.7
Axial spacing of yarns	[mm]	38	38
Cross-sectional area per yarn [1]	[mm²]	3.62 [1]	3.62 [1]
Cross-sectional area per meter [1]	[mm²/m]	95 [1]	95 [1]

[1] Filament area without epoxy-impregnation.

The bond properties of the reinforcement were analyzed in a companion investigation [69]. In contrast to non-impregnated textiles or textiles with soft impregnation, the full and hard epoxy impregnation led to form closure with longitudinal splitting of the concrete as a bond failure mechanism, rather than pull-out or jamming of the yarns. The mean length required for full anchorage was determined to be 78 mm for a concrete cover of 20 mm in the same cementitious matrix with equal tensile and compression strengths, as used in the present paper [69]. A free length of 50 mm behind the supports on both ends of the specimens proved to be sufficient to anchor the respective forces from shear and bending. All layers of reinforcement continued to the end of the specimen. Up to the point of ultimate failure, no longitudinal cracks ran up to the supports or up to the ends of the specimen. This allowed for the conclusion that no anchorage failure occurred in this study.

2.4. Cementitious Matrix

The cementitious matrix utilized in this study was specifically designed within the C^3 project to meet the requirements of textile reinforced concrete [70]. The mixture was based on [70], but adapted with locally available aggregates. Details can be found in Table 3. The maximum diameter of the crushed quartz aggregate (4 mm) matched the size of the grid openings. The high content of fine particles in the cementitious binder compound and the fine sand paired with the high-performance superplasticizer led to self-compacting properties of the fresh mix. During production, no external or internal compaction was required to achieve a dense matrix without cavities or gravel pockets.

According to DIN EN 206 [71], the mixture was no standard concrete due to its small aggregate size and its high content of fine particles. As it was produced and applied just like concrete as matrix, the term "concrete" is used in this paper and generally in the context of textile reinforced concrete for new constructions. This allows a distinction to be made from repair and retrofitting, where the term mortar is more common.

The hardened concrete exhibited high strength, both in compression and in tension. The bending tensile strength was determined on prism specimens (40 × 40 × 160 mm), according to the standard test method for mortar [72]. The mean value of all experiments (age 27 to 32 days) was 15.1 MPa with a coefficient of variation (COV) of 21.6%. The mean modulus of elasticity of the cementitious matrix was tested on cylindrical specimens (d/h = 150/300 mm) to 44595 MPa (COV 2.1%) with the method described in [73]. The mean compressive strength reached 127.6 MPa (COV 4.2%) for 150 mm cubes [74], 105.4 MPa (COV 4.7%) for cylinders (d/h = 150/300 mm, [74]), and 122.5 MPa (COV 5.1%) for the prism halves [72]. Due to the high strength of the cementitious matrix, cracks usually ran through the aggregates. The ultimate compressive strain of this concrete has been determined on two cylinders with external concrete strain gauges to 2.92‰ at the age of 28 days.

Table 3. Mix design of cementitious matrix for HF-2-165-4 (mix design adapted from [70]).

Substance	Density	Content
	kg/m^3	kg/m^3
Cementitious binder compound CEM II/C-M Deuna	2962	707
Fine quartz sand F38 S	2650	294
Quartz sand 0.1–0.5 mm	2630	243.2
Quartz sand 0.5–1.0 mm	2630	201.4
Quartz sand 1.0–2.0 mm	2630	148.9
Quartz sand 2.0–4.0 mm	2630	593.5
Superplasticizer (polycarboxylatether-basis) MC-VP-16-0205-02	1070	15
Water	1000	165

3. Results

3.1. Failure Mechanisms

In the experimental program, two main different failure mechanisms were observed, bending and shear failure. The smallest specimens with a cross-sectional height of about ~60 mm and an effective depth of ~40 mm failed in bending by rupture of one or several yarns. Figure 3a shows three representative examples for the three different load-support-distances (span lengths). For some specimens, a diagonal crack originating from the layer of reinforcement occurred prior to failure. However, this crack was not the ultimate reason for failure. Three specimens (C3-1-4-1, C3-3-4-1, and C3-3-4-3) showed significant crack formation along the layer of reinforcement prior and subsequent to rupture of one yarn. Their failure mechanism is therefore described as bending failure with subsequent shear failure.

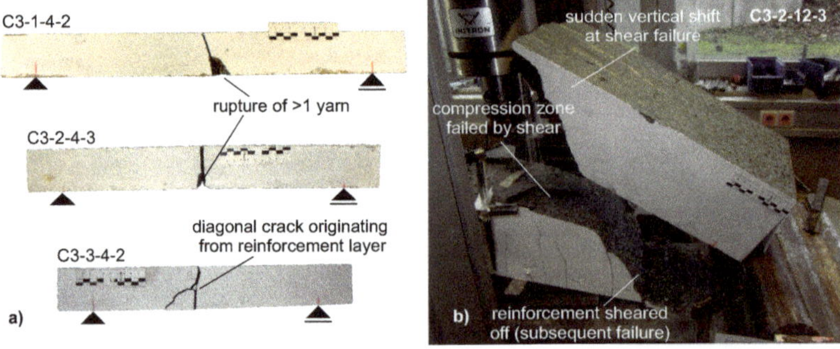

Figure 3. Failure mechanisms observed in experimental study (**a**) bending failure; (**b**) shear-compression failure.

All specimens with an effective depth of 80 and 120 mm failed in shear. Because their compression zone was further constricted by the critical shear crack prior to failure, the term shear compression failure is utilized. Figure 3b illustrates a typical failure phenomenon which occurred for several specimens. The sudden propagation of the shear crack into the compression zone led to the brittle failure. The sudden release of stored energy resulted in a mutual sliding and a vertical shift of one half of the specimen. Note that the reinforcement in the tension zone was sheared at the crack in all layers without adding significant resistance and ductility to the failure mechanism. This is typical for the anisotropic fiber-reinforcement material, but even more pronounced for textile reinforcement due to the relatively low transversal stiffness of the individual yarns.

For all specimens, one can observe that the residual compression zone in mid-span is only few millimeters deep. This can be explained by the high compressive resistance of the concrete and the good compaction of the concrete resulting in a dense and tough uppermost cementitious layer. The shear cracks passed through the aggregate grains. At failure, the constricted compression zone often buckled in the vertical direction. The high utilization of the concrete in this zone is shown in Figure 4 exemplary for three specimens, all having an effective depth of 80 mm but with different a/d-ratios.

All specimen showed a linear-elastic branch in their load-deflection curve up to a high first-cracking load. This high load and the severe drop after first cracking are not surprising for the high-strength concrete with its corresponding stiffness. Subsequent bending cracks are clearly visible in the diagram. The end of the test was marked by a sudden drop of the load without residual capacity. At this point, the compressive strain directly below the load introduction at mid-span reached or even exceeded the ultimate strain of the concrete taken from uniaxial compression tests on cylinders. Next to the observations of the crack pattern after failure, this additionally confirms the hypothesis of shear compression failure. The highest compressive strains were reached for the smallest shear slenderness ($a/d = 4.1$ for C3-3-8-2 in Figure 4b). One explanation for this observation is the superposition of compressive stress from beam action and from direct stress fields. This is a first indication towards an influence of increased arch action for the smaller shear slenderness.

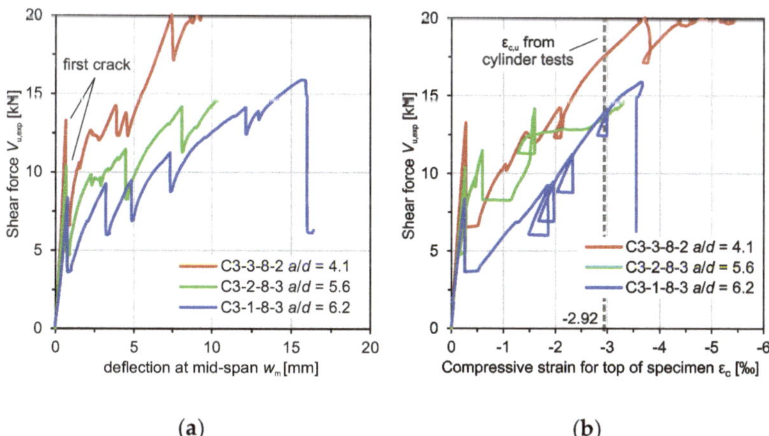

(a)   (b)

Figure 4. Experimental results for three typical specimens with $d = 8$ cm. (**a**) Load-deflection diagram; (**b**) load-compressive strain diagram (strain gauge at the top of the specimen in mid-span).

3.2. Crack Pattern and Critical Shear Crack

The characteristic crack pattern of specimens without shear reinforcement failing in shear is highly relevant for the assessment of their failure mode and their ultimate resistance. In Figures 5 and 6, the crack patterns after failure are shown for the specimens with 120 mm and 80 mm effective depth, respectively. The critical shear crack is highlighted with a bold black line. The series with 40 mm is not shown, as there was usually only one bending crack, sometimes with a single longitudinal or diagonal crack originating from the layer of reinforcement.

If a full separation of the specimens' halves at shear failure occurred, both halves were digitally rejoined for better comparison of the shear crack form. The dark-grey zones indicate concrete spalling, which occurred both in the compression zone (e.g., C3-1-12-1, C3-3-12-2, C3-1-8-3) and in the tension zone (e.g., C3-1-12-3, C3-2-12-2, C3-1-8-2). For almost all specimens, the critical shear crack propagated from a bending crack up to the area of introduction of the concentrated load. Table 4 gives detailed information on all experiments. The comparatively high scatter of the three results with an effective depth of $d = 12$ and a shear slenderness of $a/d = 4$ should be highlighted at this point. The low inclination of the critical shear crack for C3-2-12-1 and C3-2-12-2 indicates a dominating direct compressive strut towards the support. In contrast to this, the identically reinforced C3-2-12-3 has a steeper critical shear crack and thus a reduced direct load transfer. Due to the significantly lower ultimate force, fewer bending cracks and fewer longitudinal cracks are visible for this last specimen (Figure 5).

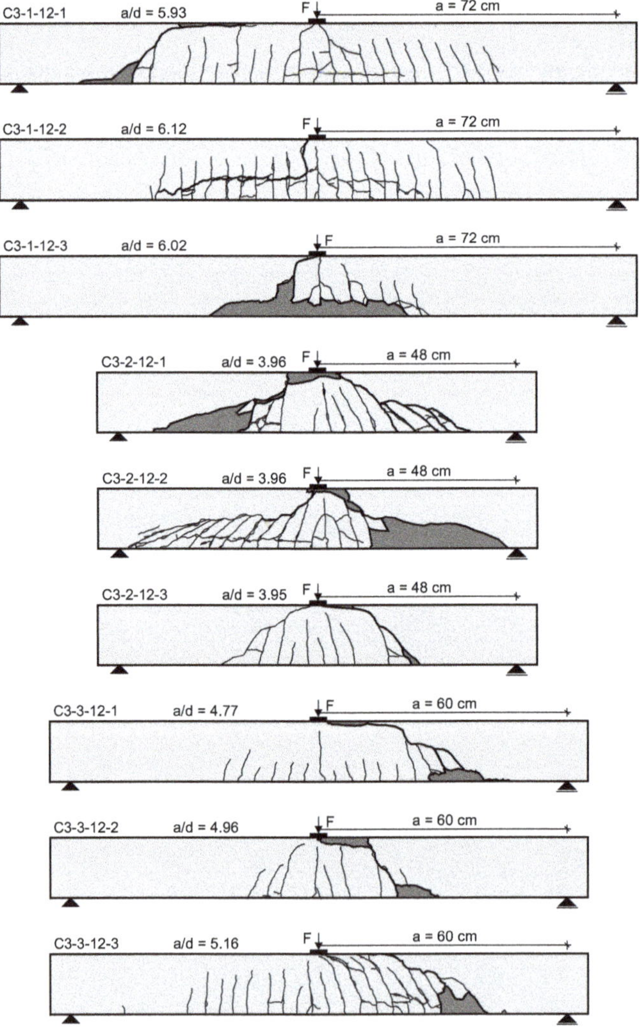

Figure 5. Crack pattern and failure mechanism for $d = 12$ cm shear test series.

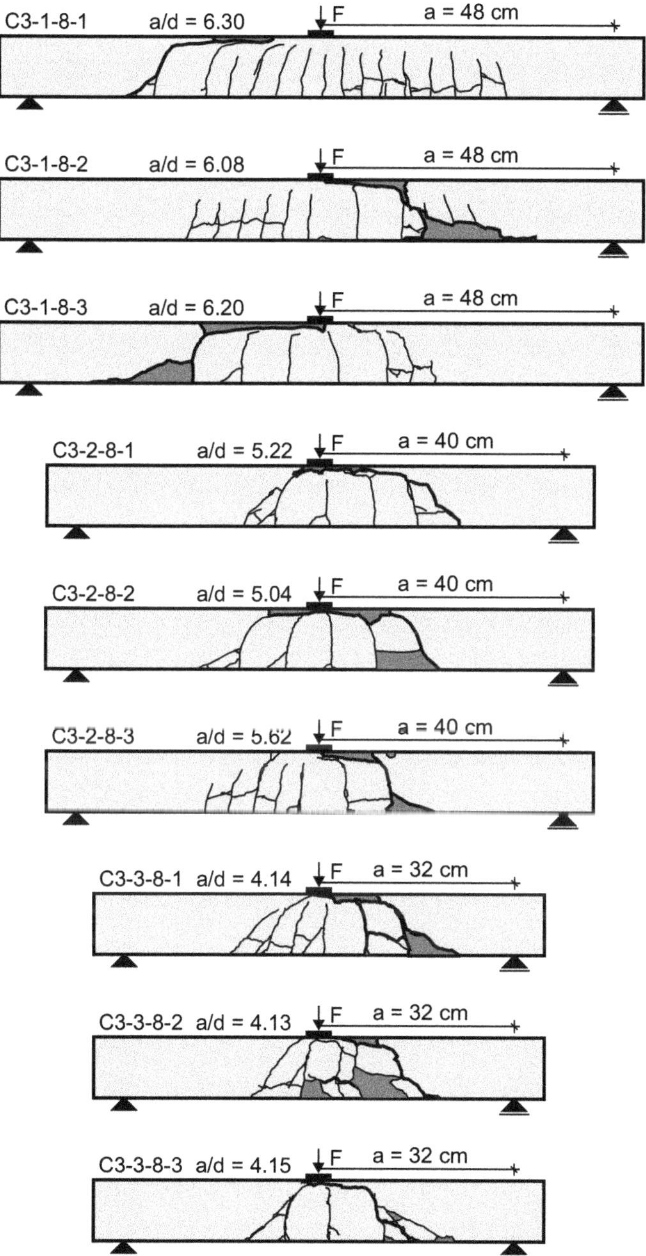

Figure 6. Crack pattern and failure mechanism for $d = 8$ cm shear test series.

Table 4. Experimental results of shear tests.

Specimen	Total Length	Span Length	Width	Height	Effective Depth	Shear Span	Shear Slenderness	No. Layers	No. Yarns	Reinforcement Area	Ratio	age	Concrete Characteristics					Initial Crack Load	Ultimate Shear Load	Ultimate Bending Moment	Failure Mode
	l_{tot}	l	b	h	d	a	a/d	n_{layer}	n_{yarn}	A_{nm}	ρ_l	age	$f_{cm,cube}$	$f_{cm,cyl}$	$f_{cm,pris}$	$f_{ctm,fl}$	E_{cm}	V_{crack}	V_u	M_u	
	[mm]	[mm]	[mm]	[mm]	[mm]	[mm]	[-]	[-]	[-]	[mm^2]	[%]	[d]	[MPa]	[MPa]	[MPa]	[MPa]	[MPa]	[kN]	[kN]	[kNm]	
C3-1-12-1	1540	1440	198	145	121	720	5.93	3	15	54.3	0.226	28	127.1	106.2	119.1	16.4	44,010	10.03	21.91	15.77	Shear compression
C3-1-12-2	1540	1440	200	145	118	720	6.12	3	15	54.3	0.231	28	127.1	106.2	119.1	16.4	44,010	10.85	21.03	15.13	Shear compression
C3-1-12-3	1540	1440	199	144	120	720	6.02	3	15	54.3	0.228	28	127.1	106.2	119.1	16.4	44,010	10.32	14.55	10.47	Shear compression
C3-2-12-1	1060	960	199	145	121	480	3.96	3	15	54.3	0.225	28	126.1	101.1	120.3	18.1	45,346	16.07	27.30	12.98	Shear compression
C3-2-12-2	1060	960	200	145	121	480	3.96	3	15	54.3	0.224	28	126.1	101.1	120.3	18.1	45,346	17.50	35.19	16.89	Shear compression
C3-2-12-3	1060	960	200	146	122	480	3.95	3	15	54.3	0.223	28	126.1	101.1	120.3	18.1	45,346	17.15	20.59	9.88	Shear compression
C3-3-12-1	1300	1200	199	147	126	600	4.77	3	15	54.3	0.217	28	129.6	108.9	126.6	15.5	44,429	12.70	20.73	12.43	Shear compression
C3-3-12-2	1300	1200	200	143	121	600	4.96	3	15	54.3	0.224	28	129.6	108.9	126.6	15.5	44,429	12.74	16.56	9.93	Shear compression
C3-3-12-3	1300	1200	200	145	116	600	5.16	3	15	54.3	0.234	28	129.6	108.9	126.6	15.5	44,429	13.75	21.43	12.85	Shear compression
C3-1-8-1	1060	960	200	99	76	480	6.30	2	10	36.2	0.238	29	127.1	106.2	116.9	15.7	44,010	8.75	16.86	8.09	Shear compression
C3-1-8-2	1060	960	200	100	79	480	6.08	2	10	36.2	0.229	29	127.1	106.2	116.9	15.7	44,010	8.78	14.80	7.10	Shear compression
C3-1-8-3	1060	960	201	100	77	480	6.20	2	10	36.2	0.233	29	127.1	106.2	116.9	15.7	44,010	8.32	15.87	7.62	Shear compression
C3-2-8-1	900	800	200	99	77	400	5.22	2	10	36.2	0.236	29	126.1	101.1	113.0	15.1	45,346	10.42	14.43	5.77	Shear compression
C3-2-8-2	900	800	199	100	79	400	5.04	2	10	36.2	0.229	29	126.1	101.1	113.0	15.1	45,346	11.01	15.89	6.35	Shear compression
C3-2-8-3	900	800	199	100	71	400	5.62	2	10	36.2	0.256	29	126.1	101.1	113.0	15.1	45,346	10.38	14.60	5.84	Shear compression
C3-3-8-1	740	640	200	100	77	320	4.14	2	10	36.2	0.234	27	129.6	108.9	126.7	11.9	44,429	13.33	17.60	5.63	Shear compression
C3-3-8-2	740	640	200	99	77	320	4.13	2	10	36.2	0.234	27	129.6	108.9	126.7	11.9	44,429	13.29	20.04	6.41	Shear compression
C3-3-8-3	740	640	198	99	77	320	4.15	2	10	36.2	0.237	27	129.6	108.9	126.7	11.9	44,429	12.42	20.38	6.52	Shear compression
C3-1-4-1	580	480	198	60	35	240	6.92	1	5	18.1	0.264	30	127.1	106.2	106.7	15.7	44,010	7.58	7.15	1.71	Bending (subseq. shear)
C3-1-4-2	580	480	198	62	41	240	5.79	1	5	18.1	0.220	30	127.1	106.2	106.7	15.7	44,010	7.98	8.89	2.13	Bending (yarn rupture)
C3-1-4-3	580	480	198	59	34	240	6.99	1	5	18.1	0.267	30	127.1	106.2	106.7	15.7	44,010	6.87	7.67	1.84	Bending (yarn rupture)
C3-2-4-1	500	400	198	60	39	200	5.12	1	5	18.1	0.234	32	126.1	101.1	127.6	17.2	45,346	9.57	9.73	1.94	Bending (yarn rupture)
C3-2-4-2	500	400	198	61	36	200	5.57	1	5	18.1	0.254	32	126.1	101.1	127.6	17.2	45,346	10.41	8.56	1.71	Bending (yarn rupture)
C3-2-4-3	500	400	198	61	41	200	4.91	1	5	18.1	0.224	32	126.1	101.1	127.6	17.2	45,346	10.01	9.27	1.85	Bending (yarn rupture)
C3-3-4-1	420	320	197	61	41	160	3.93	1	5	18.1	0.226	29	129.6	108.9	124.2	10.8	44,429	9.48	12.72	2.03	Bending (subseq. shear)
C3-3-4-2	420	320	198	61	36	160	4.44	1	5	18.1	0.254	29	129.6	108.9	124.2	10.8	44,429	10.09	9.93	1.59	Bending (yarn rupture)
C3-3-4-3	420	320	198	60	36	160	4.45	1	5	18.1	0.255	29	129.6	108.9	124.2	10.8	44,429	9.99	11.69	1.87	Bending (subseq. shear)

4. Discussion

Although the observations allow for discussion of several phenomena of shear behavior of TRC, only the influence of shear span and size effect are briefly discussed in this paper. Furthermore, the prediction of selected current design provisions for shear capacity of FRP or steel reinforced concrete are compared to the own experiments.

4.1. Effect of Shear Span Length

The question of an influence by the shear slenderness on shear capacity is of high interest for both researchers and for design in practice. Researchers need to design future experiments with an appropriate load-to-support distance to avoid overestimation of the shear capacity. On the other hand, engineers in practice exploit the direct stress transfer of concentrated loads near supports by reduction of the design shear force according current design provisions.

Figure 7a shows the ultimate shear force from all experiments in relation to shear slenderness. The shear force caused by self-weight of the specimens is neglected, because it differs along the shear span and along the critical shear crack. All results are displayed regardless of their failure mechanisms (a usual procedure, see for example in [75]). As discussed before, the specimens with $d = 4$ cm failed in bending. These results are therefore to be considered and compared with caution.

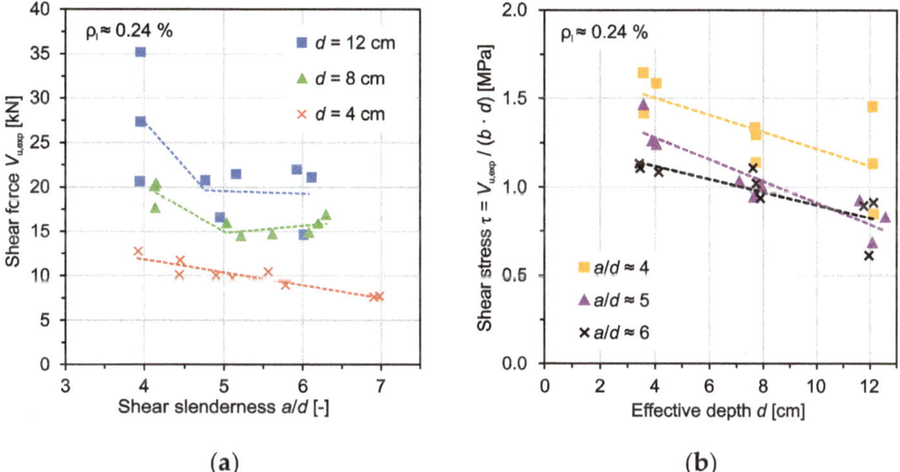

Figure 7. Results of experimental test program: (**a**) Influence of the ratio of shear span to effective depth on shear force; (**b**) Influence of effective depth on shear stress (size effect).

The highest shear forces in this test series occurred for an a/d-ratio of 4. It is noteworthy that there, the scatter for the three identical beams with the largest effective depth is highest. With increasing a/d, the shear resistance diminishes. This general observation is not surprising and well-known from steel-reinforced concrete [75] and concrete with FRP-reinforcement [76,77]. However, it should be highlighted that usually, an influence of a direct compression field is assumed up to $a/d = 2.5$–3. Here, the significant difference of shear resistance between $a/d = 4$ and 5 (while all other experimental parameters are kept the same) indicates a direct load transfer between load introduction and support, at least up to the shear slenderness of 4. There is no significant difference between $a/d = 5$ and 6, as indicated by the red and green dashed trend lines. The red dashed linear trend line for an effective depth of 4 cm is significantly influenced by the three values at $a/d = 7$, where earlier bending failure due to the higher moment governed. One possible explanation for the phenomenon of increased direct load transfer might be the influence of longitudinal cracking in the layer of reinforcement. The crack patterns in Figures 5 and 6 clearly show longitudinal cracks (resulting from high local bond stress introduced by the reinforcement) at the level of one layer of reinforcement, which connect individual bending cracks. Those longitudinal cracks lead to a reduction or even a total loss of local bond between reinforcement and matrix. Yet, the end-anchorage was sufficient, as the longitudinal cracks did not propagate to or beyond the supports. In consequence, the tension stress in the reinforcement is constant in the center part along the beam length, resulting in a more efficient tied arch action. The analysis of the crack pattern of the beams with the highest loads (C3-2-12-1, 27.3 kN and C3-2-12-2, 35.2 kN) underlines this hypothesis. There, longitudinal cracks are present almost all the way to the support, whereas for C3-2-12-2 with its 71% lower resistance (20.6 kN), fewer longitudinal cracks were visible. The first two specimens' shear resistance seems to be significantly influenced by tied arch action rather than beam action.

4.2. Size Effect

An influence of the effective depth on the shear resistance (size effect) can be derived from Figure 7b. A clear trend of diminishing shear resistance for increasing effective depths is visible. Once again, all results are presented regardless of their failure mechanism. Note that despite bending failure, the specimens with a d of ~40 mm bear the highest shear stresses. The size effect has been described extensively in literature by various researchers, of whom Bazant is arguably the most renown (see the extensive compilation in [78]). Especially for shear, current design provisions consider the size effect either directly (by a reduction factor as Eurocode 2 [79]), incorporated in the strain (e.g., in the Modified Compression Field Theory [80] and the Critical Shear Crack Theory [22,23]), or in a combined factor as in [27].

Bazant and Kim [81] describe the structural size effect according to Equation (1)

$$\phi(d) = \frac{1}{\sqrt{1+\frac{d}{\lambda_0 \cdot d_a}}} \tag{1}$$

where d = effective depth, λ_0 = empirical constant, and d_a = maximum aggregate size. For steel reinforced concrete beams, an empirical value of 25 was determined from a large set of experiments [81]. For the typical lower limit for the maximum aggregate size in normal concrete of 8 mm, the term $\lambda_0 \cdot d_a$ leads to a constant value of 200, which is often used in shear prediction models (e.g., [26]). Although the data set in this study is too small to properly adjust an empirical constant for TRC, the linear trend lines in Figure 8 indicate that the first approach with $\lambda_0 = 25$ and $d_a = 4$ mm already led to satisfying results. It should be mentioned that in the present study, the concrete cracks passed through the aggregate grains rather than around them and thus the influence of aggregate size is debatable.

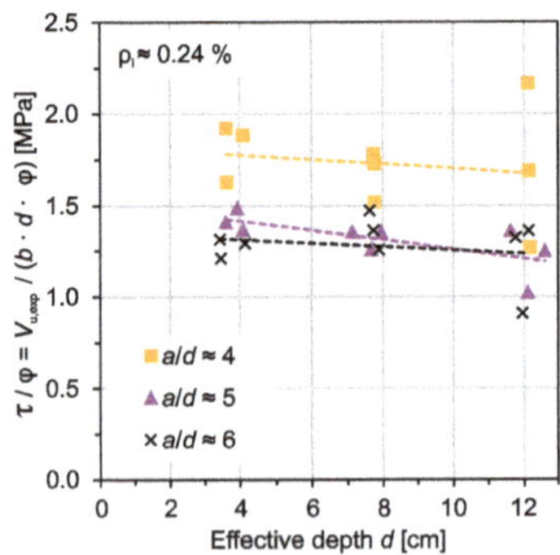

Figure 8. Relative shear stress considering the size effect factor according to Bazant [81] with $\lambda_0 = 25$ and $d_a = 4$ mm.

In contrast to the size effect law in (1), Eurocode 2 [79] limits the influence of size effect to a lower boundary of d = 200 mm. This derives both from the typical slab or beam dimensions and the typical minimum aggregate sizes of 8 mm. For TRC, smaller aggregate sizes as well as reduced depth are typical and intended. Thus, a lower limit is considered critical and more general approaches should be used for TRC. For future research, it could be advisable to additionally test specimens with doubled effective depth in logarithmic scale (10 times the current depth) in order to validate the fit of chosen size effect factors. Alternatively, nonlinear finite element modeling can be used to validate the size effect laws (e.g., [82]).

4.3. Comparison to Existing Models and Current Design Provisions

For beams and slabs with FRP-longitudinal reinforcement without shear reinforcement, engineering models and design formulas have been derived and validated by various researchers. Whether these models are directly applicable to the non-metallic grid-like carbon textile reinforcement in slabs or slab segments is discussed in this section. Due to the limited variation in key parameters such as reinforcement ratio, reinforcement ultimate strength, reinforcement modulus of elasticity, and concrete strength, no generalizable statement on transferability is possible. However, the comparison of the experimental results to predicted values enables a first assessment of the applicability of existing models and thus prepares future work.

Two models and two design provisions were chosen for calculation. The model of Mari et al. [46] has been developed specifically for FRP reinforcement based on a mechanical approach in combination with evaluation of an extensive database with genetic programming. An even more general (and more recent) model for steel-reinforced members by the same authors, the Compression Chord Capacity Model [27], is based on similar assumptions but differs in the calculation of the size effect factor. Here, however, the original model for FRP reinforcement has been used. The second model is the simplified shear design approach by Cavagnis, Fernandez Ruiz, and Muttoni [83,84] based on the Critical Shear Crack Theory [22,23]. This model is currently part of the discussion for the upcoming revision of Eurocode 2 [85] and can be used for FRP reinforcement by considering the modulus of elasticity of the longitudinal reinforcement. Keeping in mind that using similar approaches towards the shear design procedures for conventional reinforcement and non-metallic reinforcement is of major interest for practice, the application of the model in this paper is justified.

With the American Concrete Institute (ACI) code 440.1R-15 [54] and the Canadian Standards Association (CSA) Code S806-12 [53], American codes provide established models for shear design of FRP-reinforced concrete members. In contrast to European standardization, a longer experience for the use of FRP especially for bridges exist. Both codes follow the respective shear design tradition for reinforced concrete.

In Table 5, the formulae and variables necessary for calculation of the selected models are summarized. The last column gives comments and explanations regarding how the various parameters were specifically set for prediction of the experimental results of this paper.

Table 5. Summary of design provisions and models used in the study.

Code/Model	Shear Strength Prediction	Comments
CSCT/EC2 D3 [83–85]	$V_R = b_w \cdot d \cdot \tau_{Rd,c} \geq b_w \cdot d \cdot \tau_{Rdc,min}$ $\tau_{Rd,c} = 0.6 \cdot \left(100 \cdot \rho_l \cdot \frac{E_s}{200,000} \cdot f_c \cdot \frac{d_{dg}}{d}\right)^{1/3}$ $\rho_l = \frac{A_s}{b \cdot d}$, A_s : Area of longitudinal reinforcement b : web width, d : effective depth E_s : Modulus of elasticity of the longitudinal reinforcement $d_{dg} = \begin{cases} 16 + D_{lower} \leq 40 \text{ [mm] } for\ f_c \leq 60 \text{ MPa} \\ 16 + D_{lower} \cdot \left(\frac{60}{f_c}\right)^2 \leq 40 \text{ [mm] } for\ f_c > 60 \text{ MPa} \end{cases}$ D_{lower} : The smallest value of D_{max} (coarsest fraction of aggregates) $d = \begin{cases} a_v = \sqrt{\frac{a_{cs}}{4} \cdot d} & if\ a_{cs} \leq 4 \cdot d \\ d & if\ a_{cs} > 4 \cdot d \end{cases}$ $with\ a_{cs} = \left\| \frac{M_{Ed}}{V_{Ed}} \right\| \geq d$ $\tau_{Rdc,min} = 0.021 \sqrt{E_s \cdot \frac{f_c}{f_y} \cdot \frac{d_{dg}}{d}}$ f_y: Yield strength or strength that has been assumed for the flexural design of the cross-section	with $\gamma_c = 1.0$ f_c taken as $f_{cm,cyl}$ f_c taken as $f_{cm,cyl}$ Here, D_{lower} is taken as 4 mm Here, $a_v = d/2$. For 3-point loaded single span beams, $\|M_{Ed}/V_{Ed}\| = a$ Here, f_y is taken as mean ultimate reinforcement stress (3221 MPa, Table 2).
Mari et al. [46]	$V_{ut} = f_{ct,m} \cdot \zeta \cdot b \cdot d \cdot \left((1.072 - 0.01 \cdot \alpha) \cdot \frac{c}{d} + 0.036\right)$ $f_{ct,m} = \begin{cases} 0.3 \cdot f_c^{\left(\frac{2}{3}\right)} & if\ f_c \leq 50 \text{ N/mm}^2 \\ 2.12 \cdot \ln\left(1 + \frac{f_c}{10}\right) & if\ f_c > 50 \text{ N/mm}^2 \end{cases}$ $\zeta = 1.20 - 0.20 \cdot \frac{a}{d} \cdot d$ $\alpha = \frac{E_r}{E_c}$; E_r, E_c : modulus of elasticity of reinforcement and concrete $\frac{c}{d} = \alpha \cdot \rho_r \cdot \left(1 + \sqrt{1 + \left(\frac{2}{\alpha \cdot \rho_r}\right)}\right)$ $\rho_r = \frac{A_s}{b \cdot d}$, A_s : Area of longitudinal reinforcement b and d: Web width and effective depth, respectively	Calculation of $f_{ct,m}$ according to EC2, f_c taken as $f_{cm,cyl}$ a and d in m E_r, E_c taken from experimental data (see Table 4).
CSA S806-12 [53]	$V_c = 0.05 \cdot \lambda \cdot k_m \cdot k_r \cdot k_s \cdot (f_c')^{\left(\frac{1}{3}\right)} \cdot b_w \cdot d_v$ $0.11 \cdot \sqrt{f_c'} \cdot b_w \cdot d_v < V_c \leq 0.22 \cdot \sqrt{f_c'} \cdot b_w \cdot d_v$ $\lambda = 1.0$ for normal concrete $k_m = \sqrt{\frac{V_f \cdot d}{M_f}}$ V_f, M_f : Acting shear force and moment at the control section $k_r = 1 + \left(E_F \cdot \rho_{Ff}\right)^{\frac{1}{3}}$ E_F : Modulus of elasticity of longitudinal reinforcement $\rho_{Ff} = \frac{A_F}{b \cdot d}$; A_F : Area of longitudinal reinforcement $k_s = \frac{750}{450 + d} \leq 1.0$ $f_c' \leq 60$ MPa $d_v = \max(0.9 \cdot d; 0.72 \cdot h)$ b_w, d, h: Web width, effective depth and member height, respectively	with $\phi_c = 1.0$ For 3-point loaded single span beams, $V_f/M_f = 1/a$ d in mm f'_c taken as $f_{cm,cyl} < 60 = 60$ MPa
ACI 440.1R-15 [54]	$V_c = \frac{2}{5} \sqrt{f_c'} \cdot b_w \cdot k \cdot d$ $f_c' \leq 69$ MPa $k = \sqrt{2 \cdot \rho_f \cdot n_f + \left(\rho_f \cdot n_f\right)^2} - \rho_f \cdot n_f$ ρ_f : FRP reinforcement ratio $= \frac{A_f}{b_w \cdot d}$ $n_f = \frac{E_f}{E_c}$; E_f, E_c : Modulus of elasticity of reinforcement and concrete b_w and d: Web width and effective depth, respectively	Limitation of f'_c to 10,000 Psi according to [86]

The results of the comparison for all four models are shown in Figure 9. The mean value of the experimental to theoretical shear force ratio V_{exp}/V_{calc} is an indicator of accuracy, while the COV is used as a measure of precision. Generally, all models except ACI 440.1R-15 show promising results. The best mean value is obtained by the simplified approach based on critical shear crack theory with $V_{exp}/V_{calc} = 0.95$, while the CSA S806-12-model leads to the lowest COV with 18.9%.

The ACI 440.1R-15 leads to conservative results, with a mean ratio $V_{exp}/V_{calc} = 2.3$. Furthermore, this model shows the lowest COV with 22.3%. This observation is consistent with those of other researchers, e.g., [46]. For all models, the high scatter of ultimate shear load for the three largest plate segments (C3-1-12) significantly influences the prediction results. The models cannot represent the direct strut action (tied arch action) adequately. However, these first results indicate that the prediction of shear capacity for TRC with suitable existing models for FRP reinforced concrete or TRC is possible. In order to generalize this finding, comparisons to larger experimental data sets are required. Therefore, especially the variation of the modulus of elasticity of the reinforcement, e.g., by the use of impregnated glass textiles, should be focused in future experimental studies.

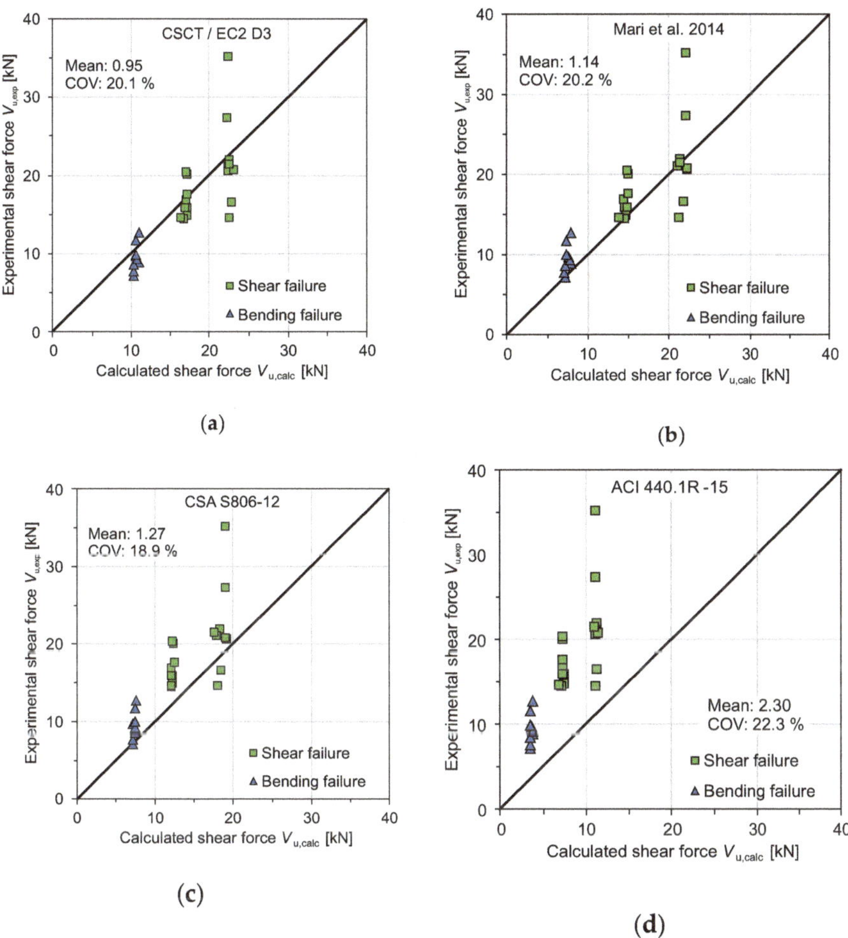

Figure 9. Comparison of experimental results to predicted shear force: (**a**) Critical shear crack theory (CSCT) [83–85]; (**b**) Model by Mari et al. [46]; (**c**) Design formula from CSA S8-06-12 [53]; (**d**) Design formula from ACI 440.1R-15 [54].

5. Conclusions

While numerous studies of FRP-reinforced concrete exist, experimental data and systematic studies on shear behavior of TRC are sparse. Despite similar raw materials (glass, carbon), grid-like textile reinforcement features certain differences compared to FRP bars, e.g. the smaller size and thus lower bending stiffness, the fixed transversal yarns, and the possible variation of cross-sectional dimensions in the longitudinal direction. The use of TRC aims at reducing the member sizes compared to conventional reinforced concrete. This leads to a typically reduced thickness in planar members (e.g., slabs), for which fewer experiments exist both in FRP and steel reinforced concrete.

Despite these differences, the results from an experimental program on 27 slab segments show clear parallels in shear behavior. The formation of the critical shear crack from bending cracks and the similar failure mechanism (shear compression failure) are evident. The formation of longitudinal cracks in the layer of reinforcement combined with concrete spalling can be seen as reason for high scatter in otherwise identical members.

By variation of shear span to depth ratio, the influence of direct load transfer from concentrated loads was investigated. It can be concluded that here, with a shear slenderness larger $a/d = 5$, no significant reduction of shear capacity occurs. However, a significant difference between $a/d = 4$ and 5 can be seen, which is a significant difference to reinforced concrete where an influence up to $a/d < 2.5$–3 is typical. This might be a result from the bond characteristics of the textile reinforcement with its significant longitudinal crack formation in the layer of reinforcement.

The variation of member height in the experimental program showed a reduction of relative shear capacity by increasing effective depth. A first approach to transfer models for consideration of size effect indicates that existing models can be used in TRC shear design.

The comparison of the experimental results to predictions from existing models for shear resistance of FRP-reinforced concrete indicates promising results; several current, readily available models are able to predict the ultimate shear force obtained in the experiments presented here with sufficient accuracy. However, this observation is yet to be validated by comparison to larger data sets in future works. There, a systematic variation of the type of reinforcement (grid opening, yarn spacing), the modulus of elasticity and strength of the reinforcement, the reinforcement ratio, and the maximum grain size and compressive strength of the concrete is necessary.

Author Contributions: Conceptualization, J.B. and V.A.; experiments: J.B.; formal analysis, J.B. and M.C.; writing—original draft preparation, J.B.; writing—review and editing, V.A., J.H. and M.C.; visualization, J.B. and V.A.; supervision, J.H. and M.C.; project administration, J.H.; funding acquisition, J.B. and J.H.

Funding: This research was funded by the German Federal Ministry of Education and Research (BMBF), grant number 03ZZ0304I. The APC was funded by German Federal Ministry of Education and Research (BMBF), grant number 16PGF0147.

Acknowledgments: The authors give their thanks to solidian GmbH for donation of the carbon grid reinforcement used for the experiments. The authors are particularly thankful for the work of Mr. Alexander Böning in preparing and testing the specimens.

References

1. Hegger, J.; Curbach, M.; Stark, A.; Wilhelm, S.; Farwig, K. Innovative design concepts: Application of textile reinforced concrete to shell structures. *Struct. Concr.* **2018**, *19*, 637–646. [CrossRef]
2. Tysmans, T.; Adriaenssens, S.; Cuypers, H.; Wastiels, J. Structural analysis of small span textile reinforced concrete shells with double curvature. *Compos. Sci. Technol.* **2009**, *69*, 1790–1796. [CrossRef]
3. Kromoser, B.; Preinstorfer, P.; Kollegger, J. Building lightweight structures with carbon-fiber-reinforced polymer-reinforced ultra-high-performance concrete: Research approach, construction materials, and conceptual design of three building components. *Struct. Concr.* **2018**, *112*, 106. [CrossRef]
4. Hegger, J.; Voss, S. Investigations on the bearing behaviour and application potential of textile reinforced concrete. *Eng. Struct.* **2008**, *30*, 2050–2056. [CrossRef]
5. Sharei, E.; Scholzen, A.; Hegger, J.; Chudoba, R. Structural behavior of a lightweight, textile-reinforced concrete barrel vault shell. *Compos. Struct.* **2017**, *171*, 505–514. [CrossRef]
6. De Munck, M.; Tysmans, T.; Verbruggen, S.; Vervloet, J.; El Kadi, M.; Wastiels, J.; Remy, O. Influence of Weathering Conditions on TRC Sandwich Renovation Panels. In *Strain-Hardening Cement-Based Composites, SHCC4*; Mechtcherine, V., Slowik, V., Kabele, P., Eds.; Springer: Berlin/Heidelberg, Germany, 2017; pp. 659–667, ISBN 978-94-024-1193-5.
7. Scheerer, S.; Chudoba, R.; Garibaldi, M.P.; Curbach, M. Shells Made of Textile Reinforced Concrete—Applications in Germany. *J. IASS* **2017**, *58*, 79–93. [CrossRef]
8. De Sutter, S.; Remy, O.; Tysmans, T.; Wastiels, J. Development and experimental validation of a lightweight Stay-in-Place composite formwork for concrete beams. *Constr. Build. Mater.* **2014**, *63*, 33–39. [CrossRef]

9. May, S.; Michler, H.; Schladitz, F.; Curbach, M. Lightweight ceiling system made of carbon reinforced concrete. *Struct. Concr.* **2018**, *19*, 1862–1872. [CrossRef]
10. Bielak, J.; Hegger, J.; Chudoba, R. Towards Standardization: Testing and Design of Carbon Concrete Composites. In *High Tech Concrete: Where Technology and Engineering Meet, Proceedings of the 2017 fib Symposium, Maastricht, The Netherlands, 12–14 June 2017*; Hordijk, D.A., Luković, M., Eds.; Springer International Publishing: Cham, Switzerland, 2017; pp. 313–320, ISBN 3319594710.
11. Hegger, J.; Kulas, C.; Horstmann, M. Realization of TRC Façades with Impregnated AR-Glass Textiles. *KEM* **2011**, *466*, 121–130. [CrossRef]
12. Kulas, C.; Schneider, M.; Will, N.; Grebe, R. Hinterlüftete Vorhangfassaden aus Textilbeton. *Bautechnik* **2011**, *88*, 271–280. [CrossRef]
13. Hegger, J.; Horstmann, M.; Shams, A. Load-carrying behavior of sandwich panels at ultimate limite state. In Proceedings of the fib Symposium Prague 2011, Praha, Czech Republic, 8–10 June 2011.
14. Shams, A.; Horstmann, M.; Hegger, J. Experimental investigations on Textile-Reinforced Concrete (TRC) sandwich sections. *Compos. Struct.* **2014**, *118*, 643–653. [CrossRef]
15. Hering Bau GmbH & Co. KG. Hering Architectural Concrete: betoshell®. Available online: https://www.heringinternational.com/en/products-services/architectural-concrete/material-concepts/textile-reinforced-concrete/ (accessed on 15 February 2019).
16. Rempel, S.; Kulas, C.; Will, N.; Bielak, J. Extremely Light and Slender Precast Pedestrian-Bridge Made Out of Textile-Reinforced Concrete (TRC). In *High Tech Concrete: Where Technology and Engineering Meet, Proceedings of the 2017 fib Symposium, Maastricht, The Netherlands, 12–14 June 2017*; Hordijk, D.A., Luković, M., Eds.; Springer International Publishing: Cham, Switzerland, 2017; pp. 2530–2537, ISBN 3319594710.
17. Helbig, T.; Unterer, K.; Kulas, C.; Rempel, S.; Hegger, J. Fuß- und Radwegbrücke aus Carbonbeton in Albstadt-Ebingen. *Beton- und Stahlbetonbau* **2016**, *111*, 676–685. [CrossRef]
18. Hegger, J.; Goralski, C.; Kulas, C. Schlanke Fußgängerbrücke aus Textilbeton. *Beton- und Stahlbetonbau* **2011**, *106*, 64–71. [CrossRef]
19. Hegger, J.; Kulas, C.; Raupach, M.; Büttner, T. Tragverhalten und Dauerhaftigkeit einer schlanken Textilbetonbrücke. *Beton- und Stahlbetonbau* **2011**, *106*, 72–80. [CrossRef]
20. Rempel, S. Erste Straßenbrücke aus Carbonbeton. In *Tagungsband der 9. Carbon- und Textilbetontage*; Tudalit e.V. und C3—Carbon Concrete Composite e.V.: Dresden, Germany, 2017; pp. 40–41.
21. Schumann, A.; Michler, H.; Schladitz, F.; Curbach, M. Parking slabs made of carbon reinforced concrete. *Struct. Concr.* **2018**, *19*, 647–655. [CrossRef]
22. Cavagnis, F.; Fernández Ruiz, M.; Muttoni, A. A mechanical model for failures in shear of members without transverse reinforcement based on development of a critical shear crack. *Eng. Struct.* **2018**, *157*, 300–315. [CrossRef]
23. Cavagnis, F.; Fernández Ruiz, M.; Muttoni, A. Shear failures in reinforced concrete members without transverse reinforcement: An analysis of the critical shear crack development on the basis of test results. *Eng. Struct.* **2015**, *103*, 157–173. [CrossRef]
24. Herbrand, M.; Kueres, D.; Claßen, M.; Hegger, J. Einheitliches Querkraftmodell zur Bemessung von Stahl- und Spannbetonbrücken im Bestand. *Beton- und Stahlbetonbau* **2016**, *111*, 58–67. [CrossRef]
25. Tue, N.V.; Theiler, W.; Tung, N.D. Schubverhalten von Biegebauteilen ohne Querkraftbewehrung. *Beton- und Stahlbetonbau* **2014**, *109*, 666–677. [CrossRef]
26. Herbrand, M. Shear Strength Models for Reinforced and Prestressed Concrete Members. Ph.D. Thesis, RWTH Aachen University, Aachen, Germany, 2017.
27. Cladera, A.; Marí, A.; Bairán, J.M.; Ribas, C.; Oller, E.; Duarte, N. The compression chord capacity model for the shear design and assessment of reinforced and prestressed concrete beams. *Struct. Concr.* **2016**, *17*, 1017–1032. [CrossRef]
28. Adam, V.; Herbrand, M.; Claßen, M. Experimentelle Untersuchungen zum Einfluss der Bauteilbreite und der Schubschlankheit auf die Querkrafttragfähigkeit von Stahlbetonplatten ohne Querkraftbewehrung. *Bauingenieur* **2018**, *93*, 37–45.
29. Herbrand, M.; Classen, M. Shear tests on continuous prestressed concrete beams with external prestressing. *Struct. Concr.* **2015**, *16*, 428–437. [CrossRef]
30. Herbrand, M.; Classen, M.; Adam, V. Querkraftversuche an Spannbetondurchlaufträgern mit Rechteck- und I-Querschnitt. *Bauingenieur* **2017**, *92*, 465–473.

31. Fischer, O.; Schramm, N.; Gehrlein, S. Labor- und Feldversuche zur realitätsnahen Beurteilung der Querkrafttragfähigkeit von bestehenden Spannbetonbrücken. *Bauingenieur* **2017**, *92*, 465–473.

32. Rombach, G.; Henze, L. Querkrafttragfähigkeit von Stahlbetonplatten ohne Querkraftbewehrung unter konzentrierten Einzellasten. *Beton- und Stahlbetonbau* **2017**, *112*, 568–578. [CrossRef]

33. Gleich, P.; Maurer, R. Querkraftversuche an Spannbetondurchlaufträgern mit Plattenbalkenquerschnitt. *Bauingenieur* **2018**, *93*, 68–72.

34. Reissen, K.; Classen, M.; Hegger, J. Shear in reinforced concrete slabs—Experimental investigations in the effective shear width of one-way slabs under concentrated loads and with different degrees of rotational restraint. *Struct. Concr.* **2018**, *19*, 36–48. [CrossRef]

35. Huber, P.; Kromoser, B.; Huber, T.; Kollegger, J. Experimentelle Untersuchung zum Querkrafttragverhalten von Spannbetonträgern mit geringer Schubbewehrung. *Bauingenieur* **2016**, *91*, 238–247.

36. Huber, P.; Kromoser, B.; Huber, T.; Kollegger, J. Berechnungsansatz zur Ermittlung der Schubtragfähigkeit bestehender Spannbetonbrückenträger mit geringem Querkraftbewehrungsgrad. *Bauingenieur* **2016**, *91*, 227–237.

37. Javidmehr, S.; Oettel, V.; Empelmann, M. Schrägrissbildung von Stahlbetonbalken unter Querkraftbeanspruchung. *Bauingenieur* **2018**, *93*, 248–254.

38. Alam, M.; Hussein, A. Shear Strength of concrete beams reinforced with Glass Fibre Reinforced Polymer (GFRP) Bars. In *Fibre-Reinforced Polymer Reinforcement for Concrete Structures, Proceedings of the 9th International Conference (FRPRCS-9), Sydney, Australia, 13–15 July 2009*; Oehlers, D.J., Griffith, M.C., Seracino, R., Eds.; University of Adelaide: Adelaide, Australia, 2009; ISBN 9780980675504.

39. Olivito, R.S.; Zuccarello, F.A. On the Shear Behaviour of Concrete Beams Reinforced by Carbon Fibre-Reinforced Polymer Bars: An Experimental Investigation by Means of Acoustic Emission Technique. *Strain* **2010**, *46*, 470–481. [CrossRef]

40. El-Sayed, A.K.; El-Salakawy, E.F.; Benmokrane, B. Shear strength of fibre-reinforced polymer reinforced concrete deep beams without web reinforcement. *Can. J. Civ. Eng.* **2012**, *39*, 546–555. [CrossRef]

41. Razaqpur, A.G.; Isgor, B.O. Proposed Shear Design Method for FRP-Reinforced Concrete Members without Stirrups. *ACI Struct. J.* **2006**, *103*, 93–102. [CrossRef]

42. Kilpatrick, A.; Easden, L. shear capacity of GFRP reinforced high strength concrete slabs. In *Developments in Mechanics of Structures and Materials, Proceedings of the 18th Australasian Conference on the Mechanics Structures and Materials (ACMSM-18), Perth, Australien, Dezember 2004*; Deeks, A.J., Hao, H., Eds.; CRC Press: Boca Raton, FL, USA, 2004; ISBN 9058096599.

43. Tariq, M. Effects of Flexural Reinforcement Properties on shear Strength of Concrete Beams. Master's Thesis, Dalhousie University, Halifax, NS, Canada, 2003.

44. Alam, M.S. Influence of Different Parameters on Shear Strength of FRP Reinforced Concrete Beams without Web Reinforcement. Ph.D. Thesis, Memorial University of Newfoundland, St. John's, NL, Canada, 2010.

45. Matta, F.; Nanni, A.; Hernandez, T.; Benmokrane, B. Scaling of strength of FRP reinforced concrete beams without shear reinforcement. In *FRP Composites in Civil Engineering, Proceedings of the 4th International Conference (CICE 2008), Zürich, Switzerland, 22–24 July 2008*; Motavalli, M., Ed.; Structural Engineering Research Laboratory EMPA: Dübendorf, Switzerland, 2008; ISBN 9783905594508.

46. Marí, A.; Cladera, A.; Oller, E.; Bairán, J. Shear design of FRP reinforced concrete beams without transverse reinforcement. *Compos. Part B Eng.* **2014**, *57*, 228–241. [CrossRef]

47. Kurth, M.; Hegger, J. Zur Querkrafttragfähigkeit von Betonbauteilen mit Faserverbundkunststoff-Bewehrung—Ableitung eines Bemessungsansatzes. *Bauingenieur* **2013**, *88*, 403–411.

48. Yost, J.R.; Gross, S.P.; Dinehart, D.W. Shear Strength of Normal Strength Concrete Beams Reinforced with Deformed GFRP Bars. *J. Compos. Constr.* **2001**, *5*, 268–275. [CrossRef]

49. Ahmed, E.A.; El-Salakawy, E.; Benmokrane, B. Behavior of concrete beams reinforced with carbon FRP stirrups. In *FRP Composites in Civil Engineering, Proceedings of the 4th International Conference (CICE 2008), Zürich, Switzerland, 22–24 July 2008*; Motavalli, M., Ed.; Structural Engineering Research Laboratory EMPA: Dübendorf, Switzerland, 2008; ISBN 9783905594508.

50. Okamoto, T.; Nagasaka, T.; Tanigaki, M. Shear Capacity of Concrete Beams Using FRP Reinforcement. *Trans. AIJ* **1994**, *59*, 127–136. [CrossRef]

51. Tottori, S.; Wakui, H. Shear Capacity of RC and PC Beams Using FRP Reinforcement. *ACI Spec. Publ.* **1993**, *138*, 615–632.

52. FIB. *Fib Model Code for Concrete Structures 2010*, 1st ed.; Ernst und Sohn: Berlin, Germany, 2013; ISBN 9783433604083.

53. Canadian Standards Association. *S806-12 (R2017): Design and Construction of Building Structures with Fibre-Reinforced Polymers*; Canadian Standards Association: Mississauga, ON, Canada, 2012.

54. American Concrete Institute. *ACI 440.1R-15 Guide for the Design and Constuction of Structural Concrete Reinforced with Fiber-Reinforced Polymer (FRP) Bars*; American Concrete Institute: Farmington Hills, MI, USA, 2015.

55. Kani, G. The Riddle of Shear Failure and its Solution. *ACI Struct. J.* **1964**, *61*, 441–468.

56. Lieboldt, M.; Tietze, M.; Schladitz, F. C^3-Projekt—Erfolgreiche Partnerschaft für Innovation im Bauwesen. *Bauingenieur* **2018**, *93*, 265–273.

57. RWTH Aachen, Lehrstuhl und Institut für Massivbau; Hegger, J.; Chudoba, R.; Scholzen, A.; Bielak, J. *Carbon Conrete composite—C3. Schlussbericht C3-B3 TP9: Standardisierte Prüfkonzepte zur Bauteilprüfung unter zyklischer Belastung: Laufzeit des Vorhabens: 01.01.2015 bis 30.06.2016*; RWTH Aachen University, Lehrstuhl und Institut für Massivbau: Aachen, Germany, 2016.

58. Kromoser, B.; Huber, P.; Preinstorfer, P. Experimental study of the shear behaviour of thin walled CFRP reinforced UHPC structures. In *Better, Smarter, Stronger. Proceedings for the 2018 fib Congress, Melbourne, Australia, 7–11 October 2018*; Foster, S., Gilbert, I.R., Mendis, P., Al-Mahaidi, R., Millar, D., Eds.; Fédération Internationale du Béton (FIB): Lausanne, Switzerland, 2018; pp. 1744–1750.

59. El Kadi, M. Experimental Characterization, Benchmarking and Modelling of 3D Textile Reinforced Cement Composites. Ph.D. Thesis, Vrije Universiteit Brussel, Brussels, Belgium, 2019.

60. Scholzen, A.; Chudoba, R.; Hegger, J. Thin-walled shell structures made of textile-reinforced concrete: Part I. *Struct. Concr.* **2015**, *16*, 106–114. [CrossRef]

61. Hegger, J.; Will, N. Textile-reinforced concrete: Design models. In *Textile Fibre Composites in Civil Engineering*; Triantafillou, T., Ed.; Woodhead Publishing: Oxford, UK, 2016; pp. 189–207, ISBN 1782424466.

62. Schütze, E.; Bielak, J.; Scheerer, S.; Hegger, J.; Curbach, M. Einaxialer Zugversuch für Carbonbeton mit textiler Bewehrung. *Beton- und Stahlbetonbau* **2018**, *113*, 33–47. [CrossRef]

63. Meßerer, D.; Heiden, B., Bielak, J.; Holschemacher, K. Prüfverfahren zur Ermittlung des Krümmungseinflusses auf die Zugfestigkeit von Textilbeton. *Bauingenieur* **2018**, *93*, 454–462.

64. Rempel, S. Zur Zuverlässigkeit der Bemessung von Biegebeanspruchten Betonbauteilen mit Textiler Bewehrung. Ph.D. Thesis, RWTH Aachen University, Aachen, Germany, 2018.

65. Hinzen, M. Prüfmethode zur Ermittlung des Zugtragverhaltens von textiler Bewehrung für Beton. *Bauingenieur* **2017**, *92*, 289–291.

66. Vorechovský, M.; Chudoba, R. Stochastic modeling of multi-filament yarns: II. Random properties over the length and size effect. *Int. J. Solids Struct.* **2006**, *43*, 435–458. [CrossRef]

67. Vorechovsky, M.; Rypl, R.; Chudoba, R. Probabilistic crack bridge model reflecting random bond properties and elastic matrix deformation. *Compos. Part B Eng.* **2018**, *139*, 130–145. [CrossRef]

68. Rempel, S.; Ricker, M. Ermittlung der Materialkennwerte der Bewehrung für die Bemessung von textilbewehrten Bauteilen. *Bauingenieur* **2017**, *92*, 280–288.

69. Bielak, J.; Spelter, A.; Will, N.; Claßen, M. Verankerungsverhalten textiler Bewehrungen in dünnen Betonbauteilen. *Beton- und Stahlbetonbau* **2018**, *113*, 515–524. [CrossRef]

70. Schneider, K.; Butler, M.; Mechtcherine, V. Carbon Concrete Composites C^3—Nachhaltige Bindemittel und Betone für die Zukunft. *Beton- und Stahlbetonbau* **2017**, *112*, 784–794. [CrossRef]

71. DIN Deutsches Institut für Normung e.V. *DIN EN 206 Beton—Festlegung, Eigenschaften, Herstellung und Konformität; Deutsche Fassung EN 206:2013+A1:2016*; ICS 91.100.30; Beuth Verlag GmbH: Berlin, Germany, 2017.

72. DIN Deutsches Institut für Normung e.V. *Prüfverfahren für Zement—Teil 1: Bestimmung der Festigkeit*; 91.100.10 (DIN EN 196-1); Beuth Verlag GmbH: Berlin, Germany, 2005.

73. DIN Deutsches Institut für Normung e.V. *DIN EN 12390-13:2014-06: Prüfung von Festbeton—Bestimmung des Elastitzitätsmoduls unter Druckbelastung (Sekantenmodul). Deutsche Fassung EN 12390-13:2013*; 91.100.30; Beuth Verlag GmbH: Berlin, Germany, 2014.

74. DIN Deutsches Institut für Normung e.V. *Prüfung von Festbeton—Teil 3: Druckfestigkeit von Probeköpern*; 91.100.30 (DIN EN 12390-3); Beuth Verlag GmbH: Berlin, Germany, 2009.

75. Leonhardt, F.; Walther, R. *Schubversuche an Einfeldrigen Stahlbetonbalken mit und ohne Schubbewehrung zur Ermittlung der Schubtragfähigkeit und der oberen Schubspannungsgrenze*; Schriftenreihe des Deutschen Ausschusses für Stahlbeton No. 151; Ernst: Berlin, Germany, 1962.
76. Kurth, M.C. Zum Querkrafttragverhalten von Betonbauteilen mit Faserverbundkunststoff-Bewehrung. Ph.D. Thesis, RWTH Aachen University, Aachen, Germany, 2012.
77. Niewels, J. Zum Tragverhalten von Betonbauteilen mit Faserverbundkunststoff-Bewehrung. Ph.D. Thesis, RWTH Aachen University, Aachen, Germany, 2008.
78. Bazant, Z.P.; Chen, E.-P. Scaling of Structural Failure. *Appl. Mech. Rev.* **1997**, *50*, 593–627. [CrossRef]
79. Deutsches Institut für Normung e.V. *DIN EN 1992-1-1:2011-01: Eurocode 2: Bemessung und Konstruktion von Stahlbeton- und Spannbetontragwerken—Teil 1-1: Allgemeine Bemessungsregeln und Regeln für den Hochbau. Deutsche Fassung EN 1992-1-1:2004 + AC:2010*; 91.010.30; 91.080.40; Beuth Verlag GmbH: Berlin, Germany, 2011.
80. Bentz, E.C.; Veccio, F.J.; Collins, M.P. Simplified Modified Compression Field Theory for calculating shear strength of reinforced concrete members. *ACI Struct. J.* **2006**, *103*, 614–624.
81. Bazant, Z.P.; Kim, J.-K. Size Effect in Shear Failure of Longitudinally Reinforced Beams. *ACI J.* **1984**, *81*, 456–468.
82. Herbrand, M.; Kueres, D.; Stark, A.; Claßen, M. Numerische Simulation von balken- und plattenförmigen Bauteilen aus Stahlbeton und UHPC mit einem plastischen Schädigungsmodell. *Bauingenieur* **2016**, *91*, 46–56.
83. Muttoni, A.; Fernández Ruiz, M.; Cavagnis, F. From detailed test observations to mechanical models and simple shear design equations. In *Towards a Rational Understanding of Shear in Beams and Slabs: Fib Bulletin 85: Workshop in Zürich, Switzerland September 2016*; Bayrak, O., Fernández Ruiz, M., Kaufmann, W., Muttoni, A., Eds.; Fédération Internationale du Béton (FIB): Lausanne, Switzerland, 2018; pp. 17–32, ISBN 2883941254.
84. Cavagnis, F. Shear in Reinforced Concrete without Transverse Reinforcement: from Refined Experimental Measurements to Mechanical Models. Ph.D. Thesis, Ecole Polytechnique Fédérale de Lausanne, Lausanne, Switzerland, 2017.
85. CEN European Committee for Standardization. *prEN 1992-1-1:2018: Eurocode 2: Design of Concrete Structures—Part 1-1: General Rules, Rules for Buildings, Bridges and Civil Engineering Structures. Draft 3*; CEN: Brussels, Belgium, 2018.
86. ACI Committee. *Building Code Requirements for Structural Concrete (ACI 318-14)*; American Concrete Institute (ACI): Farmington Hills, MI, USA, 2014.

 *applied*
sciences

Article

Flexural Strengthening of RC Structures with TRC—Experimental Observations, Design Approach and Application

Silke Scheerer [1,*], Robert Zobel [2], Egbert Müller [1], Tilo Senckpiel-Peters [1], Angela Schmidt [1] and Manfred Curbach [1]

[1]   Institute of Concrete Structures, TU Dresden, 01069 Dresden, Germany;
      Egbert.mueller@tu-dresden.de (E.M.); Tilo.senckpiel-peters@tu-dresden.de (T.S.-P.);
      Angela.schmidt@tu-dresden.de (A.S.); Manfred.curbach@tu-dresden.de (M.C.)
[2]   IBB Ingenieurbüro Baustatik Bautechnik, TU Dresden, 01069 Dresden, Germany; Zobel@ibb-wilhelm.de
*   Correspondence: Silke.scheerer@tu-dresden.de; Tel.: +49-351-463-36527

Received: 22 February 2019; Accepted: 26 March 2019; Published: 29 March 2019

Abstract: Today, the need for structural strengthening is more important than ever. Flexural strengthening with textile reinforced concrete (TRC) is a recommendable addition to already proven methods. In order to use this strengthening method in construction practice, a design model is required. This article gives a brief overview of the basic behavior of reinforced concrete slabs strengthened with TRC in bending tests as already observed by various researchers. Based on this, a design model was developed, which is presented in the main part of the paper. In addition to the model, its assumptions and limits are discussed. The paper is supplemented by selected application examples to show the possibilities of the described strengthening method. Finally, the article will give an outlook on open questions and current research.

Keywords: textile reinforced concrete (TRC); strengthening; bending; model; design; practical application

---

1. Introduction—Strengthening of Concrete Buildings—Why and How

Particularly in the past decades, the world population has grown rapidly and will continue to do so according to unanimous forecasts. Buildings are essential for mankind, but neither the natural resources nor the available room or the costs allow us to cover our needs by new buildings alone. Buildings, roads, and bridges must be carefully maintained, renovated, or reinforced if their load-bearing capacity is no longer given or if there are deficiencies in their serviceability.

Reinforced concrete has been the world's most widely used composite building material for more than a hundred years. When properly dimensioned and constructed, it is very efficient and durable. Two things in particular can be problematic:

1. Corrosion of the steel reinforcement; it can occur because of carbonation of concrete, the choice of unfavorable materials, insufficient concrete cover, or too wide cracks.
2. Insufficient load-bearing capacity; this mostly results from the loads which have steadily increased over the decades and which must be taken into account according to the standard. One example is the increased axle loads of trucks. About 100 years ago, the total weight of commercial vehicles was about 10 tons. Today, gigaliners with total weights of up to 60 tons are under discussion worldwide; however, their traffic-legal approval is the responsibility of national authorities.

Various methods exist for the rehabilitation and strengthening of plain, reinforced, and prestressed concrete structures, see e.g., References [1,2]. The most important are briefly mentioned:

- Shotcrete with steel reinforcement, e.g., References [3–5],
- Glued or notched steel lamellas, e.g., References [6–9],
- Glued or notched lamellas/sheets made of fibre-reinforced plastics (abbreviated: FRP), e.g., References [7–16], or
- Supplementation of additional components such as external tendons, e.g., References [17,18].

Another method that has become increasingly established over the past 20 years is strengthening with textile reinforced concrete (abbreviated: TRC; also known as textile reinforced mortar, abbreviated: TRM, or fibre-reinforced cementitious matrix (abbreviated: FRCM) composites) [19–24]. Similar to shotcrete, TRC is—from the material's side—very compatible with the steel reinforced concrete of the primary building element. Because the fibres used in TRC can bear higher tensile stresses than steel and do not corrode, considerably smaller layer thicknesses can be realized compared to conventional shotcrete. Compared to steel lamellas, the requirements for corrosion protection are lower. Like FRP lamellas, TRC is a relatively light building material, which can be processed by hand. The concrete cover around the fibre reinforcement in TRC is advantageous at elevated temperatures—for the textiles themselves as well as for the steel reinforcement in the basic component.

A major disadvantage of textile reinforced concrete is—with a few exceptions, e.g., References [23,25–27]—the lack of general building authority approvals, guidelines, and standards. To date, not all details of the load-bearing behavior of TRC have been clarified. In addition, there is a constantly growing number of further and new developments in the field of fibre reinforcement. Often these reinforcements show a similar load-bearing behavior. In detail, however, there can be differences that must be taken into account when dimensioning and applying textile reinforced concrete.

TRC can be used to increase the bending, torsional, longitudinal, and shear load-bearing capacity and to improve serviceability and functionality, e.g., References [22,24]. In this article, the focus lies on a model to calculate the increase of the flexural load-carrying capacity due to an additional TRC layer. On the one hand, compared to the other TRC strengthening variants, this has been very well researched. On the other hand, in our experience, there is a great need for flexural strengthening measures in building practice compared to the other possibilities.

2. Research on Flexural Strengthening with Textile Reinforced Concrete

First research with textile fabrics made of technical endless fibres embedded in mortar and fine-grained concrete respectively has been performed since the early 1990s, e.g., References [28–30]. At our institute, we carried out first own tests on steel reinforced (abbreviated: RC) concrete plates strengthened with an additional TRC layer in the bending tensile zone in 1997 as part of a feasibility study [31,32]. The objective was to demonstrate the potential of textile reinforcements to increase the flexural load-carrying capacity of RC slabs in principle. The three plate specimens were made of normal strength concrete B25 (acc. to DIN 1048 [33], this corresponds nearly to the current concrete class C20/25 acc. to DIN EN 206 [34]), and were reinforced with 5 steel bars with a diameter of 8 mm (Figure 1a). Two of the plates were strengthened with 6 textile layers embedded in fine-grained mortar—at that time still using ISBOTON as adhesion promoter. The used multiaxial grids were made of alkali-resistant (AR) glass filaments (Figure 1b). The tensile strength of the glass fibres ranged between 1200 and 1400 MPa, the ultimate strain was 20%. The strengthening layer was applied to the whole underside of the specimen. In all following examinations, the support areas remained unstrengthened. The plates were studied in 3-point bending tests. The load was applied path controlled; deformations were recorded by strain gauges and linear variable differential transformers (LVDT). Figure 1c displays load-middle deflection curves of the tests. The blue line shows all characteristics of an RC slab subjected to bending. At a deflection of approximately 1 mm, a first crack was formed, followed by a short phase of formation of further cracks. After that, a steady, almost linear increase in deflection could be detected (state II). At a deflection of 8.5 mm, observed at a load of 20 kN, the steel reinforcement reached its yield strength. From now on, no further load increase was possible.

On account of the steel reinforcement's plastic deformation capability, only the deflection increases until a final bending failure occurred.

The two strengthened plates (red and green in Figure 1c) showed a similar behavior to each other, but a different behavior compared with plate 1. The first crack opened at circa 16 kN, and the associated load level was therefore twice as high as at plate 1. With the increase of the load (state II), the deflection then increased again linearly, but—compared to plate 1—at a higher load level resulting from the additional textile reinforcement. When a load of approx. 35 kN and a deformation of approx. 12 mm were reached, the textile reinforcement teared and the force dropped down to the level given by the yield point of the steel bars. In summary, the additional reinforcement with 6 layers of AR glass textile increased the load by 69% (average value of the two tests). Upon reaching the maximum load the deflection increased by approx. 30%. Still, at the same load level, the deflection of the reinforced components was less.

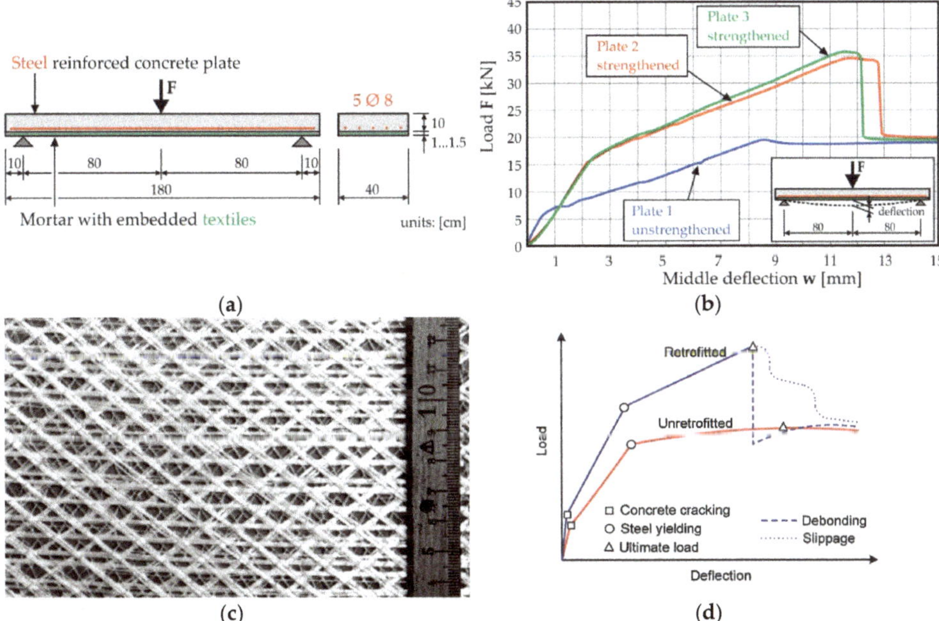

Figure 1. First test on reinforced concrete (RC) slabs strengthened with textile reinforced concrete (TRC) in Germany—(**a**) test set-up acc. to Reference, (**b**) used textile made of alkali-resistant (AR) glass, produced by the Institute for Textile and Clothing Technology of TU Dresden, (**c**) test results ((**a**–**c**) reproduced and modified with permission from [31], Institute of Concrete Structures, TU Dresden, 1997) and (**d**) principle behavior of TRC-strengthened RC plates or beams (reproduced from Koutas et al. [22] on the basis of the Creative Commons Attribution license (http://creativecommons.org/licenses/by/4.0/) under which this article open access was published, ASCE Library, 2019).

These first tests already demonstrated the basic phenomena of TRC-strengthened RC slabs or beams subjected to bending:

- Initial cracking at a higher load level,
- Reduction of deflection (at same reference load),
- Increase of the bearable load until failure.

In addition, in state II, the stiffness of TRC-strengthened components is often higher than that of unstrengthened RC elements. This is caused by the higher component's thickness and—often the decisive factor—by the increased elongation stiffness in the tensile zone (described in detail, for example, in Reference [35]).

Today a large number of research projects and publications on flexural strengthening of RC structures with TRC are known, in which comparable phenomena have been observed. Exemplary for all, we like to recommend the publication of Koutas et al. [22] and Carloni et al. [23]. The authors collected, studied, analyzed, and evaluated numerous research works from all over the world on the subject "strengthening of concrete structures with TRM" (TRC). In addition to the often considered uniaxial case, studies on two-way slabs are listed as well as experiments at elevated temperature; furthermore, various research projects on the behavior of TRC-strengthened components under impact loading are known (see for example Reference [36–39]). In References [22,23] the described basic phenomena are confirmed, based on an extensive evaluation of known experiments worldwide. Figure 1d summarizes the component's behavior in a schematic sketch by Koutas et al. [22]. The specific shape of such a load-deformation curve depends on the type and the quantity of the textile reinforcement, and, of course, on the characteristic of the RC element to be strengthened. To add a last characteristic: In general, textile reinforcements with yarn distances in the range of several millimeters up to a few centimeters cause a finer crack pattern in the bending tensile zone compared to steel reinforced concrete.

The type of failure is of particular interest for the design or practical application of a flexural strengthening with TRC. In general, the load-bearing capacity of RC components subjected to bending is reached when either the concrete fails under compression or the steel reinforcement ruptures under tension. In addition, there are failure mechanisms associated with shear (especially for beams) or detailed problems such as failure due to insufficient anchoring of the bending tensile reinforcement. If the tensile strength of the reinforcement limits the bearing capacity of a component, the application of a textile concrete layer in the bending zone can increase the bearing capacity. In doing so, it must be ensured that the "old" concrete in the pressure zone has sufficient load-bearing capacity. Otherwise, the failure mode may change. It should be noted in particular that the addition of a TRC layer with high load-bearing capacity, on the one hand, could defer crack formation in the bending tensile zone and cause a sudden concrete compression failure (see Section 3.3.3), and on the other hand (compare Reference [21] and Section 5), can cause also shear failure. Furthermore, failure mechanisms may occur which are not known from reinforced concrete construction. The following scenarios must be considered, compare e.g., References [22,24,35,40]:

- Similar to RC and steel reinforcement—exceeding the tensile strength of the textile reinforcement; this failure is indicated by increasing crack formation and deflection and is the quasi "wanted" failure form; the load-deformation behavior is essentially determined by the mechanical properties of the textile reinforcement (tensile strength, modulus of elasticity).
- Forms of failure due to the transfer of forces from the strengthening layer into the reinforced concrete base body:

  - Failure inside the old concrete (that means the concrete of the structural member to be strengthened fails first, often near the joint between old and new concrete; it is known also from other strengthening methods),
  - Delamination in the joint between old concrete and TRC layer in the end anchorage area or at opening cracks,
  - Delamination or debonding within the TRC layer (usually in the plane of the most stressed textile grid),
  - Extraction of the reinforcement from the matrix ("slippage" of the fibres through the mortar in Figure 1d).

The last two failure variants are primarily dependent on the inner bond within the yarns and on the bond between the yarn's edge filaments and the surrounding concrete matrix. The size and impregnation used have a major influence. Even though some approaches have already been developed to explain these phenomena, the topic of failure forms and bonding is still the subject of ongoing research.

In summary, much knowledge has been generated worldwide on the subject of bending retrofitting with TRC. In addition to many similarities, these results are often not directly comparable, since the known studies differ in many details. The main differences are the type of textile reinforcement, its characteristic material properties and amount, the properties of the strengthened basic RC members (geometry, material, reinforcement), and the test set-up (generated distribution of inner forces). Nevertheless, bending strengthening layers can in principle be calculated with relatively uncomplicated models. One possibility is presented and discussed in the following chapter.

3. Design Concept for Flexural Strengthening of Reinforced Concrete (RC) Components with Textile Reinforced Concrete (TRC)

3.1. General Information

There are standards on design rules and models for RC structures, e.g., in Reference [41]. In case of bending, the calculation can be done by an iterative process. A similar procedure was also chosen for the calculation of the flexural strengthening with TRC. In accordance with the design method for steel reinforced concrete, the following assumptions and simplifications are defined (compare e.g., References [41,42] and also References [23,36]):

- Cross sections remain plane (Bernoulli hypothesis).
- Strain compatibility between reinforcement and concrete is assumed.
- The concrete's tensile strength is ignored; all tensile forces are taken up by steel and textile reinforcements.
- Rigid bond between steel, concrete, and textile reinforcement may be assumed.
- The design is carried out at the ultimate limit state (ULS), i.e. at least one material (concrete, reinforcing steel, or textile reinforcement) reaches the ultimate strain.

A first model was presented by Bösche in Reference [43] in 2007. Meanwhile, it was further developed [42,44–47] and modified due to new knowledge and research findings for the later used materials. Furthermore, a similar design proposal is presented in Carloni et al. [23]. The main difference lies in the choice of the reference plane for the formation of the internal equilibrium. Compared to Reference [23], the model presented in the following has the advantage that the reference horizon is fixed by the geometry and the structural design of the component (position of the reinforcements) and does not have to be determined by stress or strain distribution.

3.2. Material Models

A standard material model for concrete was chosen [41,48,49], Figure 2a, and can be simplified into two sections—a parabolic section until the strain ε_{c2}, and a linear horizontal branch until the ultimate strain ε_{cu2} is reached. The values ε_{c2} and ε_{cu2} depend on the concrete class of the structural member that has to be strengthened.

The material model for the steel reinforcement is also taken from Reference [41,48,49], Figure 2b. It can be described as a bilinear curve, either with or without a strength increase (hardening) after reaching the yield strength.

Depending on the fibre material, there are differences in the stress–strain behavior of textile grids. The design model used here was primarily developed on the basis of tensile tests with carbon fibre reinforcements (e.g., [50]). When using material with different characteristics, the here mentioned recommendations may have to be modified. In Figure 3, different possible stress–strain relationships for carbon reinforcement are shown. The variant 1 (Figure 3a, used in Reference [25,46]) derived from material behavior observed in tension tests in former years where the investigated textile grids made of AR glass or carbon have shown a less stiff stress–strain behavior at lower load levels, e.g., due to undulation of yarns after manufacturing [45–47]. The meanwhile available carbon reinforcements show a nearly linear material behavior which led to a modification of the material model, variant 2 in Figure 3b (compare e.g., [47]) and variant 3 in Figure 3c, respectively. In Carloni et al. [23] and in Curbach et al. [47], a stress–strain curve with constant elastic modulus as in variant 3c is recommended. The different variants will be discussed in Section 3.4.1.

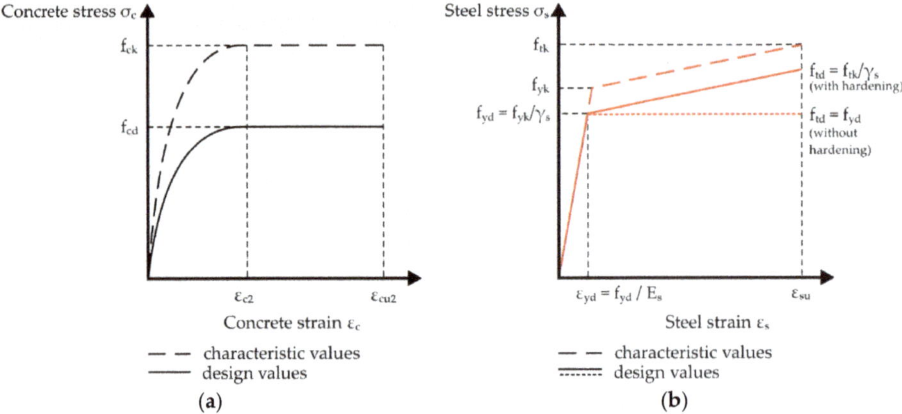

Figure 2. Stress-strain relationships of the reinforced concrete (RC) components according to European code: (**a**) parabola-rectangle diagram for concrete under compression; (**b**) bilinear diagrams for reinforcing steel (tension and compression).

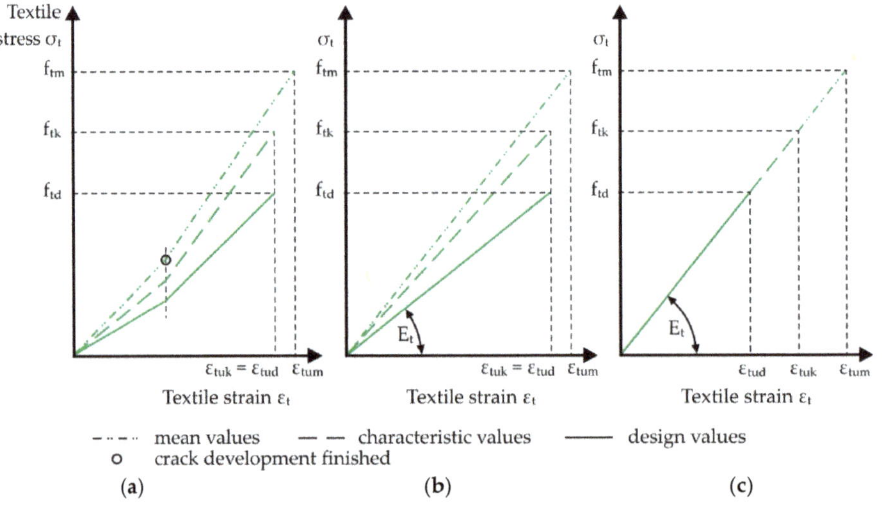

Figure 3. Possible stress-strain relationships for carbon reinforcements: (**a**) variant 1: for textiles with weaker behavior at low stress levels, e.g., because of undulated yarns; (**b**) variant 2: for textiles with linear stress–strain behavior (variable modulus of elasticity); (**c**) variant 3: for materials as in (**b**), constant modulus of elasticity.

3.3. Calculation Model

3.3.1. Iteration Process

Similar to RC, the design model is based on the equilibrium of internal and external forces. As is well known, this equilibrium cannot be solved in a closed way, one or more iterations are necessary. The iteration process is described e.g., in References [43,46,47]. In this paper, the basic formulas will be given for a cross section with forces, stresses, and strains according to Figure 4 (compare e.g., [46,47]). The definitions of the symbols are summarized in Table 1.

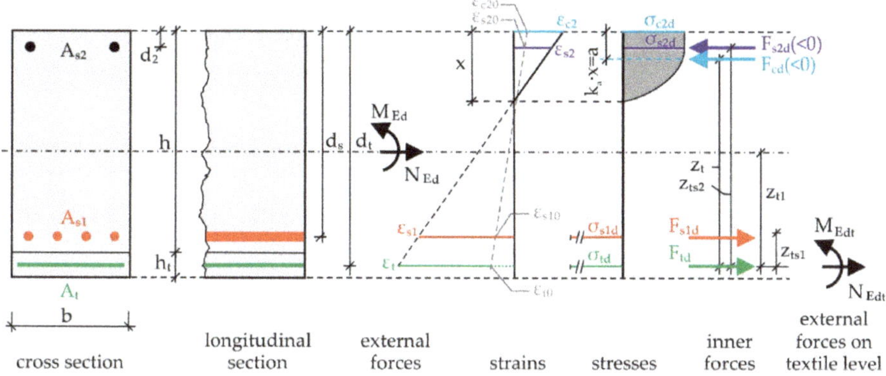

Figure 4. Principle of strengthened reinforced concrete (RC) cross section with outer and inner forces, strains and stresses.

Table 1. Definition of symbols.

Symbol	Definition
Indices	
E	external
c	concrete
s	steel
t	textile
d	design value
1	tensile stress area
2	compressive stress area
0	preload condition
Forces and Moments	
F	force
N	normal force
M	bending moment
Geometrical Values	
h	height
d	effective depth
b	width
A	reinforcement area
a	distance of compression force to top layer
x	neutral axis depth
z	inner lever arm
k_a	coefficient for distance a
α_R	block coefficient
Stresses and Strains	
σ	stress
ε	strain

The maximum resistance of the cross section is reached when at least one of the three materials reaches its ultimate capacity. The goal is to fully utilize the textile reinforcement. This is therefore the first limiting case: The textile reinforcement achieves its ultimate strain (ε_{tu}) or ultimate strength. The second possibility is a concrete failure in the compression zone ($\varepsilon_c = \varepsilon_{cu2}$). Last but not least, the steel reinforcement can fail as well, when the steel strain reaches ε_{su}, but this scenario would require an extremely large pre-deformation of the existing component prior to the strengthening measure (see "consideration of a preload", Section 3.3.3), and would apply only to exceptional cases. All other types of failure, e.g., debonding or anchorage failure, are to be excluded by separate proofs or by complying with design regulations analogous to steel reinforced concrete construction.

The first equation for the iteration is the equilibrium for horizontal forces. Irrespective of the graphic representation in Figure 4, compressive forces are introduced in equilibrium conditions under consideration of the sign (negative!).

$$N_{Ed} = F_{s1d} + F_{td} + F_{s2d} + F_{cd} \tag{1}$$

In Equation (2), the moment's equilibrium is formed in the centre of gravity of the textile reinforcement.

$$M_{Ed} - N_{Ed} \cdot z_{t1} = -F_{s2d} \cdot z_{ts2} - F_{cd} \cdot z_t - F_{s1d} \cdot z_{ts1} \tag{2}$$

In the Equations (1) and (2), there are four unknown forces (F_{s1d}, F_{td}, F_{s2d}, F_{cd}) and one unknown geometrical value z_t. Overall, there are two equations but five unknown variables. To solve this mathematical problem, more equations are necessary. For this reason, the strains will be brought into relation, according to the Bernoulli hypothesis.

$$\varepsilon_{s1} = \varepsilon_{c2} + (\varepsilon_t + \varepsilon_{t0} - \varepsilon_{c2}) \cdot \frac{d_s}{d_t} \tag{3}$$

$$\varepsilon_{s2} = \varepsilon_{c2} + (\varepsilon_{s1} - \varepsilon_{c2}) \cdot \frac{d_2}{d_s} \tag{4}$$

$$\varepsilon_{t0} = \varepsilon_{s10} + (\varepsilon_{s10} - \varepsilon_{c20}) \cdot \frac{z_{ts1}}{d_s} \tag{5}$$

For iteration, a first strain distribution must be estimated. Then, the position of the neutral axis x, including all coefficients such as k_a and α_r, and the compression force F_{cd} can be calculated. With the help of the geometric values of effective depth of the textile d_t and the centre of gravity of the resulting concrete compressive force a, the inner lever arm z_t can be determined. Due to the given amount of steel reinforcement and the according stress–strain relationship, the steel stresses ε_{s1} and ε_{s2} can be calculated. Now the equilibrium of moments can be checked. If it is fulfilled, Equation (1) can be used to determine the necessary textile reinforcement according to the material characteristics of the grid. If it is not fulfilled, the assumed strain distribution has to be modified and the calculation must be repeated.

3.3.2. Design Tables for Calculation

The iteration process described before can be time-consuming. Therefore, dimensionless design tables can be developed, e.g., References [46,47]. Analogous to RC calculation tables, specific application limits must be taken into account, e.g., the concrete strength class and the kind of steel reinforcement. In the case of flexural strengthening with TRC, the characteristic material properties of the textile reinforcement (stress–strain relation) must be considered as well. That means, for every kind of textile, special design tables have to be created. For textiles that are frequently used, however, it is worth the effort to generate such tables because overall the handling is much more efficient compared to the iteration.

For rectangular cross sections made of normal strength concrete and reinforcing steel acc. to EC2 [41,48,49] and carbon reinforcement acc. to Reference [25], (ed. 2016, characteristic ultimate strain $\varepsilon_{tuk} = 0.0075$), design tables were already developed (Frenzel [46], stress–strain relation according to Figure 3a; Zobel [47] has used the more linear stress–strain relation of modern textiles acc. to Figure 3b, as well as higher design values for the tensile strength of carbon). In Table 2, a stress–strain distribution with constant Young's modulus according to Figure 3c was used, recommended by the authors and Carloni et al. [23]. Table 2 does not include reductions in regard to durability, temperature, and permanent load when determining the design value of the carbon material strength; for discussion, see Section 3.4. In addition, in Table 2, strain hardening of the steel reinforcement was not considered (horizontal branch of σ-ε curve after reaching the yield strain); the preload ε_{t0} is assumed to be zero.

Table 2. Dimensionless design table for rectangular reinforced concrete RC member flexurally strengthened with textile reinforced concrete (TRC [1]).

μ_t	ω_t	$\xi_t = x/d_t$	$\zeta_t = z_t/d_t$	ε_{c2} (‰)	ε_t (‰)	σ_t (N/mm^2)	
0.01	0.0102	0.059	0.980	−0.37	5.95	1291.67	
0.02	0.0206	0.084	0.972	−0.54	5.95	1291.67	
0.03	0.0311	0.103	0.965	−0.68	5.95	1291.67	Rupture of textile
0.04	0.0417	0.120	0.959	−0.81	5.95	1291.67	reinforcement
0.25	0.2947	0.366	0.848	−3.43	5.95	1291.67	
0.26	0.3091	0.382	0.841	−3.50	5.67	1230.68	
0.27	0.3239	0.400	0.834	−3.50	5.25	1139.59	
0.34	0.4391	0.542	0.774	−3.50	2.95	641.40	Failure of concrete
0.35	0.4576	0.565	0.765	−3.50	2.69	584.60	

[1] stress–strain relationship acc. to Figure 3c, no preload ($\varepsilon_{t0} = 0$).

The handling is easy. First, it is assumed that the steel strains ε_{s1} and ε_{s2} reached or exceeded the yielding point. With Equation (6), the design bending moment μ_t can be calculated:

$$\mu_t = \frac{M_{Edt} + F_{s1d} \cdot z_{ts1} + F_{s2d} \cdot z_{ts2}}{b \cdot f_{cd} \cdot d_t^{\,2}} \tag{6}$$

Forces and moment can be calculated as follows:

$$F_{s1d} = A_{s1} \cdot f_{yd}; \quad F_{s2d} = A_{s2} \cdot f_{yd}; \quad M_{Edt} = M_{Ed} - N_{Ed} \cdot z_{t1}$$

The textile reinforcement ratio ω_t and the strains ε_t and ε_{c2} can be taken from Table 2. In the next step, the previous assumption for the strains ε_{s1} and ε_{s2} (yield range) must be checked with Equations (7) and (8).

$$\varepsilon_{s1} = \frac{\varepsilon_t - \varepsilon_{c2}}{d_t} \cdot d_s + \varepsilon_{c2} > \varepsilon_{yd1} \tag{7}$$

$$\varepsilon_{s2} = \left| \varepsilon_t - \frac{\varepsilon_t - \varepsilon_{c2}}{d_t} \cdot (d_t - d_2) \right| > \varepsilon_{yd2} \tag{8}$$

(a) If Equations (7) and (8) are fulfilled, the textile reinforcement area A_t can be determined on the basis of the following formula (9):

$$A_t = \frac{1}{\sigma_t} \cdot \left(\omega_t \cdot b \cdot f_{cd} \cdot d_t + N_{Ed} - F_{s1d} - F_{s2d} \right) \tag{9}$$

(b) If the yield strains are not reached, μ_t must be recalculated (Equation (6)), taking into consideration the strains calculated by (7) and (8), as listed below:

$$F_{s1d} = A_{s1} \cdot f_{yd}, \text{ if } \varepsilon_{s1} \geq \varepsilon_{yd1}; \quad F_{s2d} = -A_{s2} \cdot f_{yd}, \text{ if } |\varepsilon_{s2}| \geq \varepsilon_{yd2}$$

$$F_{s1d} = E_{s1} \cdot \varepsilon_{s1}, \text{ if } \varepsilon_{s1} < \varepsilon_{yd1}; \quad F_{s2d} = E_{s2} \cdot \varepsilon_{yd}, \text{ if } |\varepsilon_{s2}| < \varepsilon_{yd2}$$

Now, the textile reinforcement ratio ω_t and the strains ε_t and ε_{c2} can be selected again from Table 2. In the next step, the "new" defined strains $\varepsilon_{s1;new}$ and $\varepsilon_{s2;new}$ must be checked with formulas (10) and (11) as well.

$$\varepsilon_{s1;new} = \frac{\varepsilon_t - \varepsilon_{c2}}{d_t} \cdot d_s + \varepsilon_{c2} \approx \varepsilon_{s1} \tag{10}$$

$$\varepsilon_{s2;new} = \varepsilon_t - \frac{\varepsilon_t - \varepsilon_{c2}}{d_t} \cdot (d_t - d_2) \approx \varepsilon_{s2} \tag{11}$$

If the values ε_{s1} and ε_{s1} do not match, the procedure must be repeated from point (b) until the Equations (10) and (11) are fulfilled.

Finally, Equation (9) can be used to determine the required textile reinforcement area A_t.

3.3.3. Consideration of a Preload ε_{t0}

In some publications, the influence of a preload applied to the basic component before it was strengthened have been already addressed, e.g., References [22,51] with a focus on experimental results or [46,47] with regard to calculation. The thesis is that the imprinted strain state of the initial component is of particular interest for the design. Therefore, the preload influence on the design of a flexural strengthening TRC layer, already discussed in Reference [46] and further examined in Reference [47], shall be summarized.

Figure 5 displays the relation between the design bending moment μ_t and a pre-deformation ε_{t0}. ε_{t0} is a virtual size and can be determined from strain distribution of the pre-deformed component and by an assumption of the position of the textile grid in the reinforcing layer (compare Figure 4), taking a plane cross section into account (Bernoulli). Now, ε_{t0} was used as an input value for further strain analysis. In the diagram, there are two different colored areas. In the grey one, the ultimate strain of the concrete, and in the green one, the ultimate strain of the carbon reinforcement is reached in ULS. The corresponding other strain value varies from its maximum and minimum respectively to zero (compare the different colored lines in the diagram). In the green area, tensile failure of the textile occurs; the carbon reinforcement is fully utilized. In the grey zone, concrete failure will happen. The dark blue line is the limit, where concrete and carbon reinforcement fail at the same time. It can be concluded [46,47] that for the same design bending moment μ_t and increasing pre-deformation ε_{t0}, the utilization of the concrete compression zone increases. Strain distribution, type of failure, and required textile reinforcement A_t are therefore dependent on the pre-stress condition of the unstrengthened RC cross section.

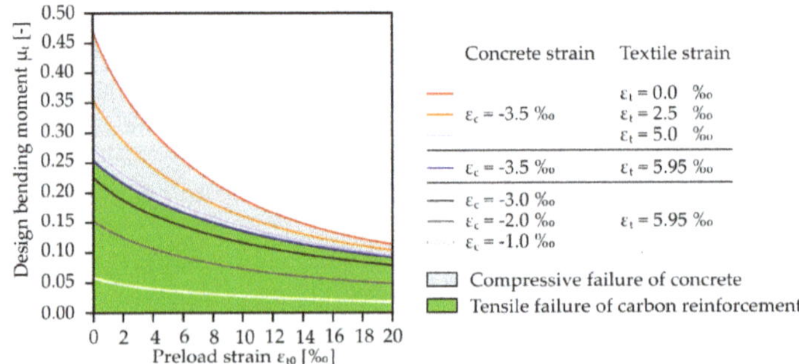

Figure 5. Influence of a pre-deformation on the strain plane and failure mode of a retrofitted reinforced concrete (RC) member as a function of the design bending moment; published by Zobel in Reference [47], modified (diagram is based on a stress-strain distribution of the carbon reinforcement acc. to Figure 3c and Reference [25]).

3.4. Special Aspects to be Taken into Account When Applying the Design Proposal

Similar to reinforced concrete construction, flexural design based on the equilibrium of internal and external forces is a reliable method. The method presented here has been tested by means of various experimental investigations. Altogether, satisfactory results were achieved. Nevertheless, there are still aspects to be discussed. These include the assumption of a stress–strain relationship for the textile reinforcement as well as the questions, which partial safety factors or which distribution function are to be used for the scattering material properties. Furthermore, not all failure phenomena observed by different researchers [22,23] so far have yet been conclusively explained. In this sense, the following sections are to be understood as a basis for discussion.

3.4.1. Influence of the Used Stress–Strain Relationship for the Textile Reinforcement

The design table presented in Section 3.3.2 uses the stress–strain curve for a carbon fibre strand shown in Figure 3b. For the reliable determination of the component's load-bearing capacity, this curve is reduced in certain values by a partial safety factor. For the material design curve currently in use, only the strength is reduced. However, this is also accompanied by a reduction in the modulus of elasticity. This corresponds to the procedure, which is also used when applying the stress–strain curve in concrete design. On the other hand, in Reference [23] or Reference [47] a further variant is presented (Figure 3c). In this variant, the strength and the elongation are reduced to the same extent, so that the modulus of elasticity does not change. This corresponds to the procedure which is also used in the design of reinforcing steel—at least for the linearly elastic, first section of the reinforcing steel stress–strain line until the yield point is reached. Here, the characteristic value of the yield strength is first reduced by the partial safety factor, after that the corresponding elongation is determined by division with the modulus of elasticity of the reinforcing steel. In the variant from Figure 3c for the carbon textile, the procedure is the same. First, the breaking stress is determined by means of single fibre tensile tests [52] or uniaxial tensile tests, e.g., according to Reference [50], and then the elongation at failure is determined by division of the strength by the modulus of elasticity. The reduction of the strength and elongation values takes place to the same extent, so that the modulus of elasticity remains the same before and after the reduction.

What effects does this changed procedure have on the load-bearing capacity design, since the value of the elongation at fracture plays an important role as one of the two limit strains in the iteration process for cross sectional design? With variant from Figure 3c, a stiffer bending component behavior is obtained by reaching the maximum textile stress earlier with smaller elongation at failure, since the elongation at failure is also reduced. This results in a faster utilization of the concrete pressure zone compared to the variant from Figure 3b in a bending component designed for the ultimate limit state. The effects investigated for the general case are explained in detail in Reference [47] and can be summarized as "lying on the safe side" since both variants do not have an intersection with the 5% or 95% quantiles of the probabilistically calculated design parameters. In Reference [47], the variant from Figure 3c was defined as the preferred variant.

3.4.2. Partial Safety Factor for Carbon Textiles

Safety factors are a crucial topic within the framework of a semi-probabilistic safety concept, which is common in the construction industry. In 2014, a proposal was already presented for the determination of partial safety factors for carbon textiles and for the characteristic and design values of the tensile strength, see Reference [25]. The distribution and the mean value of the strength were determined on the basis of 50 uniaxial tensile tests. Assuming a normally distributed quantity, a standard deviation and the 1.5% quantile were calculated, considering the low data basis for statistical purposes. The characteristic nominal strength for the investigated carbon textile acc. to [25] was specified to 1550 N/mm^2.

In Reference [53], a comparative calculation was presented using EC0 [54] on the basis of a representative bending component. The general safety level for buildings and components required by the Eurocode is verified by calibration on reinforced concrete. First, a steel reinforced concrete slab is dimensioned using the usually applied semi-probabilistic method and the required reinforcing steel reinforcement is determined. Subsequently, the distributions of the acting and resisting moment are calculated and compared using a Monte Carlo simulation under the assumption of scattering actions (dead weight and payload) and resistances (reinforcement tensile strength), so that a failure probability can be determined first for the steel reinforced concrete slab. Afterwards, the same procedure is carried out in several iteration steps for the textile reinforced concrete slab: the iterated parameter equals the partial safety factor for the textile and the iteration target equals the failure probability of the steel reinforced concrete slab, which was considered safe. With the aid of this procedure, the partial safety factor for textiles $\gamma_t = 1.2$ was introduced into the first building authority approval [25]. More recent investigations with further calculations [55], however, showed that a lower value with $\gamma_t = 1.1$ also is reasonable. These partial safety factors do not take into account reductions in regard of durability, temperature and permanent load. In Reference [53], all influences were combined to a general, and therefore significantly higher, safety factor. However, since these reductions differ for each type of material, it is recommended to adjust the corresponding characteristic value. The topic is still under research in textile concrete research, e.g., in the C^3 project [56].

4. Practical Applications

Even if there are still points open for discussion, TRC has already been used for several renovation measures. An overview on the different fields of application are given e.g., in References [22,24,57], additional examples are presented for example in References [58–60]. TRC is used for the strengthening for static or earthquake loading, but also to improve the usage properties, e.g., limiting deflections, repairing cracks, and sealing or repairing concrete surfaces. In the field of flexural strengthening, examples of applications can be found both in conventional building construction and in the field of monument protection.

An example of the first group is the renovation of a residential and commercial building in Prague (Czech Republic, 2009/2010) [59]. Ceiling slabs had to be strengthened in that project. The RC flat slabs were supported by brickwork at the edges and by point supports inside the building (column grid 12.8 m × 13.1 m). In some areas there were considerable deflections of up to 15 cm. In addition, there were problems with the flexural load-bearing capacity and the punching safety. TRC was used in the interior fields to increase the bending load-bearing capacity and the ceiling's stiffness. Up to 4 layers of carbon textile were required. After preparing the surface (roughening, pre-wetting), the first layer of fine-grained concrete (3 mm thick) was sprayed on. The carbon layer was subsequently embedded. These steps were repeated until the required number of layers was reached. The maximum thickness of the TRC layer was 2 cm. A total of approx. 3000 m^2 of carbon textile was processed.

In the field of monument protection, TRC has already been used for the renovation of several RC shell structures, where the low additional weight and the flexibility of the textile fabric are particularly advantageous. A further early application was the renovation of a hypar shell above a lecture hall at the Schweinfurt University of Applied Sciences (Germany, 2006), e.g., [61,62]. The shell, which was only 8 cm thick in the middle area, measures 38 m × 39 m. In the area of the cantilevered high points, considerable deformations occurred due to stress exceedances over the column supports. The strengthening was carried out with 3 layers of carbon textile (Figure 6a). A total of approx. 450 m^2 of carbon textile was processed. A second RC hypar shell was retrofitted with carbon reinforced concrete in Templin (Germany) in 2016. The refurbishment of a hypar shell in Magdeburg (also in Germany, Figure 6b) is currently being planned [63]. The tests to obtain an approval for individual cases based on Reference [25] were successfully completed. The strengthening with carbon textiles is scheduled for 2019.

Figure 6. Textile reinforced concrete (TRC) retrofitting of hypar shells; (**a**) application of TRC strengthening at a hypar shell in Schweinfurt (photos: Ulrich van Stipriaan and Silvio Weiland, reproduced with permission from [59], Ernst & Sohn, 2015), (**b**) hypar shell in Magdeburg before restoration (photo: Heiko Wachtel).

Until today, mostly structural engineering components have been retrofitted with textile reinforced concrete. Yet, there is also a considerable need for renovation of bridges. TRC has already been tested on concrete arch bridges (non-reinforced concrete), bridge piers, bridge caps or, as an additional concrete layer, on bridge decks (compression zone), see e.g., References [58,64–66].

5. Outlook on Current Research and Summary

Bending tests were also carried out within the BMBF-funded project 'C^3 – Carbon Concrete Composite' [56], subprojects C3-V1.2 [67] and C3-V2.7 [68] in at TU Dresden. The aim of research and development within 'C^3' is to establish carbon reinforced concrete in construction practice. Both the carbon reinforcements and the concretes used were further developed in order to improve the properties of the composite (durability, manufacture, anchoring, etc.). Load bearing mechanisms as well as design models are to be investigated and modified respectively in greater depth. In comparison to the textiles used in earlier bending tests, the now used ones are significantly stiffer and show higher tensile strengths. The main aim of the test program was therefore to demonstrate that the calculation model presented in Section 3 for strengthened RC slabs can also be applied to the new textiles. The tests were carried out on plates with the dimensions $3.3 \times 0.5 \times 0.12 \text{ m}^3$ mainly as four-point bending tests with variable distances of the load introduction points. In addition to six reference plates, a total of 18 strengthened plates were tested. For strengthening, textiles from solidian GmbH (GRID Q85/85, $a_t = 85 \text{ mm}^2/\text{m}$) and from V.Fraas Solution in Textiles GmbH (SITGrid 40, $a_t = 141 \text{ mm}^2/\text{m}$) were used. An evaluation of the experiments is currently taking place. Therefore, only the basic phenomena, observed during the experiments, are mentioned here. For an in-depth scientific analysis, please await future publications.

Depending on the configuration of the textile and the concrete base, the strengthened plates achieved 2–6 times the loads compared to the reference ones. In principle, the load-bearing behavior was in accordance with the effects described and observed earlier. The unstrengthened reference plates with low reinforcement level showed a tensile failure of the steel reinforcement; at higher reinforcement levels, a concrete compression failure occurred. The predominant type of failure of the strengthened slabs was a tensile failure of the textile reinforcement. At very high TRC strengthening levels, the formation of a critical crack could be observed, which led—after constricting the pressure zone—ultimately to a compression failure of the concrete in the pressure zone. This type of failure occurred at loads at which the range of minimum shear force resistance of the concrete base (calculated acc. to References [41,48,49]) had been nearly reached or already exceeded. In addition, in some configurations, splitting, or delamination in the plane of the textile reinforcement was observed, starting with a crack. This tendency was observed to an increasing extent with the reduction of the distance between the two individual loads. The exact reasons for the change of failure mode, however, have still to be analyzed. One assumption might be the curvature in connection with the stiffness of the textile. The observed failure types and forms lead to the conclusion that in addition to the desired flexural strength failure, other failure types can also become decisive due to the higher stiffness of the new generation of textile reinforcements. This must be taken into account in the planning and execution of strengthening measurements. In addition, proofs must be developed in order to reliably calculate these types of failure.

In summary, the article outlines the potential of flexural strengthening with TRC. A calculation model was presented to dimension a TRC strengthening. The assumptions underlying the model were explained and discussed. Critical points were identified. Thus, for the application of the model, for further research and development of TRC construction and for the practical application of TRC, the type of the textile reinforcement, the associated stress–strain curve, the bond behavior, and a preloading of the basic component must be taken into account. Hence, further research will be necessary with regard to the change of failure modes.

Author Contributions: Conceptualization, M.C.; Formal analysis, R.Z., E.M. and A.S.; Funding acquisition, M.C.; Investigation, R.Z., E.M., T.S.-P. and A.S.; Methodology, R.Z.; Project administration, M.C.; Supervision, M.C.; Validation, S.S., R.Z. and M.C.; Visualization, S.S., E.M., T.S.-P. and A.S.; Writing – original draft, S.S., E.M., T.S.-P. and A.S.; Writing – review & editing, S.S., R.Z. and M.C.

Funding: The findings presented in this paper are the result of several projects carried out over the last 20 years. The German Research Foundation (DFG) should be mentioned first and foremost. In addition to many other fundamental investigations on TRC, more than 70 flexural tests on large-format plates were examined during the DFG-CRC 528 'Textile Reinforcement for Structural Strengthening and Repair' (project number 5483454, funding period: 1999–2011), e.g., [69,70]. Within the frame of the test programme for the general building approval [25] (funded by TUDALIT e.V. [71]), further approx. 50 plate tests were carried out together with the Institute of Concrete Structures of RWTH Aachen University; in addition, the design model was improved and prepared for application in engineering offices, see [26,45,46,55,72]. Furthermore, at our institute, numerous bending tests were carried out for approvals in several individual cases for various construction measures (different funding organizations/companies, different years). Since 2014, the project consortium 'C^3—Carbon Concrete Composites' [56] with circa 170 German partners has been funded by the German Federal Ministry of Education and Research. Flexural tests on strengthened slabs are mainly carried out in the projects C3-V1.2 'Verification and testing concepts for standards and approvals' [67] (funding period: 01.2016–04.2018, grant number: 03ZZ0312A) and C3-V2.7 'Development of a general plan to strengthen an existing concrete structure with carbon reinforced concrete' [68] (funding period: 05.2017–04.2020, grant number: 03ZZ0327A).

Acknowledgments: The results presented here were generated in different projects over the past 20 years. This research is only possible with the support of our colleagues in the Otto Mohr Laboratory, where most of the experiments were carried out. We would like to express our sincere thanks to them! Further thanks go to the colleagues and partners in the various completed and ongoing research projects; above all to research group 2 of the Institute of Concrete Structures of TU Dresden and the partners of the numerous C^3 projects. We would also like to thank all the companies that have provided us with research material free of charge. Thanks also to Dajana Musiol for reviewing the English paper.

Conflicts of Interest: The authors declare no conflict of interest.

References

1. Béguin, G.H. Various Schemes for Strengthening Concrete Structures. *Pract. Period. Struct. Des. Constr.* **2018**, 23. [CrossRef]

2. Teichert, P. Die Geschichte des Spritzbetons. *Schweizer Ingenieur Architekt* **1979**, *97*, 949–960. [CrossRef]

3. ACI (Ed.) *ACI 506R-05. 2005—Guide to Shotcrete*; ACI Commitee Report; ACI: Farmington Hills, MI, USA, 2005.

4. Jesse, F.; Kaplan, F. Verfahren für Biegeverstärkungen an Stahlbetonbauteilen. *Bautechnik* **2011**, *88*, 433–442. [CrossRef]

5. Bernard, E.S. (Ed.) Shotcrete: Elements of a System. In Proceedings of the Third International Conference on Engineering Developments in Shotcrete, Queenstown, New Zealand, 15–17 March 2010; CRC Press: Boca Raton, FL, USA, 2010.

6. Swamy, R.; Jones, R.; Bloxham, J. Structural behavior of reinforced concrete beams strengthened by epoxy-bonded steel plates. *Struct. Eng.* **1987**, *65A*, 59–68.

7. Chen, J.F.; Teng, J.G. Anchorage strength models for FRP and steel plates bonded to concrete. *J. Struct. Eng.* **2001**, *127*, 784–791. [CrossRef]

8. Oehlers, D.J.; Seracino, R. *Design of FRP and Steel Plated RC Structures—Retrofitting Beams and Slabs for Strength, Stiffness and Ductility*; Elsevier: Oxford, UK, 2004.

9. DAfStb [German Association for Steel Reinforced Concrete] (Ed.) *Verstärken von Betonbauteilen mit Geklebter Bewehrung—Parts 1–4*; Technical Rule of DAfStb; Beuth: Berlin, Germany, March 2012.

10. fib (Ed.) *Externally Bonded FRP Reinforcement for RC Structures*; fib Bulletin No. 14; Technical Report; fib: Lausanne, Switzerland, 2001.

11. fib (Ed.) *Retrofitting of Concrete Structures by Externally Bonded FRPs, with Emphasis on Seismic Applications*; fib Bulletin No. 35; Technical Report; fib: Lausanne, Switzerland, 2006.

12. Hollaway, L.C.; Leeming, M. (Eds.) *Strengthening of Reinforced Concrete Structures—Using Externally-Bonded Frp Composites in Structural and Civil Engineering*; Woodhead Publ.: Cambridge, UK, 1999.

13. Bleibler, A.; Schümperli, E.; Steiner, W. Flat Strip Lamella for Reinforcing Building Components and Method for Their Production. U.S. Patent No. 6,511,727, 28 January 2003.

14. Rasheed, H.A. *Strengthening Design of Reinforced Concrete with FRP*; CRC Press: Boca Raton, FL, USA, 2014.

15. *Design Procedures for the Use of Composites in Strengthening of Reinforced Concrete Structures—State-of-the-Art Report of the RILEM Technical Committee 234-DUC 2016*; Pellegrino, C.; Sena-Cruz, J. (Eds.) Springer: Dordrecht, The Netherlands, 2016; ISBN 978-94-017-7335-5.

16. ACI Committee 440 (Ed.) *440.2R-17: Guide for the Design and Construction of Externally Bonded FRP Systems for Strengthening Concrete Structures*; ACI Committee: Farmington Hills, MI, USA, 2017.

17. Jansze, W. Strengthening of Reinforced Concrete Members in Bending by Externally Bonded Steel Plates—Design for Beam Shear and Plate Anchorage. Ph.D. Thesis, Delft University, Delft, The Netherlands, 1997. Available online: http://resolver.tudelft.nl/uuid:ca3ae758-5c6d-4da9-8866-26340c49c9e6 (accessed on 12 January 2019).

18. Pelke, E.; Scholch, U. Strengthening the major bridges on the A45 motorway: Recalculation—Immediate measures—Reinforcement concepts. In *Taller, Longer, Lighter, Proceedings of the IABSE-IASS Symposium & The 35th Annual Symposium of IABSE & The 52nd Annual Symposium of IASS and Incorporating the 6th International Conference on Space Structures, London, UK, 20–23 September 2011*; Hemming Group: London, UK, 2011; book of abstracts: p. 443, full paper 0679 on CD.

19. Scheerer, S.; Schladitz, F.; Curbach, M. Textile reinforced Concrete—From the idea to a high performance material. In Proceedings of the FERRO-11 & 3rd ICTRC (PRO 98), Aachen, Germany, 7–10 June 2015; Brameshuber, W., Ed.; S.A.R.L. Rilem Publ.: Bagneux, France, 2015; pp. 15–33.

20. Triantafillou, T.C. (Ed.) *Textile Fibre Composites in Civil Engineering*; Woodhead Publ./Elsevier: Cambridge, UK, 2016.

21. Peled, A.; Bentur, A.; Mobasher, B. *Textile Reinforced Concrete*; Series: Modern Concrete Technology No 19; CRC Press: Boca Raton, FL, USA, 2017.

22. Koutas, L.N.; Tetta, Z.; Bournas, D.A.; Triantafillou, T.C. Strengthening of Concrete Structures with Textile Reinforced Mortars: State-of-the-Art Review. *J. Compos. Constr.* **2019**, *23*. [CrossRef]

23. Carloni, C.; Bournas, D.A.; Carozzi, F.G.; D'Antino, T.; Fava, G.; Focacci, F.; Giacomin, G.; Mantegazza, G.; Pellegrino, C.; Perinelli, C.; et al. Fiber Reinforced Composites with Cementitious (Inorganic) Matrix. In *Design Procedures for the Use of Composites in Strengthening of Reinforced Concrete Structures – State-of-the-Art Report of the RILEM Technical Committee 234-DUC 2016*; Pellegrino, C., Sena-Cruz, J., Eds.; Springer: Dordrecht, The Netherlands, 2016; Chapter 9; pp. 349–392. ISBN 978-94-017-7335-5.

24. Jesse, F.; Curbach, M. Verstärken mit Textilbeton. In *BetonKalender 2010*; Bergmeister, K., Fingerloos, F., Wörner, H.D., Eds.; Ernst & Sohn: Berlin, Germany, 2010; Chapter VII; pp. 457–565.

25. DIBt. *Verfahren zur Verstärkung von Stahlbeton mit TUDALIT (Textilbewehrter Beton)*, 1st ed.; Approval no. Z-31.10-182; DIBt: Berlin, Germany, 2014; update 2016.

26. Scheerer, S.; Schütze, E.; Curbach, M. Strengthening and Repair with Carbon Concrete Composites—The First General Building Approval in Germany. In Proceedings of the SHCC4—International Conference on Strain-Hardening Cement-Based Composites, Dresden, Germany, 18–20 September 2017; Mechtcherine, V., Slowik, V., Kabele, P., Eds.; Springer: Dordrecht, The Netherlands, 2018; pp. 743–751.

27. ACI 549.4R. *Guide to Design and Construction of Externally Bonded Fabric-Reinforced Cementitious Matrix (FRCM) Systems for Repair and Strengthening Concrete and Masonry Structures*; Reported by ACI Committee 549; ACI: Farmington Hills, MI, USA, 2013.

28. Peled, A.; Bentur, A.; Yankelevsky, D. Woven fabric reinforcement of cement matrix. *Adv. Cement Based Mater.* **1991**, *1*, 216–223. [CrossRef]

29. Hayashi, R.; Yamada, K.; Inaba, S.; Tsubouchi, S. Tensile properties of carbon fiber mesh reinforced mortar with various wavings. *Proc. Jpn. Concr. Inst.* **1990**, *12*, 1043–1048.

30. Mohamed, M.H.; Bilisik, A.K. Multi-Layer Three-Dimensional Fabric and Method of Producing. U.S. Patent No. 5,465,760, 14 November 1995.

31. Curbach, M.; Fuchs, H.; Hegger, J.; Noisternig, J.F.; Offermann, P.; Reinhardt, H.-W.; Sasse, H.R.; Schorn, H.; Wörner, H.-D.; Wulfhorst, B.; et al. *Sachstandbericht zum Einsatz von Textilien im Massivbau*; Publication series of DAfStb [German Association for Steel Reinforced Concrete]; Beuth: Berlin, Germany, 1998.

32. Proske, D. textilbewehrter Beton als Verstärkung von Platten. In Proceedings of the 34th DAfStb-Forschungskolloquium 1997, Dresden, Germany, 9–10 October 1997; Institut für Tragwerke und Baustoffe of TU Dresden: Dresden, Germany, 1997; pp. 51–58.

33. DIN 1048. *Prüfverfahren für Beton*; Beuth: Berlin, Germany, June 1979.

34. DIN EN 206. *Beton—Festlegung, Eigenschaften, Herstellung und Konformität*; Deutsche Fassung EN 206:2013+A1:2016; Beuth: Berlin, Germany, January 2017.

35. Weiland, S. Interaktion von Betonstahl und textiler Bewehrung bei der Biegeverstärkung mit textilbewehrtem Beton. Ph.D. Thesis, TU Dresden, Dresden, Germany, 2009. Available online: http://nbn-resolving.de/urn: nbn:de:bsz:14-qucosa-37944 (accessed on 12 January 2019).

36. Koutas, L.N.; Bournas, D.A. Flexural Strengthening of Two-Way RC Slabs with Textile-Reinforced Mortar: Experimental Investigation and Design Equations. *J. Compos. Constr.* **2017**, *21*, 04016065. [CrossRef]

37. Bisby, L.A.; Stratford, T.J.; Hart, C.; Farren, S. Fire performance of well-anchored TRM, FRCM and FRP flexural strengthening systems. In *Advanced Composites in Construction, Proceedings of the ACIC 2013, Belfast, UK, 10–12 September 2013*; Whysell, C.J., Taylor, S.E., Eds.; Network Group for Composites in Construction: Belfast, UK, 2013; 12p.

38. Tsesarsky, M.; Katz, A.; Peled, A.; Sadot, O. Textile Reinforced Cementitious Composites for Retrofit and Strengthening of Concrete Structures under Impact Loading. *Mater. Struct.* **2015**, *48*, 471–484. [CrossRef]

39. Beckmann, B.; Hummeltenberg, A.; Weber, T.; Curbach, M. Strain Behaviour of Concrete Slabs under Impact Load. *Struct. Eng. Int.* **2012**, *22*, 562–568. [CrossRef]

40. Ortlepp, R. Untersuchungen zur Verbundverankerung textilbewehrter Feinbetonverstärkungsschichten für Betonbauteile. Ph.D. Thesis, TU Dresden, Dresden, Germany, 2007. Available online: http://nbn-resolving. de/urn:nbn:de:swb:14-1187166738380-68700 (accessed on 12 January 2019).

41. EN 1992-1-1. *Design of Concrete Structures—Part 1-1: General Rules and Rules for Buildings*; CEN: Brussels, Belgium, 2004.

42. Müller, E.; Scheerer, S.; Curbach, M. Strengthening of existing concrete structures: Design models. In *Textile Fibre Composites in Civil Engineering*; Triantafillou, T.C., Ed.; Woodhead Publishing/Elsevier: Amsterdam, The Netherlands, 2016; Chapter 14; pp. 323–359. ISBN 978-1-78242-446-8.

43. Bösche, A. Möglichkeiten zur Steigerung der Biegetragfähigkeit von Beton-und Stahlbetonbauteilen durch den Einsatz textiler Bewehrungen—Ansatz für ein Bemessungsmodell. Ph.D. Thesis, Technische Universität Dresden, Dresden, Germany, 2007.

44. Frenzel, M.; Curbach, M. Bemessungsmodell zur Berechnung der Tragfähigkeit von biegeverstärkten Stahlbetonplatten. In *Textilbeton Theorie und Praxis, Proceedings of the 6th Koll. zu textilbewehrten Tragwerken (CTRS6), Berlin, Germany, 19–20 September 2011*; Curbach, M., Ortlepp, R., Eds.; Institut für Massivbau der TU Dresden: Dresden, Germany, 2011; pp. 381–399. Available online: http://nbn-resolving.de/urn:nbn:de: bsz:14-qucosa-78041 (accessed on 12 January 2019).

45. Curbach, M.; Lorenz, E.; Schladitz, F.; Schütze, E.; Weiland, S. *Gesamtbericht der experimentellen Untersuchungen zur allgemeinen bauaufsichtlichen Zulassung für ein Verfahren zur Verstärkung von Stahlbeton mit TUDALIT (Textilbewehrter Beton)*; Unpublished Report for Deutsches Institut für Bautechnik (DIBt); Institut für Massivbau, TU Dresden: Dresden, Germany, 2014.

46. Frenzel, M. Bemessung textilbetonverstärkter Stahlbetonbauteile unter Biegebeanspruchung. *Beton-Stahlbetonbau Spezial* **2015**, *110*, 54–68. [CrossRef]

47. Curbach, M.; Schladitz, F.; Zobel, R.; Weselek, J. Eine Vision wird Realität: Der Betonbau der Zukunft ist nachhaltig, leicht, flexibel und formbar—Dank Carbon. *Der Prüfingenieur* **2017**, *51*, 20–35. Available online: https://bvpi.de/bvpi/de/aktuelles/abonnement-der-pruefingenieur.php (accessed on 22 January 2019).
48. DIN EN 1992-1-1. *Eurocode 2: Bemessung und Konstruktion von Stahlbeton- und Spannbetontragwerken—Teil 1-1: Allgemeine Bemessungsregeln und Regeln für den Hochbau*; Deutsche Fassung EN 1992-1-1:2004 + AC:2010; Beuth: Berlin, Germany, January 2011.
49. DIN EN 1992-1-1/NA. *Nationaler Anhang—National festgelegte Parameter—Eurocode 2: Bemessung und Konstruktion von Stahlbeton- und Spannbetontragwerken - Teil 1-1: Allgemeine Bemessungsregeln und Regeln für den Hochbau*; Beuth: Berlin, Germany, January 2011.
50. Schütze, E.; Bielak, J.; Scheerer, S.; Hegger, J.; Curbach, M. Einaxialer Zugversuch für Carbonbeton mit textiler Bewehrung | Uniaxial tensile test for carbon reinforced concrete with textile reinforcement. *Beton- Stahlbetonbau* **2018**, *113*, 33–47. [CrossRef]
51. Beckmann, B.; Frenzel, M.; Lorenz, E.; Schladitz, F.; Rempel, S. Biegetragverhalten von textilbetonverstärkten Platten. *Beton- Stahlbetonbau Spezial* **2015**, *110*, 47–53. [CrossRef]
52. Rempel, S.; Ricker, M. Ermittlung der Materialkennwerte der Bewehrung für die Bemessung von textilbewehrten Bauteilen. *Bauingenieur* **2017**, *92*, 280–288.
53. Just, M. Zur Sicherheit von Textilbeton. *Beton- Stahlbetonbau Spezial* **2015**, *110*, 42–46. [CrossRef]
54. DIN EN 1990. *Eurocode: Grundlagen der Tragwerksplanung*; Deutsche Fassung EN 1990:2002 + A1:2005 + A1:2005/AC:2010; Beuth: Berlin, Germany, December 2010.
55. Weselek, J.; Häussler-Combe, U. Calibrating safety factors for Carbon Concrete. *Beton- Stahlbetonbau Spezial* **2018**, *113*, 14–21. [CrossRef]
56. Homepage of C^3 Project. Available online: https://www.bauen-neu-denken.de/ (accessed on 25 January 2019).
57. Triantafillou, T.C. Strengthening of existing concrete structures: Concepts and structural behavior. In *Textile Fibre Composites in Civil Engineering*; Triantafillou, T.C., Ed.; Woodhead Publishing/Elsevier: Amsterdam, The Netherlands, 2016; Chapter 13; pp. 303–322. ISBN 978-1-78242-446-8.
58. Bournas, D. Strengthening of existing structures: Selected case studies. In *Textile Fibre Composites in Civil Engineering*; Triantafillou, T.C., Ed.; Woodhead Publishing/Elsevier: Amsterdam, The Netherlands, 2016; Chapter 17; pp. 389–411. ISBN 978-1-78242-446-8.
59. Erhard, E.; Weiland, S.; Lorenz, E.; Schladitz, F.; Beckmann, B.; Curbach, M. Anwendungsbeispiele für Textilbetonverstärkung—Instandsetzung und Verstärkung bestehender Tragwerke mit Textilbeton. *Beton- Stahlbetonbau Spezial* **2015**, *110*, 74–82. [CrossRef]
60. Scheerer, S.; Chudoba, R.; Garibaldi, M.P.; Curbach, M. Shells made of Textile Reinforced Concrete—Applications in Germany. *J. Int. Assoc. Shell Spat. Struct. J. IASS* **2017**, *58*, 79–93. [CrossRef]
61. Curbach, M.; Hauptenbuchner, B.; Ortlepp, R.; Weiland, S. Textilbewehrter Beton zur Verstärkung eines Hyparschalentragwerks in Schweinfurt. *Beton- Stahlbetonbau* **2007**, *102*, 353–361. [CrossRef]
62. Weiland, S.; Ortlepp, R.; Hauptenbuchner, B.; Curbach, M. Textile Reinforced Concrete for Flexural Strengthening of RC-Structures—Part 2: Application on a Concrete Shell. In *Design & Applications of Textile-Reinforced Concrete, Proceedings of the ACI Fall Convention, Fajardo, Puerto Rico, 14–18 October 2007*; Aldea, C.-M., Ed.; ACI: Farmington Hills, MI, USA, 2007; SP-251CD—3; pp. 41–58.
63. Hentschel, M.; Schumann, A.; Ulrich, H.; Jentzsch, S. Sanierung der Hyparschale Magdeburg. *Bautechnik* **2019**, *96*, 25–30. [CrossRef]
64. Feix, J.; Hansl, M. Zur Anwendung von Textilbeton für Verstärkungen im Brückenbau. In *Festschrift zum 60. Geburtstag von Univ.-Prof. Dr.-Ing. Manfred Keuser*; Berichte aus dem konstruktiven Ingenieurbau 12/4; Gebekken, N., Holzer, S., Keuser, M., Mangerig, I., Petersen, C., Siebert, G., Thienel, K.-Ch., Eds.; Universität der Bundeswehr München: München, Germany, 2012; pp. 289–295.
65. Al Jamous, A. Instandsetzung der historischen Betonbogenbrücke in Naila. *Beton- Stahlbetonbau* **2017**, *112*, 623–628. [CrossRef]
66. Herbrand, M.; Adam, V.; Classen, M.; Kueres, D.; Hegger, J. Strengthening of Existing Bridge Structures for Shear and Bending with Carbon Textile-Reinforced Mortar. *Materials (Basel)* **2017**, *10*, 1099. [CrossRef] [PubMed]

67. Homepage of the Project C3-V1.2. Available online: https://tu-dresden.de/bu/bauingenieurwesen/imb/forschung/Forschungsfelder/TRC-C3/C3-vorhaben/C3-V1.2?set_language=en (accessed on 25 January 2019).
68. Homepage of the Project C3-V2.7. Available online: https://tu-dresden.de/bu/bauingenieurwesen/imb/forschung/Forschungsfelder/TRC-C3/C3-vorhaben/C3-V2.7?set_language=en (accessed on 25 January 2019).
69. Homepage of Collaborative Research Centre 'SFB 528'. Available online: https://tu-dresden.de/bu/bauingenieurwesen/imb/forschung/sfb528?set_language=en (accessed on 25 January 2019).
70. Curbach, M.; Ortlepp, R. (Eds.) *Sonderforschungsbereich 528—Textile Bewehrungen zur bautechnischen Verstärkung und Instandsetzung—Abschlussbericht (gekürzte Fassung)*; Research Report; TU Dresden: Dresden, Germany, 2012; Available online: http://nbn-resolving.de/urn:nbn:de:bsz:14-qucosa-86425 (accessed on 10 January 2019).
71. Homepage of TUDALIT e.V. Available online: http://tudalit.de/ (accessed on 25 January 2019).
72. Special issue of the journal Beton- und Stahlbetonbau. Special issue of the journal Beton- und Stahlbetonbau. *Beton- Stahlbetonbau Spezial* **2015**. 110/S1—Supplement "Verstärken mit Textilbeton". Available online: https://onlinelibrary.wiley.com/toc/14371006/110/S1 (accessed on 28 March 2019).

Article

Numerical Modeling of Non-Uniformly Reinforced Carbon Concrete Lightweight Ceiling Elements

Rostislav Chudoba [1],*, Ehsan Sharei [2], Tilo Senckpiel-Peters [3] and Frank Schladitz [3]

[1]   RWTH Aachen University, 52062 Aachen, Germany
[2]   H+P Ingenieure GmbH, 52072 Aachen, Germany; esharei@huping.de
[3]   Teschnische Universität Dresden, 01069 Dresden, Germany; tilo.senckpiel-peters@tu-dresden.de (T.S.-P.); frank.schladitz@tu-dresden.de (F.S.)
*   Correspondence: rostislav.chudoba@rwth-aachen.de; Tel.: +49-241-8028150

Received: 27 March 2019; Accepted: 2 June 2019; Published: 7 June 2019

Featured Application: **The present paper contributes to the discussion on modeling methods appropriate for the structural analysis of thin-walled concrete shells, a rapidly developing field of material and structural design utilizing the high-performance cementitious composites reinforced with non-metallic reinforcement. An effective modeling support is paramount for the derivation of reliable and economic design and assessment principles in a wide range of applications.**

Abstract: The paper focuses on the specifics of macro-scale modeling of thin-walled textile-reinforced concrete shells. Application of layered shell finite elements requires systematic procedures for identification of material characteristics associated with the individual layers within the cross section. The identification of the material parameters describing the tensile behavior of a composite cross section is done using data obtained from the tensile test. Such test is usually performed only for a reference configurations with a simple layup of fabrics and a chosen thickness. The question is how to derive the strain-hardening response from the tensile test that is relevant for a changed cross-sectional configuration. We describe and discuss scaling and mixture rules that can be used to modify the material parameters for modified cross-sectional layups. The rules are examined in the context of the test results obtained on a shell that was reinforced non-uniformly, with varying types of textile fabrics and varying thickness within the shell surface.

Keywords: textile-reinforced concrete; thin-walled shells; cementitious composites; layered finite elements; mixture rules; model calibration

---

1. Introduction

Several applications of thin-walled concrete shells reinforced with high-performance textile fabrics realized in the recent decade have convincingly demonstrated the potential of the new type of composite for the design and construction of highly efficient structural members [1,2]. The combination of fine aggregate concrete matrix with textile fabric reinforcements made of carbon enabled the construction of lightweight, thin concrete shells with curved geometries [3]. Experimental investigations of structural behavior were performed for large-scale shells with strain-hardening behavior serving as roof elements [4–6], pedestrian bridges [7,8], sandwich panels and facade elements [9–11]. Because of the non-corrosive nature of the reinforcement, a high sustainability and durability of this type of structures are provided. The development of compatible material components, i.e., the cementitious matrix and the reinforcement fabrics, is continuously extending the spectrum of design and manufacturing options. This development of material and manufacturing technologies calls for an effective support in terms of efficient and validated modeling approaches delivering a correct prediction of the structural behavior of thin-walled textile-reinforced concrete (TRC) shells. Efficient and realistic models of non-uniformly reinforced fabric reinforced shells are the assumption for the development of robust design and assessment concepts [12,13]. Such modeling strategies must account for the specific features of the material and of the nonlinear structural behavior of TRC shells. The particular issues to be considered in the macro-scale modeling of TRC shells include:

- Correct reproduction of the strain-hardening response of the shell cross section for a wide range of loading conditions including the tensile or bending loads.
- Initial and damage induced anisotropy owing to the oriented crack pattern within the two-dimensional stress state in the shell plane.
- Geometrical nonlinearity in interaction with imperfections requiring the buckling analysis as documented in experimental studies [4] and numerical investigations of the effect of imperfections on the structural behavior [14,15].

The focus of the present paper is on the first item, i.e., a flexible reflection of the strain-hardening behavior in macro-scale simulations of TRC shells. The second aspect of damage-induced anisotropy is considered here as well; however, it does not have a significant influence on the structural behavior of the studied girder. The issue of geometrical nonlinearity and buckling behavior is not within the scope of the present paper.

Considering a TRC element with sufficient reinforcement ratio loaded in tension, three phases of tensile response are distinguished:

- Phase I: linear elastic behavior with the strain–stress relation characterized by the composite. elasticity modulus,
- Phase II: formation of the cracks at the level of the matrix tensile strength,
- Phase III: saturated crack state with steady matrix stress level between the cracks; the effective stiffness of the composite in this phase is approximately equal to the stiffness of the reinforcing fabrics.

Existing approaches to modeling of this tensile behavior can be roughly classified into three categories summarized in Figure 1. The theoretical descriptions based on discrete representation of the matrix cracking process and debonding between matrix and reinforcement displayed in column (a) of Figure 1 considers an elastic-brittle behavior of fabrics and matrix material components. In this model, the nonlinear strain-hardening response results from the evolution of discrete cracks emerging along the specimen with random matrix strength and nonliner bond stress—slip relation [16,17]. The ambition of this modeling approach is to predict the strain-hardening behavior of a particular cross-sectional design using the parameters of the material components and of the bond between the concrete and fabrics [18].

Even though the discrete crack modeling approaches provide a valuable insight into the process of matrix cracking and debonding, a realistic and reliable prediction of the composite response $\sigma_c(\varepsilon_c)$ is still not possible. The reason is the complexity of interaction mechanisms governing the bond behavior. Moreover, the effect of material heterogeneity and imperfections at the levels of the fiber bundles, of the fabrics and of the cross-sectional thickness introduces further challenging tasks for the development of micro- and meso-scale modeling approaches. At this level of modeling, further effort is necessary to capture the inelastic interaction effects within a multi-scale model with a broader range of validity than available so far.

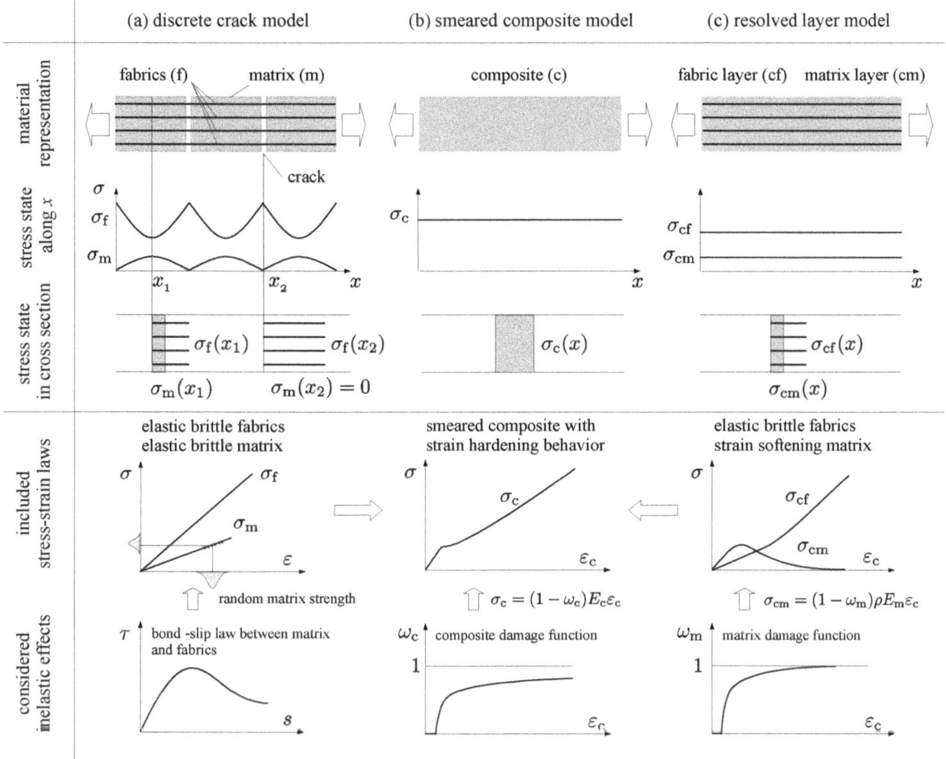

Figure 1. Classification of modeling approaches to tensile behavior of textile-reinforced concrete shells including the resolution of stresses along a uniaxial specimen, resolution of stress and strain components in a material point and examples of inelastic material characteristics.

As a result, the strain-hardening response of the composite is usually experimentally characterized using a composite tensile test. The measured response provides the input for an efficient calibration of phenomenological numerical material models. These macro-scale material models, applicable in finite element simulations, represent the strain-hardening response phenomenologically in terms of inelastic state variables, i.e., damage and plastic strains. An approach representing the composite with a uniform material behavior over the cross section depicted in column (b) of Figure 1 has been applied by the authors for the simulation of with laminated carbon TRC shells [14]. In this simulation concept, the material parameters are specific to a particular cross-sectional layup and thickness. If the cross-sectional configuration changes, the material model is not valid anymore and its material parameters must be recalibrated. Due to the plane stress state within the shell surface, the inelastic effects, i.e., damage or yielding, need to be captured only in the in-plane direction of the shell. The application of this model is limited to cross sections with uniform fabric reinforcement over their height or to structures with prevailing membrane stresses. As discussed later on, the reason is that the position of the fabrics within the cross section is not distinguished in the model so that the bending behavior of sparsely reinforced cross sections cannot be correctly reproduced.

In this paper, we focus on the modeling approach depicted in column (c) of Figure 1 with a resolved representation of the matrix and fabric layers within the cross section. In contrast to the discrete cracking models, the matrix cracking represented in a smeared way along the specimen length but is ascribed to a resolved layer of concrete within the shell cross section. Such a distinction between the effective fabrics and matrix layers allows for the rescaling of material parameters for a changed cross-sectional configuration including e.g., reinforcement ratio or layup of the fabrics within a cross section. Such scaling is needed to enhance the validity range of the macroscopic models for layered TRC shells so that a smaller amount of calibration experiments is required for the identification of the material parameters. The model classification in Figure 1 is focused on the representation of a the composite cross section. Finite element models based primarily on solid finite element discretization with an explicit resolution of material interfaces including also the localization of individual concrete cracks, presented e.g., in [19,20], have a different focus and different purpose, i.e., local stress effects in structural details of the simulation of TRC sandwich panels. For an efficient and realistic prediction of the nonlinear behavior of thin TRC shells, the layered shell models provide the appropriate dimensional reduction of the simulated boundary-value problem.

Layered shells have already been applied recently in the context of the nonuniformly reinforced TRC shells using both smeared and resolved layup within a TRC shell [21,22] in combination with damage-plasticity models available in *ABAQUS* [23]. Consistent with these studies, a generalized description of the scaling and mixture rules for elastic and inelastic material parameters reflecting modified layups of thin TRC shells is proposed in the following sections. The smeared and resolved models of the cross section are validated using the test data obtained within an experimental investigation of girder element briefly described in Section 2. The smeared representation of a composite cross section (Figure 1, column b) is realized using an anisotropic damage model characterized in Section 3. The decomposition of the tensile response for the layered cross section (Figure 1, column c) is then described in detail in Section 4 including the qualitative validation of the correct bending response. Finally, in Section 5, we present the results of the numerical simulations of the girder response performed using the calibrated material models in combination with the smeared and resolved cross-sectional idealizations.

2. Test Setup for the Carbon Concrete Girder

The experimental response of the thin-walled carbon concrete ceiling elements depicted in Figure 2 developed at the Institute of Concrete Structures at TU Dresden [24] served for the validation of the macroscopic modeling approaches (c) and (d) from Figure 1. The girders were non-uniformly reinforced with two types of fabrics of the fineness 3300 tex and 800 tex. Their thickness was non-uniform as well. The cross-sectional shape is shown in Figure 3a. As apparent from Figure 3b, two types of cementitious matrix were applied within the cross section, the outer layers were made of the Fleece Concrete Composite (FCC) and the inner textile-reinforced concrete layer (denoted as Carbon Reinforced Concrete in Figure 4) was made of a fine grain concrete with the carbon fabrics position in the middle of the 10 mm high section. The geometry, layout of fabrics within the shell and the specification of layers within a cross section of the girder is provided in Figure 4.

Figure 2. Test specimens.

(a)   (b)

Figure 3. Girder cross section: (**a**) profile of the girder with the carbon fabric reinforcement; (**b**) cross-sectional layup consisting of carbon reinforced concrete in the middle section and fleece concrete composite at the top and at the bottom.

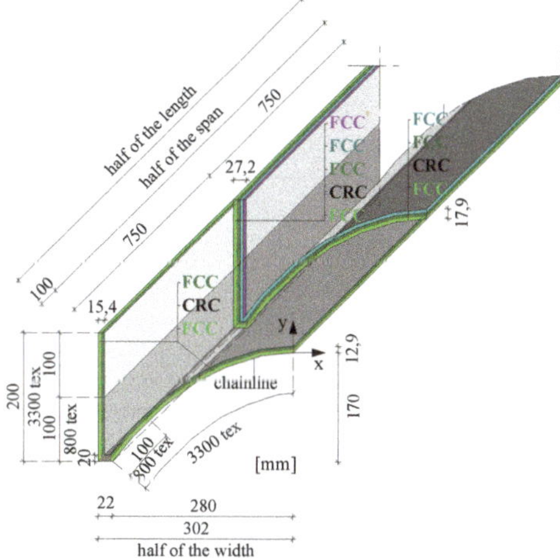

Figure 4. Cross section layups of reinforced shell sections consisting of 3300 tex and 800 tex fabrics, fleece concrete composite (FCC) and carbon reinforced concrete (CRC) displayed for a quarter section of the girder corresponding to the two planes of symmetry [24].

The girders were tested using a six-point bending test depicted in Figure 5. The load was introduced using wooden plates following the catenary curve of the vault within a girder section. The experimentally obtained load–deflection curves obtained in the test are used later on in Section 4 for the validation of the discussed modeling approaches.

Figure 5. Test setup of the carbon concrete girder.

3. Smeared Model of a Cross Section

3.1. Characterization of the Applied Anisotropic Damage Model of a Shell Layer

The classification of modeling approaches introduced in Figure 1 considers the inelastic material behavior of the composite layers in a general way. Even though a damage model is chosen in the last row of Figure 1 as an example, the classification can be used in combination with plasticity models describing the strain-hardening or strain-softening behavior as well. This classification provides the framework for the explanation of the scaling and mixture rules for inelastic material parameters presented in the sequel.

To formulate and to validate the scaling and mixture rules in combination with a particular example of an inelastic material model, a microplane damage model introduced in [25] has been chosen to reflect the nonlinear material behavior of the composite layers. Inelastic effects governing the stress–strain response are thus represented by a damage function as exemplified in the last row of Figure 1. The key idea behind the microplane models is to reflect the material state on a set of planes on a unit hemisphere or circle around a material point instead of using the usual tensorial representation. The microplanes are used to establish the mapping between the strain and stress tensors as proposed in the original model by Bažant et al. [26] sketched in Figure 6a. The macroscopic strain tensor ε is decomposed onto the microplane directions by a geometric projection, delivering the microplane strain vectors e. A constitutive law at the microplane level defines the relation between the apparent and effective microplane strain and stress vectors. Finally, the macroscopic stress tensor σ is obtained using the condition of energetic equivalence between the microplane discretization and the tensorial representation of stress at the material point. As the original version was detected to be thermodynamically inconsistent, several refinements have been proposed as documented e.g., in [27]. Currently, the microplane discretization of a material state provides a well established framework for sound formulation of anisotropic inelastic material models [28].

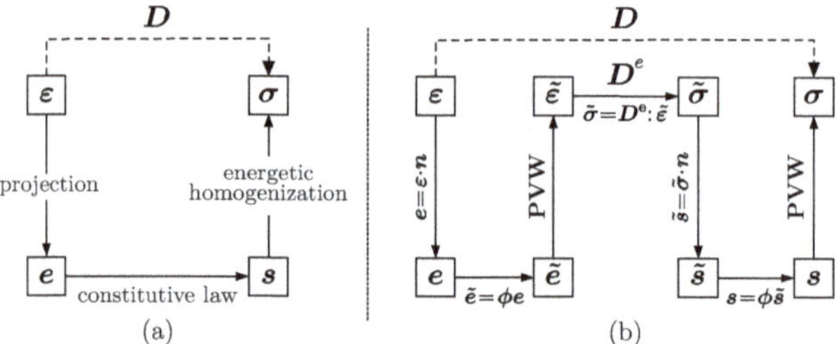

Figure 6. Algebraic structure of the microplane models (a) basic principle of microplane model; (b) constitutive stress–strain relation in the microplane damage model.

In the microplane damage model used here [25], each microplane is associated with a projected strain and damage state variables. Its important property is the separation between the apparent and effective stresses and strains. As indicated in the algebraic structure of the microplane damage model depicted in Figure 6b, the apparent stress tensor ε is first projected onto the microplanes rendering the microplane strain vectors e. Then, the effective microplane strains $\tilde{e}$ is obtained with the help of the damage/integrity function. By employing the condition of energetic equivalence, i.e., the principle of virtual work, the corresponding effective strain tensor $\tilde{\varepsilon}$ is obtained by integrating the strains over the microplane discretization. The relation between the effective strain tensor $\tilde{\varepsilon}$ and the effective stress tensor $\tilde{\sigma}$ is provided explicitly using the elasticity tensor $\boldsymbol{D}^{e}$. Analogous to the described mapping of strain, the mapping between the effective stress tensor $\tilde{\sigma}$ and apparent stress tensor σ is performed with the help of the microplane state discretization governed by the prescribed integrity function $\phi = 1 - \omega$. In the applied implementation of the model, only the normal component of the microplane strain and stress is considered. The model is thermodynamically consistent and implicitly captures the Poisson's effect of the undamaged, still an effective skeleton of a material structure in a material point.

The briefly characterized microplane damage model falls into the category of anisotropic damage models. Its application to the simulation of TRC shells has been described in [14]. The reason for choosing this model in this paper is twofold: the damage function can be automatically calibrated for both strain-hardening and strain-softening behavior, and it can reflect the anisotropy of damage due to the development of oriented cracks in the in-plane directions of the shell. The applied version of the model specialized for shells uses microplanes arranged around a unit circle centered at a material point. By aligning the stress and strain tensors in a material point to the shell geometry, the in-plane (ξ, η) and out-of-plane (ζ), the evolution of in-plane damage at a material point can be related to a polar discretization of the material state in a material point displayed in Figure 7. In this state representation, the degree of damage depends on the orientation within the shell surface. This fact introduces the damage anisotropy reflecting the evolution of oriented, fine crack pattern.

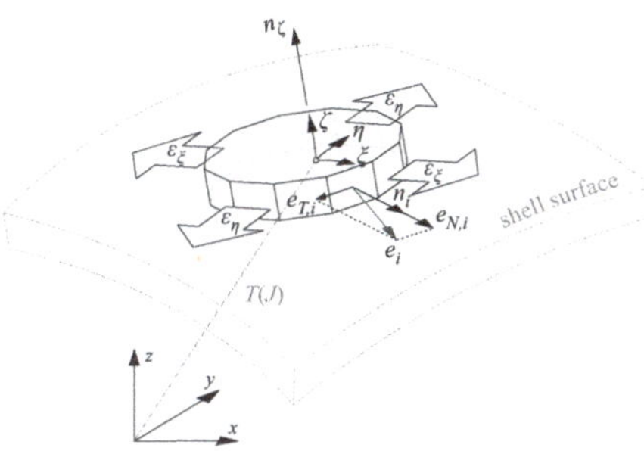

Figure 7. Microplane model within a single layer of a shell cross section.

Further details of the mathematical formulation of the microplane damage model for TRC shells would go beyond the scope of the present paper. The complete model description including the calibration procedure, elementary verification studies and model validation using three-point bending test and slab tests are provided in [14]. The formulated material model was implemented by the first two authors as a user subroutine in the finite element code *ABAQUS* and in the in-house research software *BMCS*. It is used in the following studies both in smeared and in resolved versions of the shell model to simulate the strain-hardening and strain-softening material behavior, respectively.

3.2. Calibration of a Smeared Cross Section Model Using a Tensile Test

To identify the damage functions with the same effective stress–strain behavior assumed in all material points of a cross section, the data from the tensile test following the RILEM recommendation [29] was used as input. The cross sections with similar fabric layups as those used in the tested girder were tested in two series. The specimens were reinforced with a single layer of fabrics, one with 3300 tex and one with 800 tex. Figure 8 shows the stress–strain response obtained in four tests for each cross section as gray dashed lines, and the average response curves as black solid lines. Given the cross-sectional thickness d^{test} and area of the textile fabrics per unit width a_{f}, the reinforcement ratio ρ^{test} reads

$$\rho^{\text{test}} = \frac{a_{\text{f}}}{d^{\text{test}}}.$$ (1)

The fabric area per unit length and the corresponding reinforcement ratios for the tested specimens reinforced with both types of fabrics are listed in Table 1.

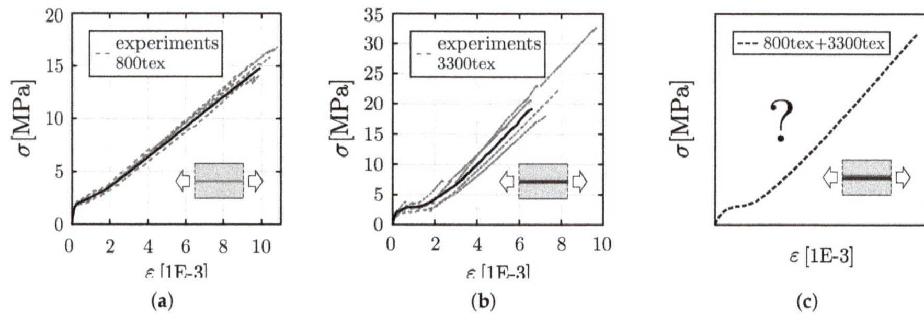

Figure 8. Tensile response of the cross sections with (**a**) 800 tex, (**b**) 3300+tex reinforcement and (**c**) of the hybrid cross section 3300+800 tex.

Table 1. Cross section characteristics; parameters of performed tests in boldface font, derived values in normal font.

fabric fineness		[tex]	**800**	**3300**	3300 + 800
fabric area	a_f	[m^2/m]	**0.616 · 10^{-4}**	**1.713 · 10^{-4}**	2.329 · 10^{-4}
specimen thickness	d	[m]	**0.01**	**0.01**	0.01
reinforcement ratio	$\rho = a_f/\bar{d}$	[-]	0.616 · 10^{-2}	1.713 · 10^{-2}	2.329 · 10^{-2}
hybrid section ratio	η (Equation (9))	[-]	0.264	0.735	1.000
final composite stiffness	E_{cf}	[GPa]	1.48	3.22	4.70
effective fabric stiffness	E_f	[GPa]	250	188	202

The parameters of the microplane damage model include the Young's modulus ($E_m = 28$ GPa) and Poisson's ratio (0.2) and the function $\phi(e) = 1 - \omega(e)$ defining the diminishing integrity evolution for an increasing normal microplane strain. The integrity functions obtained using the incremental calibration procedure are plotted in Figure 9a,b for the two tested cross sections. As observed for the cross section with 800 tex fabrics in Figure 9a, the integrity does not drop to zero but remains at a constant level corresponding to the final elastic stage of the strain-hardening behavior at which no further matrix cracks appear. In case of the 3300 tex specimens, the shape of the damage function shown in Figure 9b reveals an increasing integrity in the range of strains $0.001 < \varepsilon < 0.002$. This effect is owing to the delayed activation of filaments within the yarn cross section, usually referred to as slack, of stretched filaments in this range of strains.

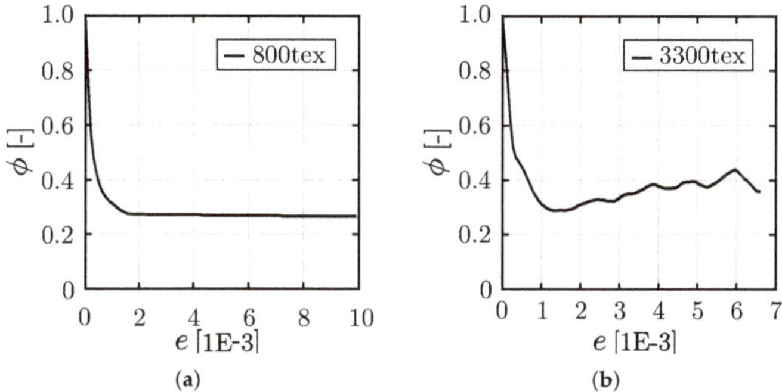

Figure 9. Calibration of integrity functions $\phi = (1 - \omega)$ using experimentally obtained stress–strain curves on 10 mm thick specimens reinforced with one layer of (**a**) 800 tex and (**b**) 3300 tex carbon fabrics.

3.3. Limitations of the Smeared Cross Section Model

The calibrated functions are valid only for the cross sections that were used in the test. Any change in the cross-sectional configuration would require conducting a new tensile test with a subsequent calibration of inelastic parameters. Such a requirement would certainly make the design, calibration, structural analysis and assessment of shells uneconomic. Thus, scaling procedures for strain-hardening response are required to extend the validity range of numerical models for a wider range of cross-sectional designs.

Another limitation of the smeared model results from the assumption of uniform material behavior in all layers of the cross section. Figure 10 shows the strain and stress profiles over the cross-sectional height that occurs upon bending load. Since every material point in the cross section follows the same strain-hardening curve, there is no possibility to reflect the effect of the position of the fabrics in the cross section. As a result, the smeared model cannot correctly capture the effect of the lever arm on the bending response in sparsely reinforced cross sections. As we shall document later on, this fact can result in an overestimation of stiffness if the loading induces non-negligible amount of bending stresses within an analyzed shell.

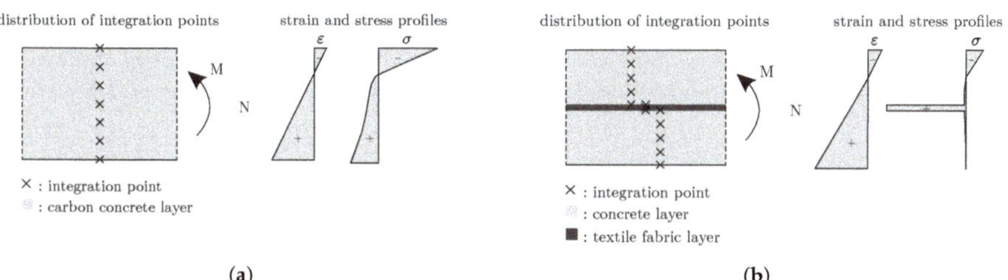

Figure 10. Stress and stress and profiles for the considered representations of the composite cross section with the distributions of integration points over the height: (**a**) smeared idealization with uniform material in each material point; (**b**) resolved cross-sectional idealization with different material behavior in plain concrete and fabric layers

4. Resolved Model of a Cross Section

To extend the range of applications to shells with a non-uniform distribution of fabrics within the cross section, the following enhancements of a material model describing the layer behavior are required:

- Scaling of the strain-hardening response for a modified reinforcement ratio related to a varying shell thickness or to the area of fabrics.
- Mixture rule to identify the strain-hardening response of a cross section layer combining several types of fabrics.
- Definition of a layered discretization of a shell cross section based on the scaling and mixture rules.

As already mentioned, these enhancements are not specific to the underlying microplane damage but are also valid for other types of inelastic material models, e.g., damage-plasticity models.

4.1. Decomposition of the Composite Stress

To derive the constitutive parameters related to a resolved, layered representation of a cross section within a shell finite element, a simple idealization of the cross section consisting of two parallel uniform material components with nonlinear, behavior is used. By idealizing the cross section as two nonlinear, springs with a stiffness defined in accordance with area fractions of the matrix and fabrics, the stress–strain measured in the reference tests are decomposed into matrix and fabrics stresses as follows:

$$\sigma_c^{\text{test}} = \sigma_{cf}^{\text{test}} + \sigma_{cm}^{\text{test}} = \rho^{\text{test}} \sigma_f + \left(1 - \rho^{\text{test}}\right) \sigma_m, \tag{2}$$

where $\sigma_{cf}^{\text{test}}$ and $\sigma_{cm}^{\text{test}}$ are related to the unit area of the composite and σ_f and σ_m to the area of the material of fabrics and matrix, respectively.

Since the stress–strain response of the studied carbon fabric material is linear elastic and brittle, a natural choice for the approximation of the fabric stress is $\sigma_f = E_f^{\text{yarn}} \varepsilon_c$, leading to the fraction of composite stress transmitted by fabrics

$$\sigma_{cf}^{\text{test}}(\varepsilon_c) = \rho^{\text{test}} E_f^{\text{yarn}} \varepsilon_c, \tag{3}$$

with E_f^{yarn} determined in a yarn tensile test.

However, the effective behavior of the fabrics in the composite may be significantly different from the one measured in the tensile test of a yarn. The major two sources of the difference are

- the nonuniform fabric strain along the composite tensile test specimen, and
- the nonuniform strain profile within a thick fiber bundle cross section.

The former effect of matrix fragmentation leading to a fluctuating stress transfer to and from the concrete matrix between the cracks indicated in column (a) of Figure 1 can be explained and visualized using the meso-scale models explicitly reflecting the multiple-cracking and debonding during the loading process [18].

The latter issue considers the non-uniformity of the stress within the yarn with roughly 50,000 filaments within the bundle as is the case for the 3300 tex carbon yarns used within the cross section. Even though these yarns are penetrated with styrene-butadiene material, their effective stiffness is much lower than in case of the 800 tex fabrics [30].

Moreover, another structural effect within the bundle that makes the transformation of material characteristics non-trivial can be recognized in Figure 11b. Fabrics with a large yarn cross-sectional area containing up to 50,000 filaments can exhibit a phenomenon called slack, meaning that the reinforcement reaches its full stiffness only after some initial level of loading. This effect occurs when a non-negligible amount of filaments is not aligned with the direction of loading so that they must be stretched before they can start to contribute to the load transfer.

The interaction of the inelastic effects, i.e., slack, multiple-cracking and debonding, explain why the material stiffness E_f^{yarn} determined in yarn or fabric tensile tests is different from the effective stiffness in the composite. Therefore, a pragmatic approach to the identification of the effective fabric stiffness sketched in Figure 11 is used as a starting point in the decomposition of the composite stress.

This approach exploits the fact that, in the saturated state of cracking, the contribution of the matrix segments between the cracks to the composite stiffness is negligible, i.e., $E_c \approx E_{cf}$. Thus, the effective fabric stiffness E_f can be directly determined from the final branch stress–strain curve as documented in Figure 11a for the test with 800 tex fabrics. Then, the effective stiffness of fabrics E_f^{test} in the composite is evaluated as

$$E_f^{test} = E_{cf}^{test} / \rho^{test}.$$

In case of 800 tex fabrics, the effective stiffness $E_{cf}^{800} = 250$ GPa approaches a level measured in a yarn tensile test. On the other hand, in the case of the 3300 tex fabrics, the value $E_{cf}^{800} = 188$ GPa reveals that a large fraction of filaments within the cross section was not activated within the tensile test—see Table 1.

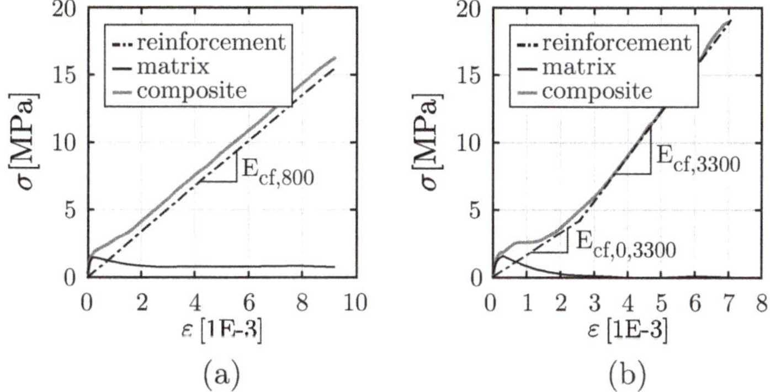

Figure 11. Stress–strain curves of the matrix (black solid lines), reinforcement (dashed lines) and composite (grey solid lines) for cross sections with (**a**) 800 tex and (**b**) 3300 tex reinforcements.

The effect of slack observed for the tests with 3300 tex fabrics has been treated by assuming a bilinear approximation of the fabric stress $\sigma_{cf}(\varepsilon_c)$ within the composite as shown in Figure 11b.

Using the obtained approximations of the effective fabric stress σ_{cf}^{test}, the corresponding fraction of matrix stress during the whole loading history is obtained as the difference between the composite and fabric stresses:

$$\sigma_{cm}^{test}(\varepsilon_c) = \sigma_c^{test}(\varepsilon_c) - \sigma_{cf}^{test}(\varepsilon_c). \tag{4}$$

The matrix stresses shown in Figure 11 grow up to the peak stress at the level of the first crack and start to diminish. The shape of the curves is similar to the stress–strain curve obtained for plain concrete in a three-point bending test. However, its mechanical interpretation is different. It does not reflect the strain-softening in the fracture process zone of a localizing macroscopic crack. It rather represents a smeared process of multiple cracking along a zone of a tensile specimen. In the former case, a stable crack growth with damage localization to a macroscopic crack is considered, while, in the latter case, multiple cracks emerge suddenly across a whole cross section in an unstable way.

4.2. Mixture Rule for Hybrid Fabric Reinforcement

In the girder member described in Section 2, four regions were reinforced with both types of fabrics i.e., 3300 + 800 tex. However, no tensile test that could be used for calibration of such a cross section has been conducted. Thus, to derive the stress–strain curve of such a cross section, a mixture rule must be used to extract the theoretical tensile behavior from the separately tested 3300 tex and 800 text cross sections.

Consider a hybrid cross section combining several types of fabrics $j = 1 \ldots m$ with a cross-sectional area per unit length $a_f^{(j)}$. For each of these fabrics, a composite tensile test has been conducted delivering the stress–strain curves $\sigma_c^{(j)}$. To combine the behavior of the fabrics in a hybrid cross section, the composite stress of each test j must first be decomposed using the procedure described in Section 4.1 into the fractions associated with fabrics and matrix $(\sigma_f^{(j)}, \sigma_m^{(j)})$. Then, the fabrics stresses $\sigma_f^{(j)}$ can be mixed into a single hybrid reinforcement layer using the area fractions of each reinforcement type in the total reinforcement area $\bar{a}_f = \sum_j^m a_f^{(j)}$ with the weight factors

$$\eta^{(j)} = \frac{a_f^{(j)}}{\bar{a}_f}.$$ (5)

The effective fabric stress within the hybrid fabric cross section is then given as

$$\bar{\sigma}_f = \sum_i^m \eta^{(j)} \sigma_f^{(j)}.$$ (6)

The matrix stress is obtained by averaging the contributions determined in the individual tests

$$\bar{\sigma}_m = \frac{1}{m} \sum_i^m \sigma_m^{(j)}.$$ (7)

Then, the composite stress in a cross section of thickness $\bar{d}$ reinforced with a hybrid fabric reads

$$\bar{\sigma}_c = (1 - \bar{\rho})\, \bar{\sigma}_m + \bar{\rho}\, \bar{\sigma}_f$$ (8)

with the reinforcement ratio

$$\bar{\rho} = \bar{a}_f / \bar{d}.$$

In reality, the usage of hybrid fabrics can introduce additional damage effects that are not reflected by this simple scaling of one-dimensional springs. In particular, a finer grid structure of overlapping fabrics reduces the contact area between the lower and upper concrete layers and can thus lead to surface delamination in the fabric plane at a low level of loading. However, in the case of the reinforcement ratio applied in the studied girder, the validity of the scaling can be assumed.

To relate the mixture rule to the studied girder, let us rewrite Equations (5)–(8) considering the case of hybrid reinforcement layer consisting of 3300 + 800 tex fabrics. Using cross-sectional characteristics of the test specimens summarized in boldface font in Table 1, the weighting factors of the fabric mixture rule read

$$\eta^{800} = \frac{a_f^{800}}{\bar{a}_f}, \quad \eta^{3300} = \frac{a_f^{3300}}{\bar{a}_f}, \quad \bar{a}_f = a_f^{800} + a_f^{3300}.$$ (9)

The fractions of fabric stress and of the matrix stress within a unit cross section are expressed as

$$\sigma_f^{3300+800} = \eta^{800} \sigma_f^{800} + \eta^{3300} \sigma_f^{3300},$$ (10)

$$\sigma_m^{3300+800} = \frac{1}{2}(\sigma_m^{3300} + \sigma_m^{800})$$ (11)

and the corresponding stress–strain curve of the composite cross section reads

$$\bar{\sigma}_c^{3300+800} = \left(1 - \rho^{3300+800}\right) \sigma_m^{3300+800}$$
$$+ \rho^{3300+800} \sigma_f^{3300+800}.$$ (12)

The result of this mixture rule is plotted in Figure 12 both as the composite stress fraction ascribed to the fabrics σ_{cf} (a) and the composite stress σ_c including also the matrix stress fraction σ_{cm} (b). To provide a comparison of effective stiffness measured in the tests with 3300 tex and 800 tex fabrics, the effective stiffness of the hybrid carbon textile fabric reinforcement in the final branch of the stress–strain curve is quantified in Table 1.

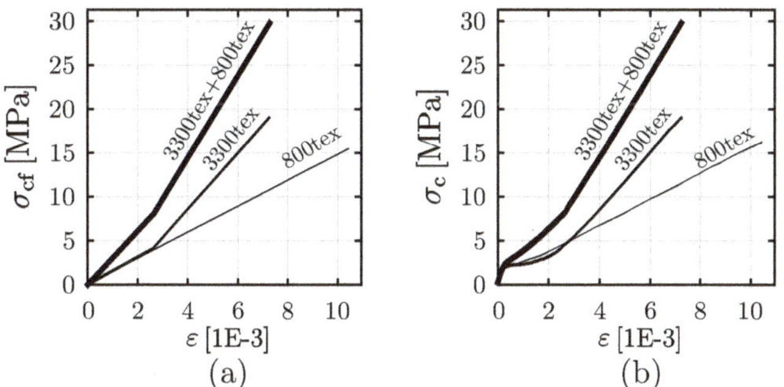

Figure 12. Reinforcement and composite stress–strain curves for cross sections with (**a**) single 800 tex and (**b**) 3300 tex reinforcements and the combined cross section.

4.3. Scaling of Composite Response for a Layer of a Shell Element

To identify the amount of stress corresponding to a layer of a finite element shell with a given layer thickness and its reinforcement ratio, the scaling formula can be used:

$$\sigma_{\mathrm{c}}^{(i)} = \frac{1 - \rho^{(i)}}{1 - \rho^{\text{test}}}\sigma_{\text{cm}}^{\text{test}} + \frac{\rho^{(i)}}{\rho^{\text{test}}}\sigma_{\text{cf}}^{\text{test}}.$$

(13)

This formula uses the decomposed stresses $\sigma_{\text{cm}}^{\text{test}}$ and $\sigma_{\text{cf}}^{\text{test}}$ evaluated from the composite stress–strain curve valid for the reinforcement ratio ρ^{test}. The source stress–strain curve can be either the original curve measured in the test or it can be a result of the mixture rule described in Section 4.2. The decomposition into the matrix and fabric stresses is performed using the procedure described in Section 4.1.

To verify this scaling formula, let us consider the limiting cases of a composite layer with the reinforcement ratios 1 and 0 and substitute it into Equation (13) to obtain the stress in a layer that represents either the reinforcement or matrix materials, respectively as

$$\rho^{(1)} = 1 \implies \sigma_{\mathrm{c}}^{(1)} = \frac{\sigma_{\text{cf}}^{\text{test}}}{\rho^{\text{test}}} = \sigma_{\mathrm{f}},$$

(14)

$$\rho^{(2)} = 0 \implies \sigma_{\mathrm{c}}^{(2)} = \frac{\sigma_{\text{cm}}^{\text{test}}}{1 - \rho^{\text{test}}} = \sigma_{\mathrm{m}}.$$

Thus, a layer corresponding to either plain matrix or plain fabric material is correctly recovered. By composing two layers such as these into a single cross section with the reinforcement ratio ρ^{test}, the original composite stress obtained in the test can be recovered using Equation (2).

4.4. Calibration for Resolved Cross-Sectional Idealization

The last step needed for the characterization of an arbitrary composition of layered cross section consistently reproducing the tensile strain-hardening response is the calibration of the material model describing the effective strain-softening behavior of the concrete matrix layers. This nonlinear relation $\sigma_{\text{cm}}(\varepsilon_{\mathrm{c}})$ is shown for the tested specimens in Figure 11. To reflect this behavior in the applied microplane damage model, the calibration of integrity functions ϕ representing the inelastic material parameters has been performed for the tested cross sections. The results are displayed in Figure 13.

As previously specified in (7), the contribution of the matrix stress $\bar{\sigma}_m$ in a mixed cross section is introduced as an average of contributions $\sigma_m^{(j)}$ determined from the tensile tests characterizing the matrix behavior in combination with the applied types of fabric reinforcement j. With regard to this fact, the contribution of the matrix stress to the cross section is nearly proportional to the area of matrix within the cross section. Thus, when composing a layup in a shell finite element with varying thickness, it is possible to simply add an effective matrix layer of a corresponding thickness and material behavior $\bar{\sigma}_m(\varepsilon_c)$ without the need to recalibrate the material parameters of the layers in the cross section.

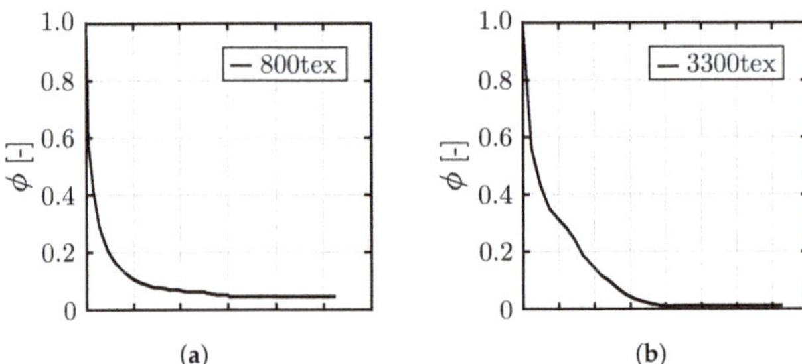

(a)   (b)

Figure 13. Damage functions describing the matrix cracking calibrated using the matrix stress–strain curve in Figure 11 extracted from the composite tensile tests with (**a**) 800 tex and (**b**) 3300 tex.

4.5. Verification of the Resolved Approach for Bending

As mentioned in Section 3.3, a correct prediction of the bending response is one of the requirements that leads to the usage of resolved, layered representation of the shell cross section. To test the ability of the resolved model to correctly represent the development of an effective lever arm within a cross section, a parametric study of a three-point bending test with the span of 500 mm and width of 100 mm was performed with three different positions of the fabric within the cross sections. A discrete load was applied at the mid-span of the beam. The cross section with the height of 10 mm was reinforced with one layer of carbon fabric with 3300 tex. The behavior of the fabrics was assumed isotropic within the reinforcement layer with the stress–strain curve $\sigma_f^{3300+800}(\varepsilon_c)$ and a layer thickness adjusted such that it matches the area of the fabric $a_f^{3300+800}$ within the cross section. The studied configurations (i), (ii) and (iii) are shown in Figure 14 together with the corresponding load–deflection curves obtained from the simulation. The deflection was recorded using the midspan section of the girder.

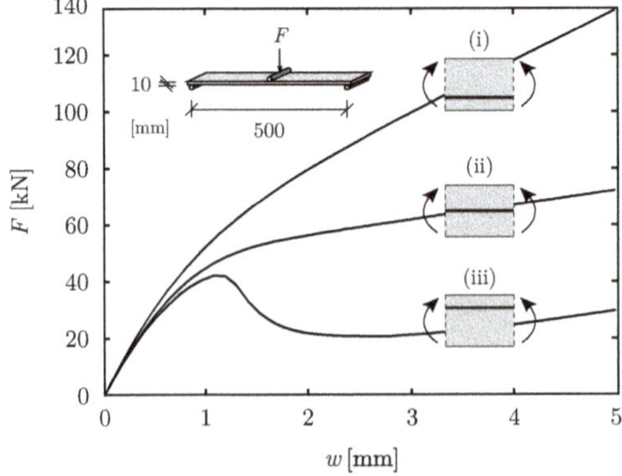

Figure 14. Load displacement curves of the simulation of carbon concrete beam with variation of the reinforcement position in the cross section subjected to bending action.

As apparent from the figure, the effect of the placement of the fabric is reproduced plausibly. The placement of the fabric at the bottom of the cross section in the tensile zone leads to a large lever arm and, thus, to an activation of a higher amount of compressive stresses within the cross section. As a result, this study shows the structural response with high stiffness and ductility (iii). The qualitative effect of reduced lever arm for configurations (i) and (ii) is reproduced correctly. For illustration, the development of the stress over the cross-sectional thickness for three selected levels of load is shown in Figure 15.

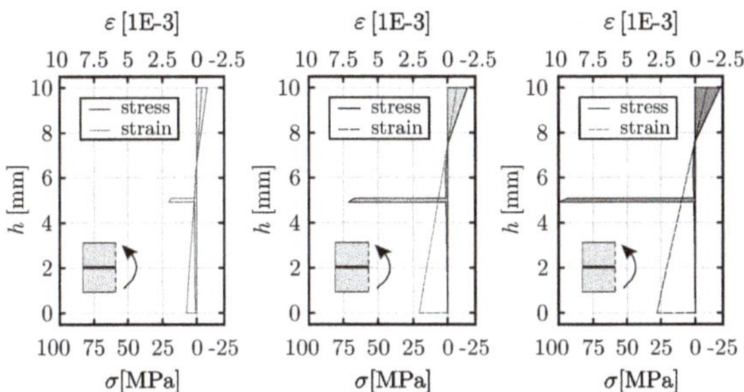

Figure 15. Development of strain and strain profiles of the carbon concrete beam with cross section (ii) according to Figure 14.

Let us note that, in case of a smeared cross-sectional model, the effect of fabric placement within the cross section cannot be captured. Considering the case of the studied girder with the fabric placed in the middle of the cross section, the smeared approach necessarily leads to an overestimation of the lever arm as can be recognized by comparing the stress profiles in a smeared and resolved cross section models depicted in Figure 10.

5. Finite Element Simulation of the Carbon Concrete Girder

Finite element model of the tested girder described in Section 1 has been decomposed into twelve zones to account for the different cross-sectional layups as shown in Figure 16. Sections 4–12 correspond to the reinforced sections displayed previously in the technical drawing in Figure 4. In addition to the reinforced sections, the finite element model also includes the non-reinforced boundary sections 1–3 positioned behind the supports.

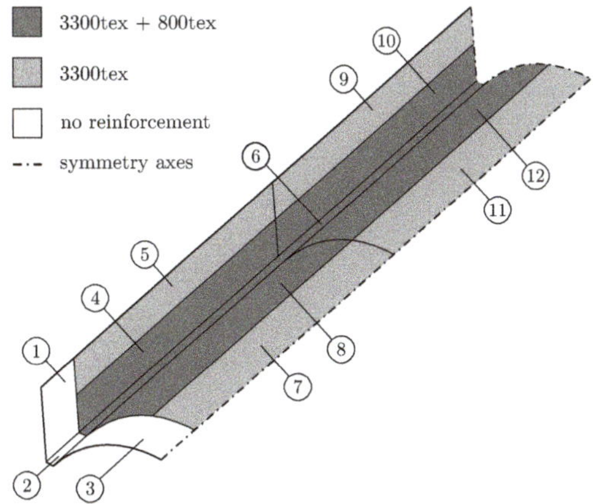

Figure 16. Cross sections of the girder with the corresponding layouts described in Tables 2 and 3.

The finite element discretization with the twelve sections is displayed in Figure 17. The finite element type was a bilinear quadrilateral layered shell element with three translational and two rotational degrees of freedom at each node. In the applied *ABAQUS* code, this element is referred to as a *conventional* shell element [31]. This type of element provides the possibility to define several layers within the cross section that can be associated with different material models and material parameters. This feature has been utilized to combine the microplane damage model for the concrete layers and the explicitly defined bilinear stress–strain curves for the fabric layers in a single cross section.

The symmetry of the shell was exploited to reduce the size of the simulated structure to a half of the girder. The load was introduced using displacement-control using stiff plates at positions indicated in Figure 17. To introduce the load from the plates to the shell consistently with the test setup, a frictionless contact was defined between the plates and the girder surface. The deflection of the girder was measured at the mid-span on the bottom side of the girder.

The described finite element model was used to study the effect of the cross section representation using four versions of cross section representation (i–iv) displayed in Figure 18. The simulated load–deflection curves corresponding to the four versions of the model are plotted in Figure 19 together with the test result. The obtained response curves are discussed in detail in the following sections.

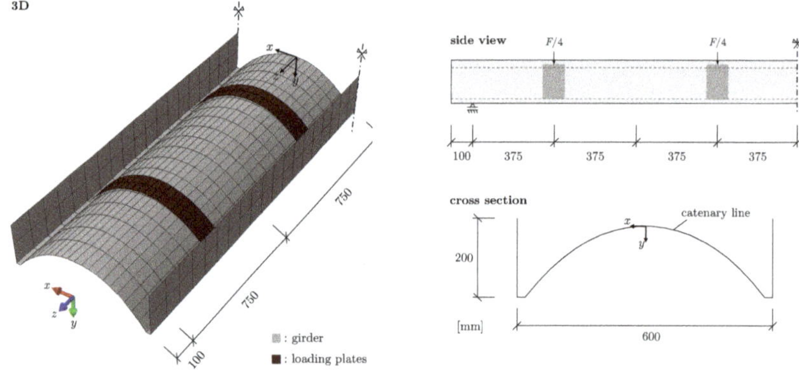

Figure 17. Finite element model of the lightweight girder: (**a**) finite element mesh and (**b**) dimensions of the girder cross section.

model thickness	fabrics smeared in composite	fabrics resolved in layers
uniform consistent with tensile test	(i) 10 mm	(iii) 10 mm
nonuniform following the girder design	(ii) 10 mm	(iv) 10 mm

Figure 18. Studied types of cross-sectional representations (i–iv).

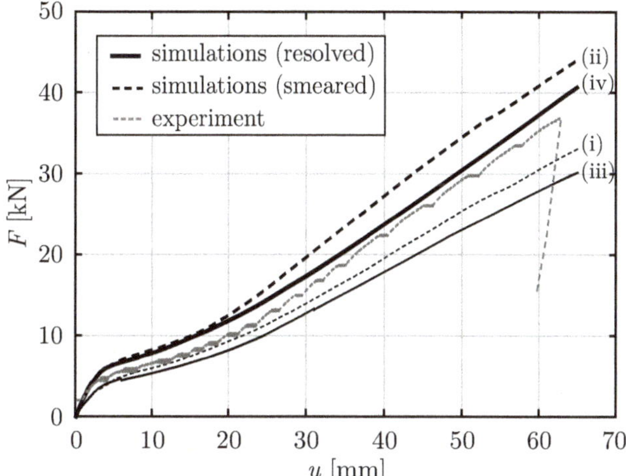

Figure 19. Simulated load–deflections curves in comparison with the test results.

5.1. Constant Thickness and Smeared Cross Section

To provide a reference example of the shell simulation with a simple non-scaled version of the cross-sectional parameters, a girder model with a constant thickness of 10 mm was used first. This cross-sectional model shown in Figure 18 (i) corresponded to the original cross section of the tensile test specimens used to identify the integrity functions.

Three calibrated integrity functions were used in this simulation to reflect the varying cross section within the shell. The zones 5, 7, 9, 11 reinforced with 3300 tex were associated with the strain-hardening behavior $\sigma_c^{3300}(\varepsilon_c)$ given in Figure 9d. The zones 4, 6, 8, 12 reinforced with hybrid fabric 3300 + 800 tex were associated with integrity functions calibrated using the stress–strain curve $\sigma_c^{3300+800}(\varepsilon_c)$ given in Figure 12b. The non-reinforced zones 1, 2, 3 were associated with a plain concrete strain-softening behavior $\bar{\sigma}_m(\varepsilon_c)$.

The obtained load–deflection curve (i) shown in Figure 19 has a lower final stiffness than the measured response of the girder test. The reason for the lower stiffness is the fact that the tested girder had additional layers of plain concrete (FCC) in zones 4–12 that were not included in this version of the model.

5.2. Varying Thickness and Smeared Cross Section

To evaluate the effect of additional concrete layers denoted as FCC in Figure 4, additional layers with strain-softening behavior $\bar{\sigma}_m(\varepsilon_c)$ have been added to the 10 mm thick shell model as indicated in Figure 18 (ii). The thickness parameters of the sections 1–12 are summarized in Table 2. The meaning of the parameters representing the added thickness in combination with the material model used for each layer is visualized in the cross-sectional layup depicted in Figure 20.

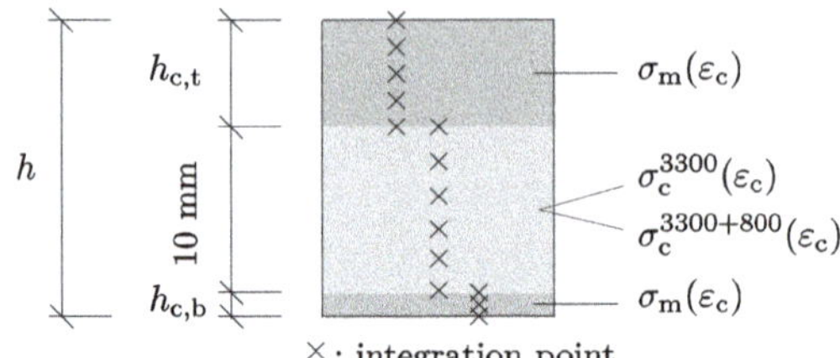

Figure 20. Cross-section with an effective strain-hardening layers either σ_c^{3300} or $\sigma_c^{3300+800}$ representing the tested cross section and strain-softening layers σ_m for add-on plain concrete layers.

Table 2. Cross section layups corresponding to the curve (ii) in Figure 16.

Cross Section	$h_{c,b}$ [mm]	$h_{m,f}$ [mm]	$h_{c,t}$ [mm]	h [mm]
1	-	-	-	15.4
2	-	-	-	20.0
3	-	-	-	12.9
4	2.7	10	2.7	15.4
5	2.7	10	2.7	15.4
6	5.0	10	5.0	20.0
7	1.45	10	1.45	12.9
8	1.45	10	1.45	12.9
9	2.7	10	15.1	27.8
10	2.7	10	15.1	27.8
11	1.45	10	6.45	17.9
12	1.45	10	6.45	17.9

The load–deflection curve obtained using the model version (ii) is shown Figure 19. Compared to version (i), the additional concrete layers increase the structural stiffness. However, now the stiffness is even higher than the real stiffness of the tested girder.

Let us remark that, if there were no bending stresses in the girder, the smeared cross-sectional representation correctly reflecting the distribution of thickness and reinforcement ration in the shell surface should able to correctly predict the structural response of the girder. Thus, we can postulate that the amount of cross-sectional bending stresses in the shell was non-negligible and that the stiffness overestimation only reveals the deficit of the smeared model mentioned at the end of Section 4.5, i.e., that it is not able to reflect the effect of placement of the fabric within the cross section. In the studied girder, the fabrics were placed nearly in the middle of the cross section. Then, the smeared strain-hardening model leads to an overestimation of the lever arm. The situation is shown in Figure 10a, ascribing the same strain-hardening behavior to all integration points within the cross section. In such a case, the material points in the bottom region of the cross section cannot reflect the effect of cracks emerging in the tensile zone that would lead to a reduction of the lever arm.

5.3. Constant Thickness and Resolved Cross Section

To show the isolated effect of the lever arm on the load–deflection response, the girder was simulated with a constant thickness of 10 mm as done in version (i). The layered cross section with the fabrics placed in the middle is shown in Figure 18 (iii). As expected, the resulting load deflection curve (iii) of a thin girder with a constant thickness shown in Figure 19 exhibits a significantly lower stiffness in the post-cracking regime than the model (i) with the smeared cross section with the same thickness.

5.4. Varying Thickness and Resolved Cross Section

Finally, the resolved cross section model was combined with the additional layers of the plain-concrete $\bar{\sigma}_m(\varepsilon_c)$ similarly to the model version (ii). The parameters of the cross-sectional layup corresponding to the 12 sections are specified in Table 3. The association of the layer thicknesses to the individual layers is depicted in Figure 21.

The calculated load–deflection curve (iv) reflects the stiffening effect of the additional plain concrete layers with respect to the curve (iii). At the same time, it shows the effect of an improved representation of the lever arm in the cross section compared to the model (ii) as documented by more realistic prediction of the real girder behavior as the level of force at the post-cracking stage and the final stiffness are predicted reasonably well.

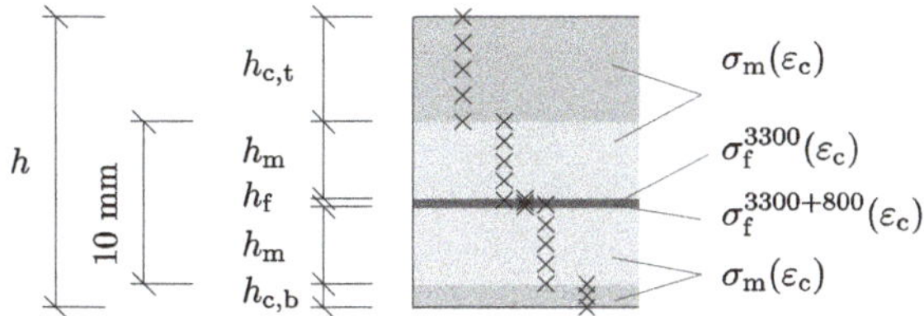

Figure 21. Cross section with a resolved layer of fabrics and with strain-softening layers σ_f^{3300} or $\sigma_f^{3300+800}$ and strain-softening layers σ_m for add-on plain concrete layers.

Table 3. Cross section layups corresponding to the curve (iv) in Figure 21.

Cross Section	$h_{c,b}$ [mm]	h_m [mm]	h_f [mm]	$h_{c,t}$ [mm]	h [mm]
1	-	-	-	-	15.4
2	-	-	-	-	20.0
3	-	-	-	-	12.9
4	2.7	4.8835	0.233	2.7	15.4
5	2.7	4.9135	0.173	2.7	15.4
6	5.0	4.8835	0.233	5.0	20.0
7	1.45	4.9135	0.173	1.45	12.9
8	1.45	4.8835	0.233	1.45	12.9
9	2.7	4.9135	0.173	15.1	27.8
10	2.7	4.8835	0.233	15.1	27.8
11	1.45	4.9135	0.173	6.45	17.9
12	1.45	4.8835	0.233	6.45	17.9

Apparently, the force level in the stage of the multiple cracking is slightly overestimated. The possible reason is that the applied scaling rule does not account for the effect of the reinforcement ratio on the crack localization process in the matrix. An application of meso-scale models [17,18,32,33] with discrete crack representation that would additionally capture the interaction between damage localization in concrete and the stress transfer due to the finely distributed fiber and fabric reinforcement might help to analyze this effect in more detail. In addition, the interaction with the additional concrete layers made of a different material (FCC) has not been included in the model due to the lack of experimental data. A refinement of the scaling rules based on the meso-scale models and additional tests could significantly improve the prediction of the structural behavior in the service state.

Still, even in the present, pragmatic form, the applied modeling approach provides the possibility of deeper interpretation of the test results in terms of the visualized damage propagation through the thin shell. In Figure 22, the deformed shapes of the girder are plotted with the distribution of the maximum damage values in each material point. The plot visualizes the damage variable ω defined within the microplane damage model as

$$\omega = 1 - \phi_{\mathrm{min}},\tag{15}$$

with ϕ_{min} denoting the lowest microplane integrity level at a material point. The propagation of the damage through the shell is depicted in Figure 22 separately for the top and the bottom surfaces at three levels of loading corresponding to deflections $u_y = 20\,\mathrm{mm}$, $u_y = 40\,\mathrm{mm}$ and $u_y = 60\,\mathrm{mm}$.

By comparing the bottom and top views on the shell at the three selected levels of load, the simulation results exhibit a difference between the damage values at the top and the bottom surface of the shell. Such a difference within a cross section is due to a significant amount of bending stresses within the cross section along the top middle section of the vault. This observation confirms the importance of the correct reflection of the changing lever arm during the multiple cracking phase of the test.

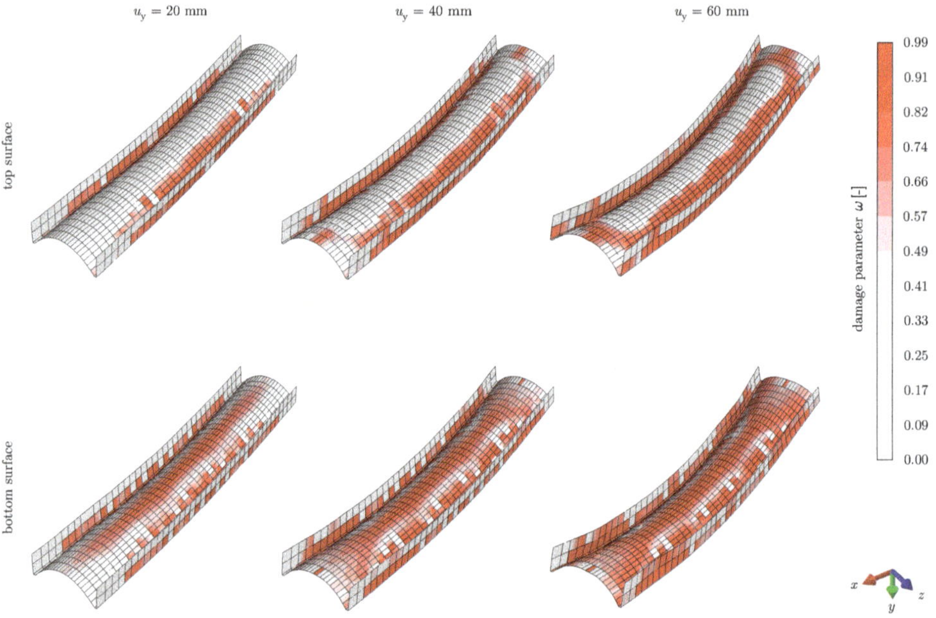

Figure 22. Damage propagation in the surface of the girder for top and bottom surfaces.

Furthermore, the damage distribution plots reveal the local effect of load transfer through the wooden plates. Even though the load was introduced via a frictionless contact model combined with the free displacement of the shell in the in-plane direction, local bending could be observed with the applied relatively coarse size of finite element. However, no significant influence on the overall structural behavior could be observed.

6. Conclusions

The strain-hardening behavior of textile-reinforced concrete has been modeled at the level of a cross section using both smeared and resolved representation. By decomposing the composite stress measured in the tensile test into the fractions associated with the fabric layer and with the matrix, it was possible to define the mixture and scaling rules that allowed us to derive cross-sectional material parameters corresponding to changed layup and thickness.

The derived characteristics were applied in the simulation of the structural response of a girder element and compared with the experimentally obtained results. The systematically performed parametric studies confirmed the fact that the smeared representation of cross section is not directly applicable to shells with a nonuniform layup and variable thickness. Still, the smeared cross-sectional representation can be effectively used for layups with laminated fabrics with thin layers of cementitious matrix. This approach was used for non-penetrated fabrics uniformly distributed over the cross-sectional thickness in the applications presented e.g., in [34].

The resolved cross section model has shown a good agreement with the experimentally observed behavior. As such, it has the potential to provide valuable insight into the stress redistribution process during loading and can be used for further improvements to the geometrical design of the shell to achieve high structural quasi-ductility. Besides the validation of the resolved model using the girder test, the study was used to describe a general mixture and scaling procedure that can be applied to derive the wider range of cross-sectional characteristics from elementary tensile tests with a single layer fabric reinforcement.

The described simulation was primarily focused on the damage process inducing stress redistribution both along the thickness direction and in the in-plane direction of the shell surface. The ultimate failure was predicted reasonably well based on the strength measured in the tensile test specimens. Such a good prediction is due to the fact that critical cross section of the shell was primarily loaded in tension and the amount of bending at this particular location was negligible. As discussed in [13], the effective strength of the fabrics in cross sections exposed to bending can be significantly higher so that a refined ultimate-state criterion is needed that explicitly distinguishes the tensile strength of a cross section loaded in bending [35].

Author Contributions: Conceptualization,R.C.; methodology, R.C.; Software, E.S.; validation, E.S.; investigation, E.S. and T.S.-R.; data curation, T.S.-R.; writing—original draft preparation, R.C., writing—review and editing, R.C. and E.S.; visualization, E.S. and T.S.-R.; supervision, R.C. and F.S.; project administration, F.S.

Funding: The theoretical part of this research was performed in Aachen and funded by the German Federal Ministry of Education and Research (BMDF, Grant No. 03ZZ0312C). The experimental part of the present work was conducted in Dresden and funded by the German Federal Ministry of Economic Affairs and Energy (BMWi, Grant No. KF2505611KI3).

Conflicts of Interest: The authors declare no conflict of interest.

References

1. Scholzen, A.; Chudoba, R.; Hegger, J. Thin-walled shell structure made of Textile Reinforced Concrete; Part I: Structural design and construction. *Struct. Concr.* **2015**, *16*, 106–114. [CrossRef]
2. Scholzen, A.; Chudoba, R.; Hegger, J.; Will, N. Light-weight precast shell elements made of textile reinforced concrete: Production, experimental investigations and application potential. *Beton- und Stahlbetonbau* **2016**, *111*, 663–675. [CrossRef]
3. Scheerer, S.; Chudoba, R.; Garibaldi, M.; Curbach, M. Shells Made of Textile Reinforced Concrete—Applications in Germany. *J. Int. Assoc. Shell Spat. Struct.* **2017**, *58*, 79–93. [CrossRef]
4. Verwimp, E.; Tysmans, T.; Mollaert, M.; Berg, S. Experimental and numerical buckling analysis of a thin TRC dome. *Thin-Walled Struct.* **2015**, *94*, 89–97. [CrossRef]
5. Petzoldt, C.; Gelbrich, S.; Röhrkohl, M.; Müller, C.; Freund, J.; Kroll, L. Textile Reinforced Lightweight Shells. *Mater. Sci. Forum* **2015**, *825–826*, 319–327. [CrossRef]
6. May, S.; Michler, H.; Schladitz, F.; Curbach, M. Lightweight ceiling system made of carbon reinforced concrete. *Struct. Concr.* **2018**, *19*, 1862–1872. [CrossRef]
7. Scheerer, S.; Michler, H. Freie Formen mit Textilbeton. *Beton- und Stahlbetonbau* **2015**, *110*, 94–100. [CrossRef]

8. Rempel, S.; Kulas, C.; Will, N.; Bielak, J. Extremely Light and Slender Precast Pedestrian-Bridge Made Out of Textile-Reinforced Concrete (TRC). In *High Tech Concrete: Where Technology and Engineering Meet, Proceedings of the 2017 Fib Symposium, Maastricht, The Netherlands, 12–14 June 2017*; Hordijk, D., Luković, M., Eds.; Springer: Berlin, Germany, 2017.

9. Finzel, J.; Häußler-Combe, U. Textile Reinforced Concrete sandwich panels: Bending tests and numerical analyses. In *Computational Modelling of Concrete Structures (EURO-C 2010)*; Bićanić, N., de Borst, R., Mang, H., Meschke, G., Eds.; CRC Press: Boca Raton, FL, USA, 2010; pp. 789–795, doi:10.1201/b10546-95.

10. Shams, A.; Horstmann, M.; Hegger, J. Experimental investigations on Textile-Reinforced Concrete (TRC) sandwich sections. *Compos. Struct.* **2014**, *118*, 643–653. [CrossRef]

11. Shams, A.; Stark, A.; Hoogen, F.; Hegger, J.; Schneider, H. Innovative sandwich structures made of high performance concrete and foamed polyurethane. *Compos. Struct.* **2015**, *121*, 271–279. [CrossRef]

12. Hawkins, W.; Orr, J.; Ibell, T.; Shepherd, P. An analytical failure envelope for the design of textile reinforced concrete shells. *Structures* **2018**, *15*, 56–65. [CrossRef]

13. Scholzen, A.; Chudoba, R.; Hegger, J. Ultimate limit state assessment of TRC structures with combined normal and bending loading. In *11th International Symposium on Ferrocement (FERRO-11) and 3rd International Conference on Textile Reinforced Concrete (ICTRC-3), Aachen, Germany, 7–10 June 2015*; Brameshuber, W., Ed.; RILEM Publications: Paris, France, 2015; pp. 159–166.

14. Sharei, E.; Scholzen, A.; Hegger, J.; Chudoba, R. Structural behavior of a lightweight, textile-reinforced concrete barrel vault shell. *Compos. Struct.* **2017**, *171*, 505–514. [CrossRef]

15. Verwimp, E.; Tysmans, T.; Mollaert, M.; Wozniak, M. Prediction of the buckling behaviour of thin cement composite shells: Parameter study. *Thin-Walled Struct.* **2016**, *108*, 20–29. [CrossRef]

16. Naaman, A. Tensile strain-hardening FRC composites: Historical evolution since the 1960. In *Advances in Construction Materials 2007*; Grosse, C.U., Ed.; Springer: Berlin/Heidelberg, Germany, 2007; pp. 181–202.

17. Fantilli, A.; Mihashi, H.; Vallini, P. Multiple cracking and strain hardening in fiber reinforced concrete under uniaxial tension. *Cement Concr. Res.* **2009**, *39*, 1217–1229. [CrossRef]

18. Li, Y.; Chudoba, R.; Bielak, J.; Hegger, J. A Modelling Framework for the Tensile Behavior of Multiple Cracking Composite. In *Strain-Hardening Cement-Based Composites (SHCC 2017)*; Mechtcherine, V., Slowik, V., Kabele, P., Eds.; Springer: Berlin, Germany, 2018; Volume 15, pp. 418–426, doi:10.1007/978-94-024-1194-2_49.

19. Djamai, Z.I.; Bahrar, M.; Salvatore, F.; Larbi, A.S.; El Mankibi, M. Textile reinforced concrete multiscale mechanical modelling: Application to TRC sandwich panels. *Finite Elem. Anal. Des.* **2017**, *135*, 22–35. [CrossRef]

20. Colombo, I.G.; Colombo, M.; di Prisco, M.; Pouyaei, F. Analytical and numerical prediction of the bending behaviour of textile reinforced concrete sandwich beams. *J. Build. Eng.* **2018**, *17*, 183–195. [CrossRef]

21. Senckpiel-Peters, T.; Häußler-Combe, U.; Metschies, H.; Eisewicht, P. Entwicklung eines neuartigen leichten Deckenträgers aus Carbonbeton: Experimentelle und numerische Untersuchungen. *Beton-und Stahlbetonbau* **2018**, *113*, 850–858. [CrossRef]

22. El Kadi, M.; Tysmans, T.; Verbruggen, S.; Vervloet, J.; De Munck, M.; Wastiels, J.; Van Hemelrijck, D. A layered-wise, composite modelling approach for fibre textile reinforced cementitious composites. *Cement Concr. Compos.* **2018**, *94*, 107–115. [CrossRef]

23. Senckpiel, T.; Häußler-Combe, U. Model comparisson for a shell structure made of textile reinforced concrete. In *Computational Modelling of Concrete Structures (EURO-C 2018)*; Meschke, G., Pichler, B., Rots, G., Eds.; CRC Press: Boca Raton, FL, USA, 2018; pp. 819–828.

24. Senckpiel, T.; Häussler-Combe, U. Experimental and computational investigations on shell structures made of carbon reinforced concrete. In Proceedings of the Annual Meeting of the International Association of Shell & Spatial Structures (IASS), Hamburg, Germany, 25–27 September 2017; pp. 1–8.

25. Jirásek, M. Comments on microplane theory. In *Mechanics of Quasibrittle Materials and Structures*; Hermes Science Publications: Paris, France, 1999; pp. 55–77.

26. Bažant, Z.P.; Prat, P.C. Microplane Model for Brittle-Plastic Material: I. Theory. *J. Eng. Mech.* **1988**, *114*, 1672–1688. [CrossRef]

27. Carol, I.; Bažant, Z.P. Damage and plasticity in microplane theory. *Int. J. Solids Struct.* **1997**, *34*, 3807–3835. [CrossRef]

28. Wu, J.Y. An alternative approach to microplane theory. *Mech. Mater.* **2009**, *41*, 87–105.

29. RILEM Technical Committee 232-TDT (Wolfgang Brameshuber). Recommendation of RILEM TC 232-TDT: Test methods and design of textile reinforced concrete: Uniaxial tensile test: Test method to determine the load bearing behavior of tensile specimens made of textile reinforced concrete. *Mater. Struct.* **2016**, *49*, 4923–4927. [CrossRef]

30. Chudoba, R.; Vořechovský, M.; Konrad, M. Stochastic modeling of multi-filament yarns. I. Random properties within the cross-section and size effect. *Int. J. Solids Struct.* **2006**, *43*, 413–434. [CrossRef]

31. *ABAQUS Analysis User's Manual, Volume IV: Elements, Version 6.11*; Dassault Systèmes Simulia Corp.: Providence, RI, USA, 2011.

32. Curtin, W. Fiber Pull-Out and Strain Localization in Ceramic Matrix Composites. *J. Mech. Phys. Solids* **1993**, *41*, 35–53.

33. Cuypers, H.; Wastiels, J. A stochastic cracking theory for the introduction of matrix multiple cracking in textile reinforced concrete under tensile loading. In *ICTRC'2006—1st International RILEM Symposium*; Hegger, J., Brameshuber, W., Will, N., Eds.; RILEM Publications SARL: Paris, France, 2006; pp. 193–202.

34. Scholzen, A.; Chudoba, R.; Hegger, J. Thin-walled shell structure made of Textile Reinforced Concrete; Part II: Experimental characterization, ultimate limit state assessment and numerical simulation. *Struct. Concr.* **2015**, *16*, 115–124. [CrossRef]

35. Sharei, E.; Chudoba, R.; Scholzen, A. Cross-sectional failure criterion combined with strain-hardening damage model for simulation of thin-walled textile-reinforced concrete shells. In *ECCOMAS Congress 2016—VII European Congress on Computational Methods in Applied Sciences and Engineering, Crete Island, Greece, 5–10 June 2016*; Papadrakakis, M., Papadopoulos, V., Stefanou, G., Plevris, V., Eds.; National Technical University of Athens (NTUA): Athens, Greece, 2016; Volume 4, pp. 6823–6831.

 *applied sciences*

Article

Damage Mechanisms of Polymer Impregnated Carbon Textiles Used as Anode Material for Cathodic Protection

Amir Asgharzadeh * and Michael Raupach

Institute of Building Materials Research at RWTH Aachen University, 52062 Aachen, Germany; raupach@ibac.rwth-aachen.de
* Correspondence: zadeh@ibac.rwth-aachen.de, Tel.: +49-241-8095140

Received: 25 November 2018; Accepted: 25 December 2018; Published: 29 December 2018

Featured Application: Conventional cathodic corrosion protection (CP) systems usually use MMO-coated titanium as anode material. Carbon textiles as anode material provide sufficient current densities for CP as well as strength for structural reinforcement. Therefore, durability is an important issue for both the structure and the CP view. This work should show the border potentials where the impregnation and sizing of carbon textiles are destroyed under anodic polarization.

Abstract: Carbon textiles as anode material for cathodic corrosion protection (CP) have been used in several reinforced steel structures. However, experience with durability is limited. To date, various influencing factors have been discovered and systematic tests on different carbon textiles with different impregnation materials in various environmental media have been carried out and considered the degradation of the impregnation materials. In this work the boundary potentials are determined at which the impregnation and sizing is destroyed under anodic polarization and the damage mechanisms are described.

Keywords: textile reinforced concrete; cathodic corrosion protection; durability

---

1. Introduction

Carbon textiles consist of thousands of bundled carbon fibres. Before carbon fibres are combined into bundles, their surface needs to be electrochemically activated (by anodic oxidation in an electrolytic solution). Each fibre is then coated by a mixture of various chemicals (sizing) to protect it from mechanical damage during production handling and improve its wetting performance [1–3]. Furthermore sizing leads to an increased chemical reactive site as well as enhanced adhesion between carbon and impregnation material [4,5]. After single carbon fibres are bundled together, the so-called carbon roving is again impregnated to improve adhesion and handling for further application. Usually, epoxy resin or styrene-butadiene rubber (SBR) based polymers are used for impregnation.

The thesis of Wetjen [6] gives an overview of the production and processing of carbon. The following statements are based on this source, unless otherwise indicated. The carbon rovings usually consist of thousands of individual combined carbon fibres, which are also called filaments and have an average thickness of 5–7 microns. Carbon fibres consist of 92% by weight of carbon. They can be produced chemically from different polymers, the most common being the production from polyacrylonitrile (PAN), since the highest strengths with high carbon yield can be achieved at comparatively low price. Graphite consists of hexagonal layers in which each carbon atom is connected to three other carbon atoms with a strong covalent bond. The individual layers are connected only by relatively weak London forces, whereby the distance of the layers in relation to the distance of the carbon atoms in the hexagons is large and the direction-dependent conductivity of graphite can be explained [6].

Compared to graphite, carbon fibres have a turbostratic structure that results from random spinning, folding, splitting, canting, branching and concatenation of the lattice structure in the manufacturing process. This inhomogeneity explains both the increased strength between the basal planes and the increased reactivity of the carbon fibre surface in comparison. In addition, the structural inhomogeneity increases the spacing between the lattice planes.

After production of the carbon fibre, surface activation of the carbon fibre follows, typically by electrochemical anodic oxidation in an electrolyte. The amount and type of functional groups formed on the surface depends inter alia on the duration of polarization, the electrolyte and the electrical potential. The surface activation is done to obtain a good wettability of the carbon fibre over a later applied sizing or other impregnating materials by a higher proportion of polar functional groups [6].

Poltavtseva et al. [7] have described the formation of surface oxides in carbons by summarizing relevant literature. The formation of surface oxides is complex because it is not possible to separate the different reactions from each other. It depends on the pH, electrode potential, pressure, temperature and humidity as well as the physical and chemical properties of the carbon compounds. Thus, a different dissolution process results for different carbon-based materials, since materials such as carbon black, graphite and carbon fibres differ in the size and orientation of the graphitic crystalline layers.

Pure graphite consists of hexagonal layers as described above. While graphite has many surface defects, carbon fibres without activation have a high degree of longitudinal orientation with fewer defects. Without these lattice defects, the graphitic crystalline layers behave inertly. The existing lattice defects, pores and edges provide for free valences, which are already functionalized under normal conditions to form surface oxides. Anodic polarization causes the dissolution of water and the generation of further oxygen-containing functional groups on the surface. First, hydroxyl ions are discharged to hydroxyl radicals by the electrolysis of water according to Equations (1) and (2).

$$C_{(s)} + H_2O \rightleftharpoons C_{(s)}OH_{ads} + H^+ + e^- \tag{1}$$

$$C_{(s)} + OH^- \rightleftharpoons C_{(s)}OH_{ads} + e^- \tag{2}$$

These radicals immediately react to form hydroxyl, carboxyl, carbonyl or lactone moieties, which are accompanied by an increase in the oxygen content at the carbon surface. The possible functional groups on the graphite surface are shown in Figure 1. Rueffer et al. [8], however, doubt the formation of hydroxyl radicals on the basis of their investigations as an intermediate step and assume a direct incorporation of oxygen into the functionalized groups. The functional groups have different oxidation states and can be further oxidized up to the elimination of carbonates and carbon dioxide [6]. According to Chung [9], these surface oxides have a lower conductivity than non-activated graphitic carbon fibres.

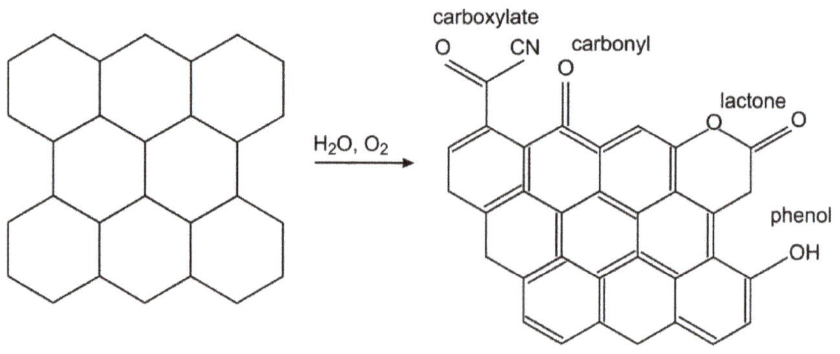

Figure 1. Formation of surface oxides in graphite [7].

After activation of the carbon fibre and cleaning of detached carbon fibre fragments, the carbon fibre is always coated in industrial applications [6]. For this, the carbon fibre is passed through a sizing bath filled with a polymer solution or dispersion. However, there are other methods of coating such as electrolytic deposition, electrolytic polymerization and plasma polymerization. As size typically polymer systems are used which have good chemical-physical compatibility for the further use of the carbon fibre. When using epoxy resins, epoxies are used, for example. Other polymer systems include polyhydroxyether, polyphenylene oxides, polysulfones, silanes and cyanamides. The sizing ensures good processability in subsequent processes, good wetting behaviour of the carbon fibre for the impregnation matrix and a strong chemical-physical interaction between these two components for good adhesion. Possible adhesion mechanisms are: mechanical interlocking, secondary bonds, such as London forces and chemical bonds, such as covalent bonds.

In the final processing step, the carbon fibres are combined into rovings and impregnated with materials such as epoxy resin matrix or styrene butadiene rubber matrix (SBR). For a good fibre-matrix interaction, a good mixing of the components carbon fibre, sizing and matrix is essential. During the application of the matrix, with "no release" of the layer, "partial release" or "complete release" of the size, there may be three different interphase formations. In the case of complete release, the sizing thoroughly mixes with the matrix, providing the best mechanical properties. The worst mechanical properties are achieved when there is no intermixing between the size and the matrix, that is no detachment.

Earlier research [10–12] has shown that carbon textile is a suitable anode-material for corrosion protection (CP). Various experiments were conducted in order to investigate the behaviour of carbon textiles under anodic polarization. Carbon textiles were tested both in saturated calcium hydroxide solution and in mortar test specimens. It has been found, that the structure of carbon textiles itself is not affected by anodic polarization. However, sizing and impregnation materials suffer considerable destruction in some cases.

Asgharzadeh et al. [12] investigated the durability of impregnated carbon textile under permanent anodic polarization while stored in saturated calcium hydroxide solution. Carbon textile was potentiodynamically polarized in saturated calcium hydroxide solution. Starting with the open circuit potential, the applied potential was increased at a constant rate of 2 mV per minute until an overall potential shift of 2200 mV in anodic direction was achieved. Based on the obtained current-density versus (compensated) potential curves, three potentials were selected, which indicate changes in the carbon textiles' electrochemical properties. Potentiostatic tests were performed with the identified potentials. After the polarization tests were completed, the test solution was collected. In order to draw conclusions about a possible decomposition of the carbon textile, the carbon content of the solution was analysed. It was found, that the solutions in which polarization of the carbon textile occurred had higher carbon contents than solutions in which unpolarized carbon textiles were stored as reference solution. Furthermore, the carbon content increases with increasing potential of potentiostatic polarization. However, the additional carbon could originate from several sources, such as the carbon textile itself, the impregnation material or the surrounding air. In order to further identify the carbon source, the experiments were repeated with unimpregnated carbon textile. In this case, the carbon content of the solution corresponded to the carbon content of the reference solution. It is therefore assumed, that the additional carbon, which was found in the solution in which polarization tests of impregnated carbon textiles were conducted, originates either from the impregnation material or the surrounding air.

Although SEM images showed a destruction of the impregnating material after polarization, the information at which potentials they are destroyed is missing, which are to be investigated in this paper. Furthermore, based on these results, the influence of polarization on various impregnation materials for carbon textiles is investigated in more detail by additional tests.

2. Materials

As an anode material, the styrene-butadiene rubber (SBR) impregnated textile S4 was selected because this material achieved the highest current densities and thus best polarization behaviour in former experiments [12]. For reasons of comparability, the epoxy impregnated textile E4 was taken which has the same mesh size as S4. In addition, carbon rovings without impregnation but with sizing (UC, see Table 1) were used to determine the destruction of the sizing under anodic polarization. For comparison purposes, carbon rovings without impregnation and without sizing (UC_WS, see Table 1) were also used. This confirmed the actual destruction of the sizing.

Table 1. Details of carbon textiles used in the investigations.

	Specimen Specification	Sizing	Impregnation Material	Mesh Size [mm/mm] 0°/90°
impregnated carbon textile	E4	yes	Epoxy	38/38
	S4	yes	SBR	38/38
unimpregnated carbon textile	UC	yes	-	-
	UC_WS	no	-	-

3. Investigations

3.1. Specimen Preparation

The carbon rovings and the counter electrode made of MMO-coated titanium mesh were attached to PVC spacers with cable ties. A reference electrode MnO_2 was attached to the spacer between the working electrode and the counter electrode and care was taken to ensure that there is no short-circuit between the reference electrode and any protruding carbon fibres.

The specimens were placed in a container with saturated calcium hydroxide solution.

3.2. Potentiodynamic Polarization

Potentiodynamic tests are used to obtain current density potential curves that allow conclusions to be made about the material behaviour.

The potential was increased from OCP to 2200 mV. The duration of the potentiodynamic test was approximately 18 h.

3.3. Potentiostatic Polarization

In order to perform the potentiostatic measurements, the potentiodynamic measurements had to be evaluated and the potential points for the tests had to be selected. The different panel areas and especially the change of the slope of the curves in Section 4.1 could be attributed to different electrochemical processes. In order to verify this, it was decided to carry out tests in these three areas, which are hereinafter referred to as potentiostatic tests.

The potentials of the potentiostatic tests are listed in Table 2. They are referred to below as levels 1, 2 and 3.

Table 2. Potentiostatic potentials on level 1, 2 and 3.

Material	Level 1	Level 2 [mV] vs. NHE	Level 3
E4	940	1490	1940
S4	490	1490	1840
UC	690	1300	1940
UC_WS	690	1300	1940

For the potentiostatic tests, specimens with the carbon fabrics E4, S4, UC and UC_WS were chosen as working electrode. Counter and reference electrode were produced as already described. The specimen setup for E4 and S4 can be seen in Figure 2. In these tests, the anodes of E4 and S4 consisted of two superimposed parts, each with one connection. The larger element had dimensions of 12 × 12 cm and the smaller element 10 × 10 cm. The test specimens with unimpregnated carbon fibres corresponded exactly to the test specimens in the presented potentiodynamic tests. Saturated calcium hydroxide solution was used as electrolyte and the test specimens were stored in the solution again 24 h before the start of the test. Before and after the test, the pH value and the resting potential (OCP) were measured.

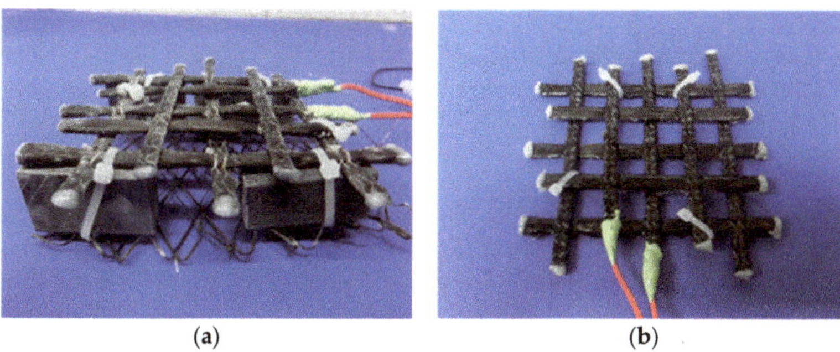

(a)   (b)

Figure 2. Specimen with S4 anode, spacers and titanium electrode (**a**) and an E4 anode (**b**).

It should be figured out whether and how the carbon surfaces change on the different levels. In addition, the time factor for destruction can be evaluated, since the holding times are higher than those of the pure potentiodynamic tests. The potentiostatic polarization was maintained for over 72 h. The polarization was started at the potential of the reference electrode, since a shift of the OCP by the polarization was to be expected.

4. Results and Discussion

4.1. Potentiodynamic Tests

The results of the potentiodynamic tests on carbon textiles are shown in Figure 3 as a current-potential curve in logarithmic scaling. The potential labels P1 to P3 indicate the potentials for the potentiostatic tests, see Section 3.3.

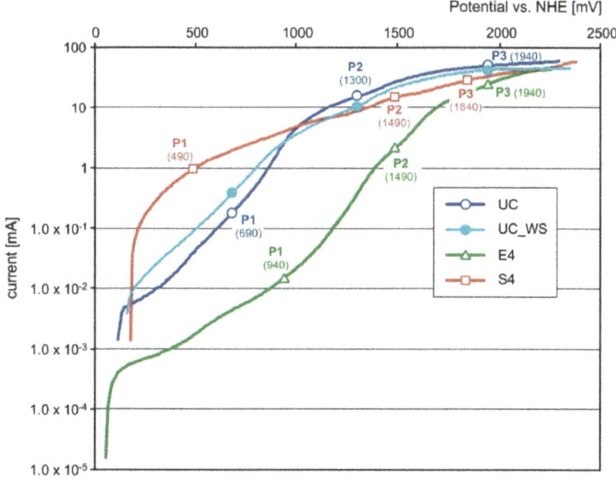

Figure 3. Current-potential curve for all materials.

Note, that current and not current density is plotted in Figure 3. This is due to the fact that the exact calculation of the anode surface of rovings without impregnation is complicated: the exact number of fibres in contact with the solution is undefined.

From Figure 3 it is seen that the polarizability of carbon impregnated with SBR-based material until 1000 mV versus NHE is better and thus delivers the highest current, while E4 achieves the lowest current. From 1100 mV versus NHE UC delivers more current than carbon impregnated with SBR-based material. The curve of the test with UC_WS is similar to UC with the difference that UC_WS can deliver more current up to 900 mV versus NHE and after that potential less current than UC.

The OCP of the non-impregnated carbon fibres amounts to about 100 mV versus NHE and is between the OCPs of E4 (60 mV versus NHE) and S4 (180 mV versus NHE). Panel area 1 is between 200 and 900 mV versus NHE for the non-impregnated carbon fibres. The transition section is small, the panel area 2 is between 950 and about 1100 mV versus NHE. For S4, panel area 1 is between about 300 and 1150 mV and panel area 2 between 1225 and 1300 mV. The transition section of E4 is not clearly defined, it could be between 950 and 1150 mV. The transition sections of all experiments are characterized by a brief reduction in the slope of current density followed by a renewed and steeper increase in current density.

The images of carbon samples UC before and after polarization are exemplarily displayed in Figure 4. Strong deposits can be seen on the polarized sample, which will be discussed later in this paper.

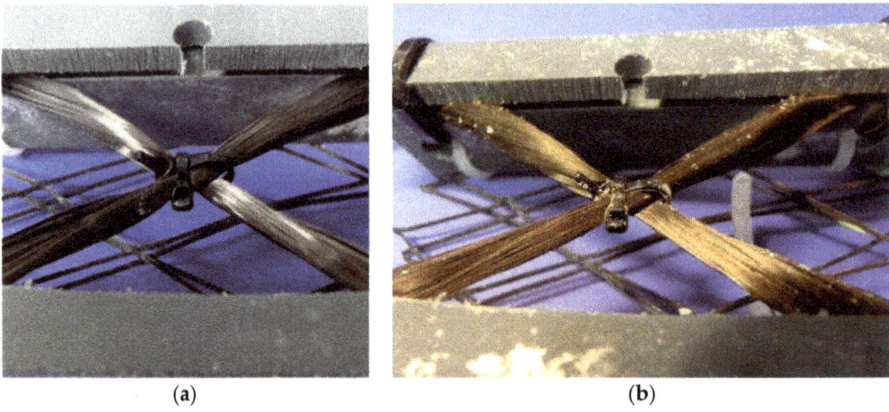

(a)   (b)

Figure 4. Unimpregnated carbon fibres before the test (**a**) and after (**b**).

A discoloration of the calcium hydroxide solution could not be observed for the potentiodynamic experiments on all experiments.

The SEM investigations of S4 and E4 before and after polarization in Asgharzadeh et al. [12] showed a destruction of the impregnation material. There was not investigated, whether the sizing under the impregnation was affected or not. To this end, the UC sample is examined in the present study. Figure 5a shows the unpolarized carbon fibre. On the right the polarized carbon fibre is shown (Figure 5b). The carbon fibres are round and have slight longitudinal grooves. This shows cracks in the sizing both in the longitudinal and in the radial direction at regular distances. The carbon fibre underneath seems to be intact. Such a clear destruction of the sizing was observed at several carbon fibres. It is concluded that the sizing deteriorates and the carbon fibre remains intact.

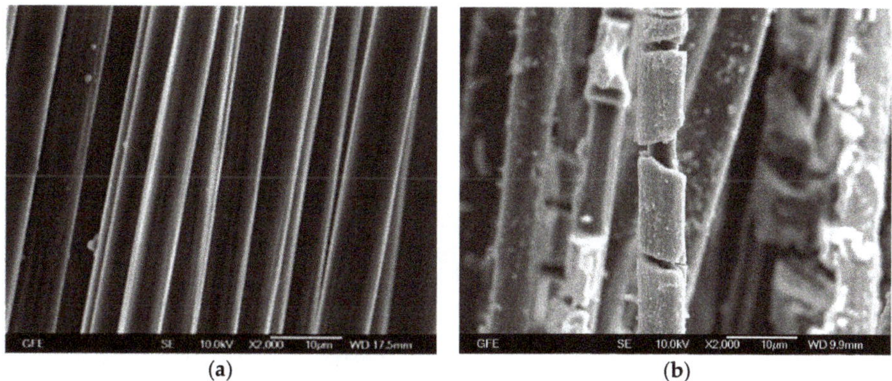

(a)   (b)

Figure 5. SEM images of unpolarized carbon fibres (**a**) and a polarized carbon fibre (**b**).

An energy dispersive X-ray spectroscopy (EDX) of the polarized sample probing the deposition shows the presence of calcium, carbon and oxygen (Figure 6).

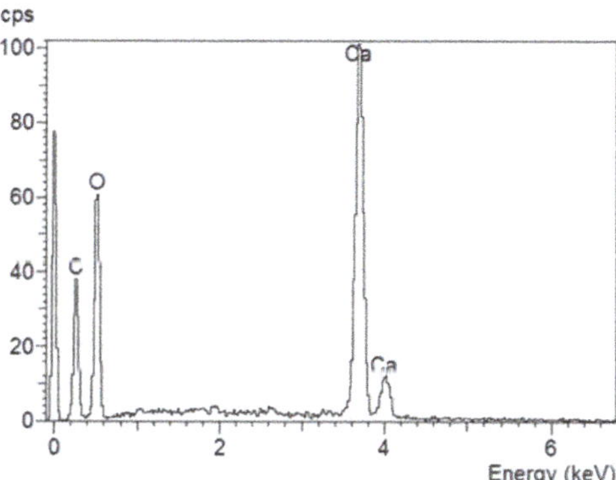

Figure 6. EDX analysis of the deposited film.

4.2. Potentiostatic Tests

4.2.1. Epoxy Impregnated Carbon (E4)

No optical changes could be detected on the carbon textile E4 after the potentiostatic polarization at level 1. The epoxy resin impregnation remained even, smooth and shiny. The specimen of level 2 shows isolated areas on which the surface is mat and rough and the structure of the carbon textile is recognizable as well as white deposits indicating the chemical bonds between carbon, calcium and oxygen. These deposits become more significant with increasing polarization, which can be seen at Figure 7 right at level 3 of potentiostatic polarization.

Figure 8 shows the SEM images of the carbon fibre E4 without polarization and after the polarization. No pores can be seen on the entire reference specimen (Figure 8a) in contrast to previous findings [12]. There were only a few of them, as well as defects. Apart from impurities and a mechanical crack, the surface is smooth due to the sample preparation. In some places the carbon fibres shine through the impregnation. The epoxy resin layer does not appear to be uniformly thick. Epoxy resin was used in various test series to interrupt conductivity [10,13–16]. Van Nguyen et al. [17] have found that the embedding of carbon anodes with epoxy resin was not successful due to insufficient conductivity.

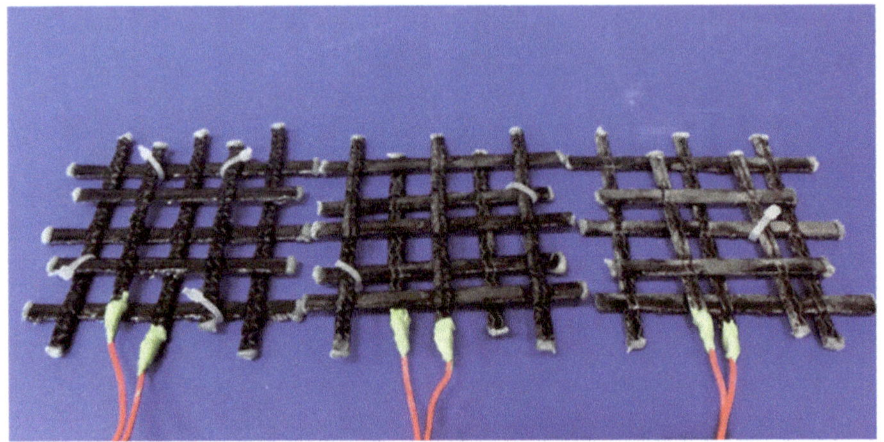

Figure 7. From left to right: Anodes E4 of levels 1 to 3 after polarization.

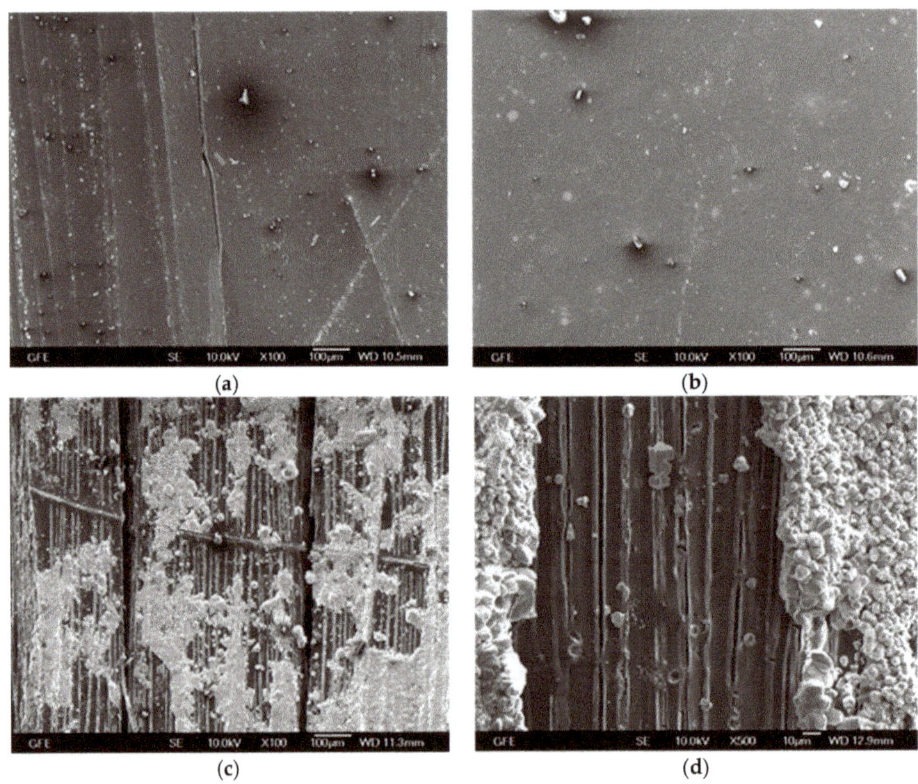

Figure 8. SEM images of E4: unpolarized reference (**a**), after polarization at level 1 (**b**), level 2 (**c**) and level 3 (**d**).

Due to the pores, defects and the different thickness of the impregnation, it can be explained why the epoxy resin-impregnated carbon fibres are polarizable at all. The potentiostatic test at level 1 did not cause any change in the surface (Figure 8b). The image of the specimen polarized at level 2 (Figure 8c) shows grooves next to two mechanical cracks and is covered approximately in half with a crystalline deposited film. Calcium, oxygen and carbon could be detected in the coating by EDX analysis. The grooves run along the carbon fibres over which the epoxy resin impregnation is probably thinnest and are due to the destruction of the impregnation. Pores have occasionally formed in the epoxy resin unlike the accumulation of pores in Asgharzadeh et al. [12]. As with the unimpregnated carbon fibres, for the carbon fibres with epoxy resin impregnation destruction in panel area 2 is used, whereas no destruction was observed in panel area 1. Thus, the start of the destruction of the epoxy resin matrix could again be in the transition range between about 1050 and 1150 mV.

The specimen of the polarization at level 3 (Figure 8d) is almost completely covered with the crystalline deposited film. There are undamaged areas that are not covered with deposited film. This and individual damages between the film indicate that most damages are obstructed by the film. Hence, it can be assumed that the damage is more pronounced than at level 2.

4.2.2. Carbon Coated with SBR-based Material (S4)

The SBR-impregnated carbon S4 has a smooth and glossy surface before polarization. Overall, the surface remains glossy but is no longer smooth but rough. The specimens of levels 2 and 3 are mostly mat and rough, there are only a few glossy surfaces left. In addition, an increased resolution of the impregnation in the area of the intersection points can be observed (Figure 9).

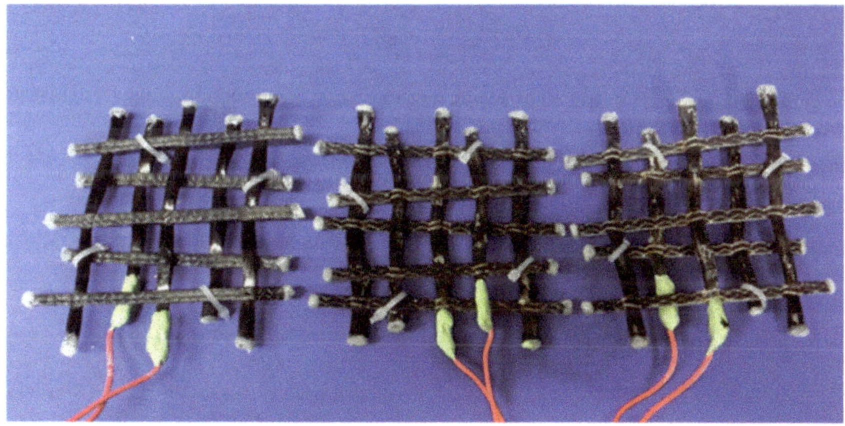

Figure 9. From left to right: S4 anodes of stages 1 to 3 after polarization.

Figure 10 shows the SEM images of the SBR-impregnated carbon fibre S4, the surface of the reference (Figure 10a) is mostly smooth apart from impurities. Sometimes the carbon fibres shine through; there are randomly arranged pores of different sizes. In the reference specimen, there are also imperfections of different sizes from 20 µm, at which the carbon fibres are exposed. Rubber is known to be non-conductive. The pores and imperfections in the impregnations as well as the poor adhesion between SBR impregnation and carbon fibre explain why the SBR-impregnated carbon fabric also shows polarizability. After the polarization at level 1, destructions are already visible at isolated points where carbon fibres are exposed (Figure 10b). Around these areas is a white coating that looks different. According to EDX analysis, this consists mainly of carbon and oxygen and low intensities of calcium and silicon can also be detected. Overall, significantly more carbon fibres are visible due to the impregnation, which suggests that the impregnation dissolves over a large area. A destruction of the SBR impregnation therefore already occurs in panel area 1.

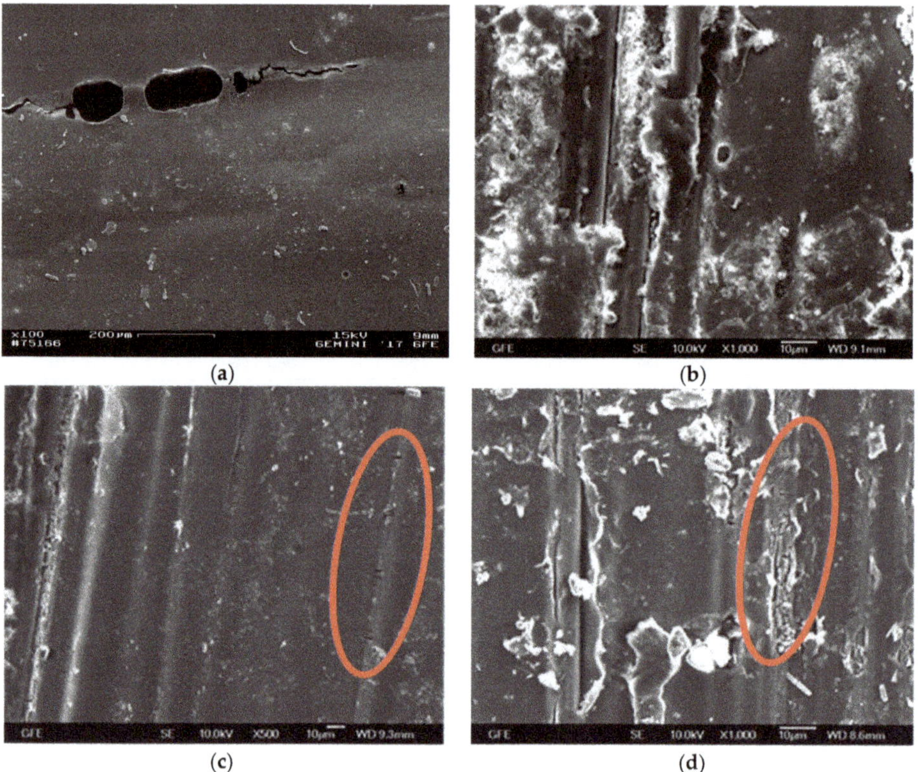

Figure 10. SEM images of S4: unpolarized reference (**a**), after polarization at level 1 (**b**), level 2 (**c**) and level 3 (**d**).

With increasing polarization both, the amount of deposited material and the removed surface of the impregnation increase. The impregnations already show imperfections and pores before polarization, so that a liquid electrolyte can enter the roving. Due to the poor adhesion between SBR and carbon fibre, this is particularly pronounced for the SBR-impregnated material and explains its good polarizability compared to carbon with epoxy resin impregnation. By destruction the impregnations, underlying carbon fibres come into contact with the electrolyte and increase the effective surface area with increasing dissolution. In Figure 10c a filament can be seen, which has become free and transverse cracks are recognizable, which can be either in impregnation or in carbon fibre. Figure 10d also shows a filament, which is free of impregnation. On the basis of this picture one can say exactly that with anodic polarization this filament is attacked here and cracks developed, which do not come from the production.

Dissolution of the carbon fibres could not be achieved by the potentiostatic tests. Before the carbon fibres are dissolved, the impregnations and the sizing are first destroyed. This can be explained by the fact that the carbon fibre consists of mainly covalent carbon compounds in contrast to the impregnations and the size and the covalent bond has a higher bonding energy. Since no optical changes have occurred with E4 and S4 despite significant decomposition of the impregnation materials, destruction of the impregnation materials to carbon dioxide and calcium carbonates can be assumed. Visible decomposition particles have only formed as a result of the decomposition of the sizing of the unimpregnated carbon fibres. However, the exact composition of these particles could not be determined, since the freeze-dried residual solution consisted mainly of calcium carbonate. Calcium carbonate is formed by the reaction of calcium hydroxide with carbon dioxide:

$$Ca(OH)_2 + CO_2 \rightarrow CaCO_3 + H_2O \tag{3}$$

However, the carbon dioxide can be a product of carbon from the dissolution reactions as well as from the air.

4.2.3. Unimpregnated Carbon with Sizing (UC)

On most specimens, optical changes could already be detected after polarization. Photos of all specimens from the potentiostatic tests with unimpregnated carbon fibres are shown in Figure 11. The unimpregnated carbon fibres were glossy and soft, comparable to a brush before the test was carried out. After polarization at level 1, the carbon fibre is still soft but less shiny. This could be explained by the deposits of calcium carbonate. The solution of the tests at levels 1 and 2 showed no difference. The solution of the experiment at level 3, on the other hand, was discoloured brown and small black particles floated in the solution. The particles deposited on the ground after a few days. These sediments were freeze-dried with a small remainder of the solution and identified as calcium carbonate by Fourier transform infrared spectroscopy. Since the substance, unlike pure calcium carbonate, was not white but grey, further components must be present. However, these could not be detected by infrared spectroscopy.

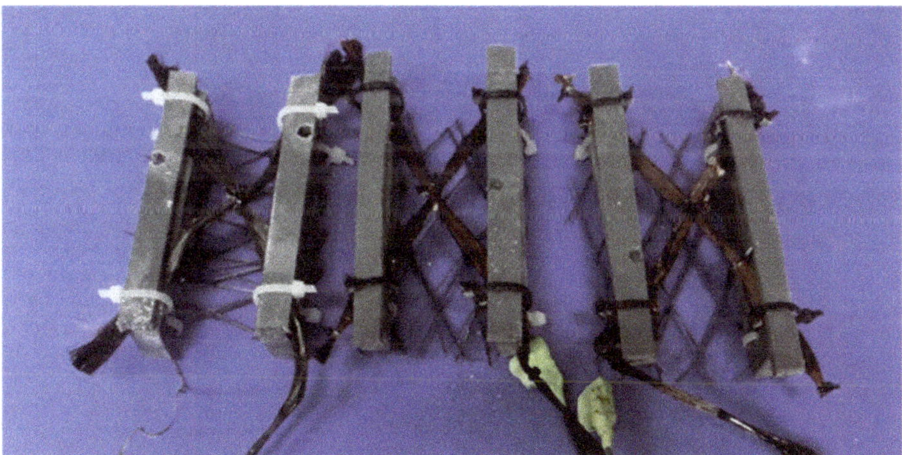

Figure 11. From left to right: specimen P1 of levels 1 to 3 after polarization.

Figure 12 shows the SEM images of UC specimens. The unimpregnated carbon fibres show no damage after polarization at level 1 (Figure 12b). Slight longitudinal grooves can already be observed at the reference (Figure 12a) and no consequence of polarization. Differences in contrast, especially in bright white areas, are caused by charges from the electron ray with insufficient surface conductivity. After polarization at level 2, the sizing of many fibres is decomposed (Figure 12c). This can be seen in most fibres through the formation of deep but narrow longitudinal furrows. Cracks in the radial direction also occur in some fibres, as they have already been detected in the potentiostatic tests. This means that an anodic polarization on level 2 decomposes the size of unimpregnated carbon fibres. On level 1 no destruction of the coating was observed. The change in the slope in the current potential diagram in the phase of the transition range at approximately 900 mV versus NHE is therefore probably due to the reaction that sets in. The sizing of most carbon fibres polarized at level 3 shows very clear destruction phenomena in the form of radial cracks (Figure 12d). The decomposition of the sizing of the unimpregnated carbon fibres also took place inside the roving. It can therefore be assumed that all carbon fibres conduct electricity and are in direct contact with the electrolyte. The effective surface area of the unimpregnated carbon fibres should raise after destruction of the sizing.

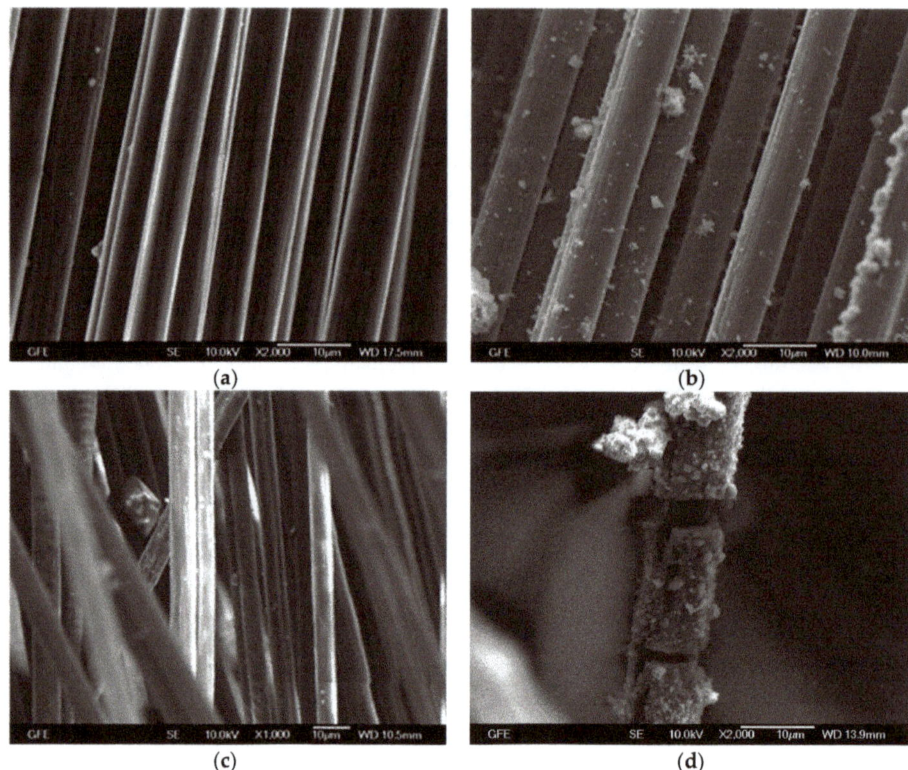

Figure 12. SEM images of UC: unpolarized reference (**a**), after polarization at level 1 (**b**), level 2 (**c**) and level 3 (**d**).

In Figure 12d it is questionable, whether this destruction is in carbon or in sizing. Because the damaged layer looks thick and the residual cross section of carbon small, this question had to be examined more closely. Three further methods were used to investigate it in more detail.

(1) Examination of the average diameter of fibres in a roving

The examination of the diameter of unimpregnated carbon showed that the fibres have a diameter between 5 and 10 µm. Figure 12d shows that the diameter of carbon, under the deposited film around the fibre, is 6 µm and therefore in the range of diameters of unpolarized samples. Hence, no destruction of the carbon fibre is observed. This has been confirmed with several samples.

(2) EDX examination of the deposited film

EDX analysis was performed on an unimpregnated carbon (UC), in a region, where the deposited film was partly flaked off. So both sides of the film could be analysed.

Figures 13 and 14 show that the inner part of the destroyed layer consists mainly of carbon, which could be part of the sizing. The outer layer indicates the presence of calcium, carbon and oxygen. These are the deposits formed due to chemical reaction between calcium of the solution and carbon from the sizing. These deposits can be found more strongly on the surface of the carbon as the polarization potential increases.

Figure 13. SEM image of unimpregnated carbon. 1 and 2 EDX analysis in the marked area.

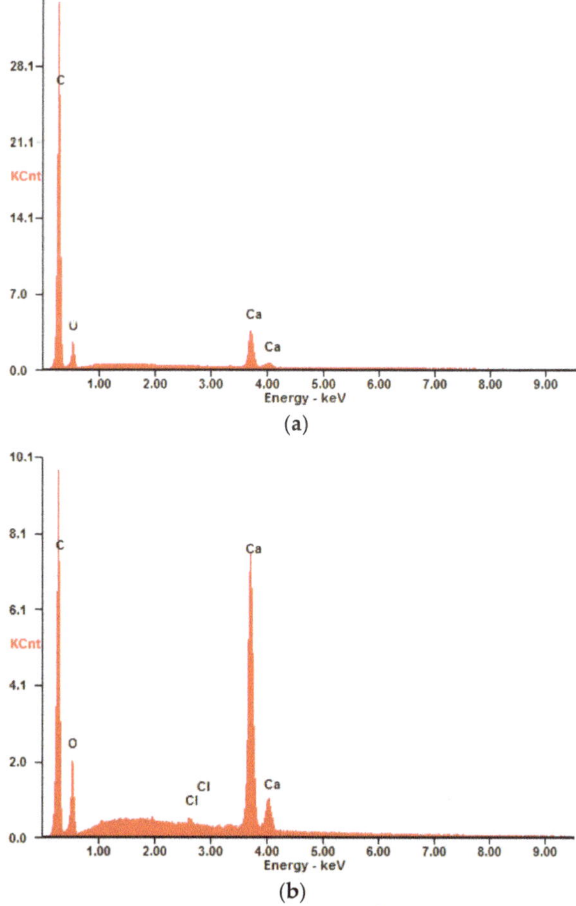

Figure 14. EDX analysis in the marked area 1 (**a**) and EDX analysis in the marked area 2 (**b**).

(3) Investigation of an unimpregnated carbon without sizing (UC_WS)

Thermal gravimetric analysis (TGA)

First, TGA measurements were performed to be sure that the two unimpregnated samples with and without sizing were different and to identify whether the sample UC really has a sizing. Since the differences cannot be seen from the SEM, TGA measurements were performed to see if the weight changes with increasing temperature. From the Figure 15 it can be seen that the sample UC_WS, which has no sizing, shows no change in mass at elevated temperature. The curve of UC shows that at 300 °C the weight decreases. This indicates that the sizing begins to dissolve. The comparison of the two curves shows that the sizing has about 1 wt% of the total weight of the carbon filament. After 450 °C, it stabilizes to 600 °C and then the curve drops again. From there the carbon textiles could start to decompose. It can be assumed that the examined UC samples with sizing really have a sizing and that the destroyed layer in Figure 12d is the sizing including deposited film.

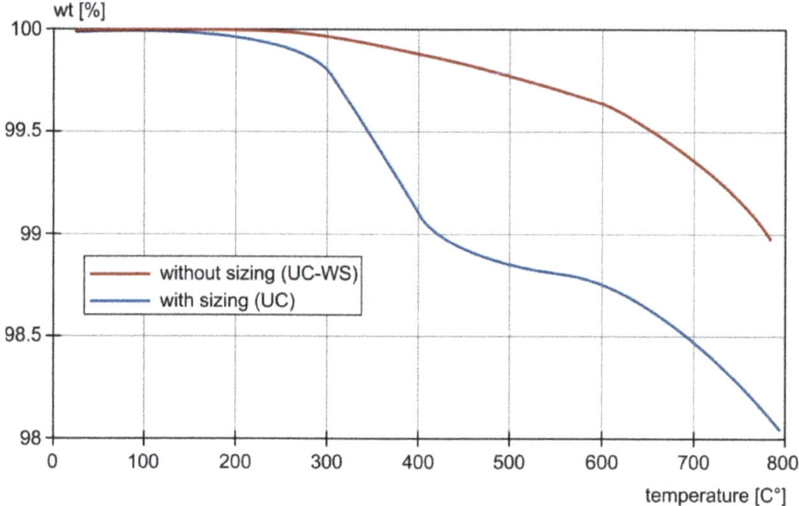

Figure 15. TGA measurement of both unimpregnated Carbons.

The decrease in the weight of sample UC-WS up to 600 °C could be due to the combustion of impurities and dirt.

4.2.4. Unimpregnated Carbon without Sizing (UC_WS)

For comparison purposes, potentiostatic measurements were also carried out on unimpregnated carbon without sizing at the same potentiostatic potentials.

The SEM pictures in the Figure 16 confirm the statement that the pictures in Figure 12 show the destruction of the sizing. The unimpregnated carbon fibres without sizing show no damage after polarization at level 1 to 3. Slight longitudinal grooves can already be observed on all samples and are not a consequence of polarization. Differences in contrast, especially bright white areas, are caused by charges from the electron beam with insufficient surface conductivity. After polarization at level 2 (Figure 16c), the deposition can be seen on the surface, which is slightly increased at level 3 (Figure 16d). But, the deposition is not as strong as with the samples with sizing. It can therefore be concluded that the deposit adheres better to sizing material than carbon. No cracks or damage can be seen on any fibres. This means that anodic polarization does not destroy carbon.

There are no other test results available in the literature, therefore the test results cannot be compared.

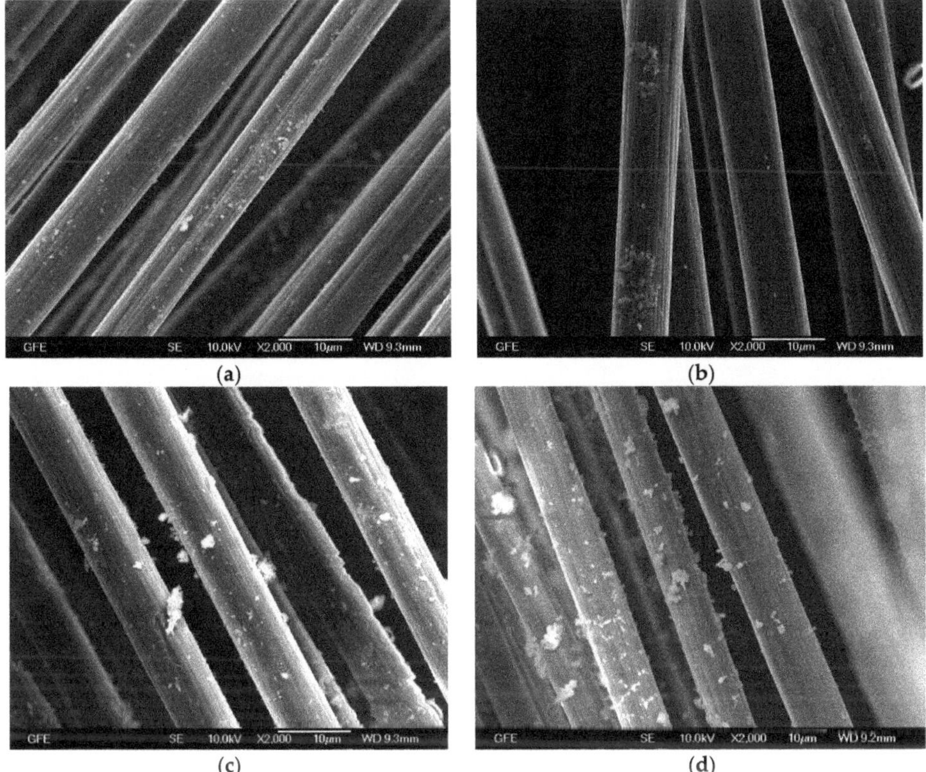

Figure 16. SEM images of UC_WS: unpolarized reference (**a**), after polarization at level 1 (**b**), level 2 (**c**) and level 3 (**d**).

5. Conclusions and Outlooks

Until now, it was unclear up to which potentials the impregnation materials and the sizing remain undamaged by anodic polarization. The results contribute to the clarification of these questions for the investigated materials.

The results show that the operation of CP with carbon textiles as anode material leads to deterioration of the polymer impregnations and sizing. Since impregnation and sizing are available to increase the strength capacity of carbon textiles, the deterioration of these materials is very interesting. With epoxy impregnated carbon it is possible to use carbon textiles as CP anode without destroying the epoxy and sizing up to the mentioned potentials.

With the increasing polarization, the strength of the deposited film on the sizing increases.

The conclusion of this work can be summarized as follows:

1. The investigations in solution have shown that CP with investigated carbon anodes up to 2200 mV versus NHE is possible
2. The Carbon filaments within the Carbon textiles as CP anode have not been destroyed up to the investigated potentials up to 2200 mV versus NHE.
3. With SBR impregnated samples, the impregnation is destroyed right from the start during polarization.
4. The sizing is destroyed at a potential of about 900 mV versus NHE.
5. Epoxy impregnation started to destroy between 1050 and 1150 mV versus NHE.

Further tests must be carried out to investigate the bond between polymer impregnated carbon textiles and mortar. It must be observed whether the destruction of the impregnation or sizing has a negative influence on the bond or durability.

Author Contributions: Conceptualization, A.A. and M.R.; Methodology, A.A.; Software, A.A.; Validation, A.A.; Formal Analysis, A.A.; Investigation, A.A.; Resources, A.A. and M.R.; Data Curation, A.A.; Writing—Original Draft Preparation, A.A.; Writing—Review & Editing, A.A. and M.R.; Visualization, A.A.; Supervision, M.R.

Funding: This research received no external funding

Conflicts of Interest: The authors declare no conflict of interest.

References

1. Gao, S.L.; Mäder, E.; Plonka, R. Nanostructured coatings of glass fibres: Improvement of alkali resistance and mechanical properties. *Acta Mater.* **2007**, *55*, 1043–1052. [CrossRef]
2. Zhang, R.L.; Huang, Y.D.; Liu, L.; Tang, Y.R.; Su, D.; Xu, L.W. Effect of emulsifier content of sizing agent on the surface of carbon fibres and interface of its composites. *Appl. Surf. Sci.* **2011**, *257*, 3519–3523. [CrossRef]
3. Lesko, J.J.; Swain, R.E.; Cartwright, J.W.; Chin, J.W.; Reifsnider, K.L.; Dillard, D.A.; Wightman, J.P. Interphases developed from fibre sizings and their chemical-structural relationship to composite performance. *J. Adhes.* **1994**, *45*, 43–57. [CrossRef]
4. Xu, L. Interfacial Engineering of the Interphase between Carbon Fibers and Vinyl Ester Resin. Ph.D. Thesis, Department of Chemical Engineering and Materials Science, Michigan State University, East Lansing, MI, USA, 2003.
5. Broyles, N.S.; Verghese, K.N.E.; Davis, S.V.; Li, H.; Davis, R.M.; Lesko, J.J.; Riffle, J.S. Fatigue performance of carbon fibre/vinyl ester composites: The effect of two dissimilar polymeric sizing agents. *Polymer* **1998**, *39*, 3417–3424. [CrossRef]
6. Wetjen, D. Wechselwirkung von Carbonfasern, Schlichte und Epoxidbasierter Polymerer Matrix in Carbonfaserverstärkten Kunststoffen. Ph.D. Thesis, Mathematisch-Naturwissenschaftlich-Technische Fakultät, Universität Augsburg, Augsburg, Germany, 2016.
7. Poltavtseva, M.; Ebell, G.; Mietz, J. Electrochemical investigations of carbon-based conductive coatings for application as anodes in ICCP systems of reinforced concrete structures. *Mater. Corros.* **2015**, *66*, 627–634. [CrossRef]
8. Rueffer, M.; Bejan, D.; Bunce, N.J. Graphite. An active or an inactive anode? *Electrochim. Acta* **2011**, *56*, 2246–2253. [CrossRef]
9. Chung, D.D.L. Review Graphite. *J. Mater. Sci.* **2002**, *37*, 1475–1489. [CrossRef]
10. Asgharzadeh, A.; Raupach, M.; Koch, D. Investigations on the Suitability of Technical Textiles for Cathodic Corrosion Protection. In Proceedings of the 4th International Conference on Concrete Repair, Rehabilitation and Retrofitting (ICCRRR), Leipzig, Germany, 5–7 October 2015.
11. Asgharzadeh, A.; Raupach, M. Development of a test Method for the Durability of Carbon Textiles under Anodic Polarisation. In *Service Life and Durability of Reinforced Concrete Structures (RILEM Book Series)*; Springer International Publishing: Basel, Switzerland, 2018; pp. 143–158.
12. Asgharzadeh, A.; Raupach, M. Durability behaviour of polymer impregnated carbon textiles in alkaline solution as CP anode. *Mater. Corros.* **2018**, 1–12. [CrossRef]
13. Bertolini, L.; Bolzoni, F.; Pastore, T.; Pedeferri, P. Effectiveness of a conductive cementitious mortar anode for cathodic protection of steel in concrete. *Cem. Concr. Res.* **2004**, *34*, 681–694. [CrossRef]
14. Chini, M.; Antonsen, R.; Vennesland, Ø.; Mork, J.H.; Arntsen, B. Polarization Behavior of Carbon Fiber as an Anodic Material in Cathodic Protection. In Proceedings of the 11DBMC International Conference on Durability of Building Materials and Components, Istanbul, Turkey, 11–14 May 2008.
15. Chini, M. Pan-Based Carbon Fiber as Anode Material in Cathodic Protection Systems for Concrete Structures. Avhandling–Norges Teknisk-Naturvitenskapelige Universitet, Trondheim. Ph.D. Thesis, Department of Structural Engineering, Faculty of Engineering Science and Technology, Norwegian University of Science and Technology, Trondheim, Norway, 2010.

16. Zhang, E.Q.; Tang, L.; Zack, T. Carbon Fiber as Anode Material for Cathodic Prevention in Cementitious Materials. In Proceedings of the 5th International Conference on the Durability of Concrete Structures, Shenzhen, China, 30 June–1 July 2016; Xing, F., Han, N., Zhu, J.-H., Eds.; Shenzhen University: Shenzhen, China; Purdue University Press: West Lafayette, IN, USA, 2016.
17. Van Nguyen, C.; Lambert, P.; Mangat, P.; O'Flaherty, F.; Jones, G. The Performance of Carbon Fibre Composites as ICCP Anodes for Reinforced Concrete Structures. *ISRN Corros.* **2012**, *2012*, 1–9. [CrossRef]

MDPI
St. Alban-Anlage 66
4052 Basel
Switzerland
Tel. +41 61 683 77 34
Fax +41 61 302 89 18
www.mdpi.com

Applied Sciences Editorial Office
E-mail: applsci@mdpi.com
www.mdpi.com/journal/applsci